Serono Symposia, USA
Norwell, Massachusetts

PROCEEDINGS IN THE SERONO SYMPOSIA, USA SERIES

OVARIAN CELL INTERACTIONS: Genes to Physiology
 Edited by Aaron J.W. Hsueh and David W. Schomberg

IN VITRO FERTILIZATION AND EMBRYO TRANSFER IN PRIMATES
 Edited by Don P. Wolf, Richard L. Stouffer, and Robert M. Brenner

CELL BIOLOGY AND BIOTECHNOLOGY: Novel Approaches to Increased Cellular Productivity
 Edited by Melvin S. Oka and Randall G. Rupp

PREIMPLANTATION EMBRYO DEVELOPMENT
 Edited by Barry D. Bavister

MOLECULAR BASIS OF REPRODUCTIVE ENDOCRINOLOGY
 Edited by Peter C.K. Leung, Aaron J.W. Hsueh, and Henry G. Friesen

MODES OF ACTION OF GnRH AND GnRH ANALOGS
 Edited by William F. Crowley, Jr., and P. Michael Conn

FOLLICLE STIMULATING HORMONE: Regulation of Secretion and Molecular Mechanisms of Action
 Edited by Mary Hunzicker-Dunn and Neena B. Schwartz

SIGNALING MECHANISMS AND GENE EXPRESSION IN THE OVARY
 Edited by Geula Gibori

GROWTH FACTORS IN REPRODUCTION
 Edited by David W. Schomberg

UTERINE CONTRACTILITY: Mechanisms of Control
 Edited by Robert E. Garfield

NEUROENDOCRINE REGULATION OF REPRODUCTION
 Edited by Samuel S.C. Yen and Wylie W. Vale

FERTILIZATION IN MAMMALS
 Edited by Barry D. Bavister, Jim Cummins, and Eduardo R.S. Roldan

GAMETE PHYSIOLOGY
 Edited by Ricardo H. Asch, Jose P. Balmaceda, and Ian Johnston

GLYCOPROTEIN HORMONES: Structure, Synthesis, and Biologic Function
 Edited by William W. Chin and Irving Boime

THE MENOPAUSE: Biological and Clinical Consequences of Ovarian Failure: Evaluation and Management
 Edited by Stanley G. Korenman

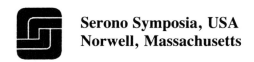

Serono Symposia, USA
Norwell, Massachusetts

Don P. Wolf   Richard L. Stouffer
Robert M. Brenner   Editors

In Vitro Fertilization and Embryo Transfer in Primates

Foreword by Gary D. Hodgen

With 76 Figures

Springer-Verlag
New York Berlin Heidelberg London Paris
Tokyo Hong Kong Barcelona Budapest

Don P. Wolf, Ph.D.
Richard L. Stouffer, Ph.D.
Robert M. Brenner, Ph.D.
Oregon Regional Primate Research Center
505 N.W. 185th Avenue
Beaverton, OR 97006
USA

Proceedings of the Symposium on In Vitro Fertilization and Embryo Transfer in Primates, sponsored by Serono Symposia, USA, held May 28 to 31, 1992, in Beaverton, Oregon.

For information on previous volumes, please contact Serono Symposia, USA.

Library of Congress Cataloging-in-Publication Data
In vitro fertilization and embryo transfer in primates / Don P. Wolf,
  Richard L. Stouffer, Robert M. Brenner, editors. — 1st ed.
    p. cm.
  "Serono Symposia, USA".
  Includes bibliographical references and index.
  ISBN 0-387-94056-1 (New York: alk. paper). — ISBN 3-540-94056-1
(Berlin: alk. paper)
  1. Primates—Embryos—Transplantation—Congresses.
  2. Fertilization in vitro—Congresses. 3. Embryology—Mammals—
Congresses. I. Wolf, Don P. II. Stouffer, Richard L.
III. Brenner, Robert M.
QL737.P9I45   1993
599.8'0433—dc20   93-28106

Printed on acid-free paper.

© 1993 Springer-Verlag New York, Inc.
All rights reserved. This work may not be translated or copied in whole or in part without the written permission of the publisher (Springer-Verlag New York, Inc., 175 Fifth Avenue, New York, NY 10010, USA), except for brief excerpts in connection with reviews or scholarly analysis. Use in connection with any form of information storage and retrieval, electronic adaptation, computer software, or by similar or dissimilar methodology now known or hereafter developed is forbidden.
The use of general descriptive names, trade names, trademarks, etc., in this publication, even if the former are not especially identified, is not to be taken as a sign that such names, as understood by the Trade Marks and Merchandise Marks Act, may accordingly be used freely by anyone.
While the advice and information in this book are believed to be true and accurate at the date of going to press, neither the authors, nor the editors, nor the publisher, nor Serono Symposia, USA, nor Serono Laboratories, Inc., can accept any legal responsibility for any errors or omissions that may be made. The publisher makes no warranty, expressed or implied, with respect to the material contained herein.
Permission to photocopy for internal or personal use, or the internal or personal use of specific clients, is granted by Springer-Verlag New York, Inc., for libraries registered with the Copyright Clearance Center (CCC), provided that the base fee of $5.00 per copy, plus $0.20 per page is paid directly to CCC, 21 Congress Street, Salem, MA 01970, USA. Special requests should be addressed directly to Springer-Verlag New York, Inc., 175 Fifth Avenue, New York, NY 10010, USA.

Production coordinated by Marilyn Morrison and managed by Francine McNeill; manufacturing supervised by Jacqui Ashri.
Typeset by Best-set Typesetter Ltd., Hong Kong.
Printed and bound by Edwards Brothers, Inc., Ann Arbor, MI.

Printed in the United States of America.

9 8 7 6 5 4 3 2 1

ISBN 0-387-94056-1 Springer-Verlag New York Berlin Heidelberg
ISBN 3-540-94056-1 Springer-Verlag Berlin Heidelberg New York

SYMPOSIUM ON IN VITRO FERTILIZATION AND EMBRYO TRANSFER IN PRIMATES

Scientific Committee

>Robert M. Brenner, Ph.D.
Oregon Regional Primate Research Center
Beaverton, Oregon

>Don P. Wolf, Ph.D.
Oregon Regional Primate Research Center
Beaverton, Oregon

>Richard L. Stouffer, Ph.D.
Oregon Regional Primate Research Center
Beaverton, Oregon

Organizing Secretary

>Bruce K. Burnett, Ph.D.
Serono Symposia, USA
100 Longwater Circle
Norwell, Massachusetts

Foreword

It is a genuine honor and a privilege of distinction to provide the foreword for the proceedings that follow. This marvelous symposium reflects the unique qualities of its two principal sponsors. Whenever the Oregon Regional Primate Research Center is involved in matters of science, we see a consistent record of exceptional quality in both the work and the people who do it. Likewise, Serono Symposia, USA is unequaled in assuring a world-class speakers' forum, utterly without taint of commercial influence. The proceedings published herein are a remarkable testimony, not only to the contributors, but to Drs. Brenner, Wolf, Stouffer, and Burnett, who have shepherded its conception, presentation, and publication.

Readers will notice immediately one of the universal strengths of this total composition; that is, the diversity of investigational interests among attendees. Presentations on the core topic of primate in vitro fertilization and embryo transfer were joined by specialist presentations in related areas, including human-assisted reproductive technologies, reproductive physiology of the great apes, sperm biology, implantation mechanisms, cryobiology, coculture systems for embryogenesis, micromanipulation technologies, and genetic diagnosis of heritable diseases in the pre-embryo. Even though pro-fertility issues understandably dominated this forum, there was significant attention given to fertility-control research, especially the role of primates in this endeavor (more on this subject below).

One dimension of the quality of these chapters is that data are presented at so many levels, ranging from use of a variety of molecular technologies to elucidation of basic information on the survival of endangered nonhuman primates whose pool of genetic diversity is severely limited. The progression of contributions also reveals that real knowledge and understanding requires comparative data from many primate species, as well as nonprimate models.

During the course of the symposium, I made a wish list of items wherein scientific progress is sorely needed, as mentioned in many chapters in this book. For example, we wish for

Useful systems of oocyte maturation in vitro
Efficacious methodology for in vitro maturation of zygotes to full blastocyst hatching
A stock of age-matched homozygous twins
Optimal environmental cues that underlie reproductive success
Avoidance of antigenicity from hormonal regimens used to achieve controlled ovarian hyperstimulation
Better yields of high-quality sperm, both fresh and cryopreserved
Determination of how much LH needs to be added to FSH for optimal ovarian response
Knowledge of optimal gas tensions for specific developmental states in vitro
Knowledge of the metabolic substrates, cofactors, and growth promoters to achieve optimal gametes and pre-embryos
Knowledge of the growth factors essential to a suitable implantation milieu
Knowledge of better coculture systems for oocyte and pre-embryo development
Among endangered species, prevention of disease transmission through stored genetic material
An understanding of the frequent basis of ectopic pregnancy in human primates versus its infrequent occurrence in nonhuman primates
Clonal systems for development of laboratory primates of minimal genetic diversity

While this wish list is in no way complete, these examples set the stage for some of the arenas of coming progress.

As is given to scientific endeavors, not infrequently, several paradoxes were illuminated during these proceedings. Two quite diverse ones seem worthy of mention: (i) that certain threatened or endangered species (primate and nonprimate) require contraception to avoid overpopulating available/affordable artificial habitat, and (ii) that with the adaptation of some technologies already applied in the livestock industry to laboratory primates commonly used in physiologic and pharmacologic studies, we may have the best of both situations in conserving their numbers; that is, genetic diversity representative of human response (the noninbred state) and production of homozygous twins or even more numerous clonal members in substantial numbers. Such an evolution would produce very high reproducibility within test groups without smothering genetic diversity via inbreeding (see Figs. A and B).

Last, along with others, I point to the *population bomb*, arguably the greatest threat to both humankind and the variety of primates worldwide that we strive so much to know and ultimately preserve. As habitat dwindles and genetic diversity diminishes, it is we, ourselves, who increasingly endanger them. Accordingly, perhaps few scientific contribu-

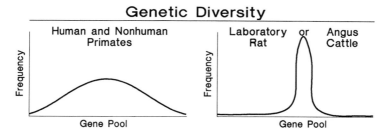

FIGURE A. Example model by which genetic diversity and clonal development of primates can enhance experimental power while conserving endangered or threatened species used in biomedical research.

	Control Group		Test Group
Primate	1a	*versus*	1b
	2a	*versus*	2b
	etc...		

FIGURE B. Assignment of clonally derived primates to minimize genetic sources of response variation without inbreeding.

tions will be more important to the survival of nonhuman primates than human population control. It is from this message that I suggest to the leadership of the Oregon Regional Primate Research Center that their next such symposium be on the topic "Primate Models for Contraceptive Technologies."

GARY D. HODGEN

Preface

Goals of the Symposium

Gametogenesis, fertilization, early development, and implantation are critical events for successful reproduction in mammals. In the past 10–15 years, the increased availability of in vitro approaches has opened these processes to experimental study. Despite the fact that in vitro fertilization–embryo transfer is now considered standard medical practice for the treatment of several categories of human infertility, a nonhuman primate model remains highly desirable for the study of reproductive and developmental processes, as well as for the establishment of improved disease models for medical research and the preservation of endangered nonhuman primate species. Clearly, such a nonhuman primate model is essential for research on early developmental processes that cannot be performed in women.

This symposium analyzed in depth our current knowledge of IVF-ET technology as applied to nonhuman primates. The combination of podium lectures, discussions, and perspectives of leading authorities offered an up-to-date summary of the field. A better understanding of the primate reproductive system is essential to facilitate a global effort to control human population and to aid the numerous childless families who are attempting to overcome their infertility. The publication of these proceedings should contribute effectively to these goals.

Don P. Wolf
Richard L. Stouffer
Robert M. Brenner

Contents

Foreword vii
Preface xi
Contributors xvii

Part I. IVF in Primates

1. State of the Art and Future Directions in Human IVF 3
   ALAN TROUNSON

2. IVF in Nonhuman Primates: Current Status and Future Directions 30
   BARRY D. BAVISTER AND DOROTHY E. BOATMAN

3. Assisted Reproduction in the Great Apes 46
   KENNETH G. GOULD AND JEREMY F. DAHL

4. Assisted Reproduction in New World Primates 73
   W. RICHARD DUKELOW

5. IVF-ET in Old World Monkeys 85
   DON P. WOLF AND RICHARD L. STOUFFER

Part II. Gamete Biology/Ovarian Physiology

6. Sperm-Zona Pellucida Interaction in Macaques 103
   JAMES W. OVERSTREET AND CATHERINE A. VANDEVOORT

7. Nonhuman Primate Oocyte Biology: Environmental Influences on Development 110
   DOROTHY E. BOATMAN, RALPH D. SCHRAMM, AND BARRY D. BAVISTER

8. Stimulation of Follicle and Oocyte Development in Macaques for IVF Procedures 124
   RICHARD L. STOUFFER, MARY B. ZELINSKI-WOOTEN, YASMIN ALADIN CHANDRASEKHER, AND DON P. WOLF

xiv   Contents

Part III.   Preimplantation/Implantation Biology

9. Overview of the Morphology of Implantation in Primates.... 145
   ALLEN C. ENDERS

10. Physiology of Implantation in Primates 158
    J.P. HEARN, P.B. SESHAGIRI, AND G.E. WEBLEY

11. Interactions Between the Embryo and Uterine Endometrium
    During Implantation and Early Pregnancy in the Baboon
    (*Papio anubis*) 169
    ASGERALLY T. FAZLEABAS, SHERI HILD-PETITO,
    KATHLEEN M. DONNELLY, PATRICIA MAVROGIANIS, AND
    HAROLD G. VERHAGE

12. Regulation of Human Cytotrophoblast Invasion 182
    KATHRYN E. BASS, IRIS ROTH, CAROLINE H. DAMSKY, AND
    SUSAN J. FISHER

13. Expression and Binding of Transforming Growth Factor βs in
    the Mouse Embryo and Uterus During the Periimplantation
    Period.. 195
    S.K. DEY, S.K. DAS, B.C. PARIA, K.C. FLANDERS, AND
    G.K. ANDREWS

Part IV.   Gamete/Embryo Manipulation

14. Cryobiology of Gametes and Embryos from Nonhuman
    Primates.. 223
    W.F. RALL

15. Application of Micromanipulation in the Human 246
    JACQUES COHEN, MINA ALIKANI, ALEXIS ADLER,
    ADRIENNE REING, TONI A. FERRARA, ELENA KISSIN, AND
    CINDY ANDERSON

16. Embryo Coculture and the Regulation of Blastocyst
    Formation In Vitro 279
    YVES MENEZO, LAURENT JANY, AND
    CHAQUE KHATCHADOURIAN

17. Proteoglycans as Modulators of Embryo-Uterine
    Interactions 290
    DANIEL D. CARSON, ANDREW L. JACOBS, JOANNE JULIAN, AND
    LARRY H. ROHDE

18. A Primatologist's Perspective on Assisted Reproduction for Nonhuman Primates 308
R.G. RAWLINS

Part V. Gamete/Embryo Applications

19. Nuclear Transfer in Mammals 317
N.L. FIRST AND M.L. LEIBFRIED-RUTLEDGE

20. Assisted Reproduction in the Propagation Management of the Endangered Lion-Tailed Macaque (*Macaca silenus*) 331
M.R. CRANFIELD, B.D. BAVISTER, D.E. BOATMAN, N.G. BERGER, N. SCHAFFER, S.E. KEMPSKE, D.M. IALEGGIO, AND J. SMART

21. Genetic Abnormalities in the Human Preimplantation Embryo 349
ALAN H. HANDYSIDE

22. Intra-Acrosomal Contraceptive Vaccine Immunogen SP-10 in Human, Macaque, and Baboon 360
JOHN C. HERR, RICHARD M. WRIGHT, CHARLES J. FLICKINGER, ALEX FREEMERMAN, KENNETH KLOTZ, JAMES FOSTER, AND JOHN SHANNON

Author Index 384
Subject Index 386

Contributors

ALEXIS ADLER, The Center for Reproductive Medicine and Infertility, Department of Obstetrics and Gynecology, The New York Hospital–Cornell University Medical College, New York, New York, USA.

MINA ALIKANI, The Center for Reproductive Medicine and Infertility, Department of Obstetrics and Gynecology, The New York Hospital–Cornell University Medical College, New York, New York, USA.

CINDY ANDERSON, The Center for Reproductive Medicine and Infertility, Department of Obstetrics and Gynecology, The New York Hospital–Cornell University Medical College, New York, New York, USA.

G.K. ANDREWS, Department of Biochemistry and Molecular Biology, University of Kansas Medical Center, Kansas City, Kansas, USA.

KATHRYN E. BASS, Department of Stomatology, University of California, San Francisco, California, USA.

BARRY D. BAVISTER, Wisconsin Regional Primate Research Center and Department of Animal Health and Biomedical Sciences, University of Wisconsin, Madison, Wisconsin, USA.

N.G. BERGER, Baltimore, Maryland, USA.

DOROTHY E. BOATMAN, Department of Animal Health and Biomedical Sciences, University of Wisconsin, Madison, Wisconsin, USA.

DANIEL D. CARSON, Department of Biochemistry and Molecular Biology, M.D. Anderson Cancer Center, Houston, Texas, USA.

YASMIN ALADIN CHANDRASEKHER, Division of Reproductive Sciences, Oregon Regional Primate Research Center, Beaverton, Oregon, and Department of Physiology, Oregon Health Sciences University, Portland, Oregon, USA.

JACQUES COHEN, The Center for Reproductive Medicine and Infertility, Department of Obstetrics and Gynecology, The New York Hospital–Cornell University Medical College, New York, New York, USA.

M.R. CRANFIELD, Baltimore Zoo, Baltimore, Maryland, USA.

JEREMY F. DAHL, Yerkes Regional Primate Research Center, Emory University, Atlanta, Georgia, USA.

CAROLINE H. DAMSKY, Departments of Anatomy and Stomatology, University of California, San Francisco, California, USA.

S.K. DAS, Departments of Obstetrics-Gynecology and Physiology, University of Kansas Medical Center, Kansas City, Kansas, USA.

S.K. DEY, Departments of Obstetrics-Gynecology and Physiology, University of Kansas Medical Center, Kansas City, Kansas, USA.

KATHLEEN M. DONNELLY, Department of Obstetrics and Gynecology, University of Illinois College of Medicine, Chicago, Illinois, USA.

W. RICHARD DUKELOW, Endocrine Research Center, Michigan State University, East Lansing, Michigan, USA.

ALLEN C. ENDERS, Department of Cell Biology and Human Anatomy, University of California School of Medicine, Davis, California, USA.

ASGERALLY T. FAZLEABAS, Department of Obstetrics and Gynecology, University of Illinois College of Medicine, Chicago, Illinois, USA.

TONI A. FERRARA, The Center for Reproductive Medicine and Infertility, Department of Obstetrics and Gynecology, The New York Hospital–Cornell University Medical College, New York, New York, USA.

N.L. FIRST, Department of Meat and Animal Science, University of Wisconsin, Madison, Wisconsin, USA.

SUSAN J. FISHER, Departments of Stomatology, Anatomy, Obstetrics, Gynecology and Reproductive Sciences, and Pharmaceutical Chemistry, University of California, San Francisco, California, USA.

K.C. FLANDERS, Laboratory of Chemoprevention, National Cancer Center, National Institutes of Health, Bethesda, Maryland, USA.

CHARLES J. FLICKINGER, Center for Recombinant Gamete Contraceptive Vaccinogens, Department of Anatomy and Cell Biology, University of Virginia Health Sciences Center, Charlottesville, Virginia, USA.

JAMES FOSTER, Center for Recombinant Gamete Contraceptive Vaccinogens, Department of Anatomy and Cell Biology, University of Virginia Health Sciences Center, Charlottesville, Virginia, USA.

ALEX FREEMERMAN, Center for Recombinant Gamete Contraceptive Vaccinogens, Department of Anatomy and Cell Biology, University of Virginia Health Sciences Center, Charlottesville, Virginia, USA.

KENNETH G. GOULD, Yerkes Regional Primate Research Center, Emory University, Atlanta, Georgia, USA.

ALAN H. HANDYSIDE, Institute of Obstetrics and Gynaecology, Royal Postgraduate Medical School, Hammersmith Hospital, London, UK.

JOHN P. HEARN, Wisconsin Regional Primate Research Center, University of Wisconsin, Madison, Wisconsin, USA.

JOHN C. HERR, Center for Recombinant Gamete Contraceptive Vaccinogens, Department of Anatomy and Cell Biology, University of Virginia Health Sciences Center, Charlottesville, Virginia, USA.

SHERI HILD-PETITO, Department of Obstetrics and Gynecology, University of Illinois College of Medicine, Chicago, Illinois, USA.

GARY D. HODGEN, The Jones Institute for Reproductive Medicine, Department of Obstetrics/Gynecology, Eastern Virginia Medical School, Norfolk, Virginia, USA.

D.M. IALEGGIO, Baltimore Zoo, Baltimore, Maryland, USA.

ANDREW L. JACOBS, Department of Biochemistry and Molecular Biology, M.D. Anderson Cancer Center, Houston, Texas, USA.

LAURENT JANY, Unité de Fécondation in vitro, Polyclinique Hotel Dieu, Clermont-Ferrand, France.

JOANNE JULIAN, Department of Biochemistry and Molecular Biology, M.D. Anderson Cancer Center, Houston, Texas, USA.

S.E. KEMPSKE, Baltimore Zoo, Baltimore, Maryland, USA.

CHAQUE KHATCHADOURIAN, INSA Laboratoire de Biologie, Villeurbanne, France.

ELENA KISSIN, The Center for Reproductive Medicine and Infertility, Department of Obstetrics and Gynecology, The New York Hospital–Cornell University Medical College, New York, New York, USA.

KENNETH KLOTZ, Center for Recombinant Gamete Contraceptive Vaccinogens, Department of Anatomy and Cell Biology, University of Virginia Health Sciences Center, Charlottesville, Virginia, USA.

M.L. LEIBFRIED-RUTLEDGE, Department of Meat and Animal Science, University of Wisconsin, Madison, Wisconsin, USA.

PATRICIA MAVROGIANIS, Department of Obstetrics and Gynecology, University of Illinois College of Medicine, Chicago, Illinois, USA.

YVES MENEZO, IRH/Fondation Merieux, Bron, and INSA Laboratoire de Biologie, Villeurbanne, France.

JAMES W. OVERSTREET, Division of Reproductive Biology and Medicine, Department of Obstetrics and Gynecology, University of California, Davis, California, USA.

B.C. PARIA, Departments of Obstetrics-Gynecology and Physiology, University of Kansas Medical Center, Kansas City, Kansas, USA.

WILLIAM F. RALL, Department of Animal Health, National Zoological Park, Smithsonian Institution, Washington, District of Columbia, USA.

R.G. RAWLINS, Department of Obstetrics and Gynecology, Section of Reproductive Endocrinology and Infertility, Rush-Presbyterian-St. Luke's Medical Center, Chicago, Illinois, USA.

ADRIENNE REING, The Center for Reproductive Medicine and Infertility, Department of Obstetrics and Gynecology, The New York Hospital–Cornell University Medical College, New York, New York, USA.

LARRY H. ROHDE, Department of Biochemistry and Molecular Biology, M.D. Anderson Cancer Center, Houston, Texas, USA.

IRIS ROTH, Departments of Anatomy and Stomatology, University of California, San Francisco, California, USA.

N. SCHAFFER, Lincoln Park Zoological Gardens, Chicago, Illinois, USA.

RALPH D. SCHRAMM, Wisconsin Regional Primate Research Center, Madison, Wisconsin, USA.

POLANI B. SESHAGIRI, Wisconsin Regional Primate Research Center, Department of Veterinary Science, University of Wisconsin, Madison, Wisconsin, USA.

JOHN SHANNON, Protein and Nucleic Acid Sequencing Center, Department of Anatomy and Cell Biology, University of Virginia Health Sciences Center, Charlottesville, Virginia, USA.

J. SMART, Department of Obstetrics and Gynecology, Johns Hopkins University, School of Medicine, Baltimore, Maryland, USA.

RICHARD L. STOUFFER, Division of Reproductive Sciences, Oregon Regional Primate Research Center, Beaverton, Oregon, and Department of Physiology, Oregon Health Sciences University, Portland, Oregon, USA.

ALAN TROUNSON, Centre for Early Human Development, Institute of Reproduction and Development, Monash University, Clayton, Victoria, Australia.

CATHERINE A. VANDEVOORT, California Regional Primate Research Center, University of California, Davis, California, USA.

HAROLD G. VERHAGE, Department of Obstetrics and Gynecology, University of Illinois College of Medicine, Chicago, Illinois, USA.

GEORGINA E. WEBLEY, The Zoological Society of London, The Institute of Zoology, London, UK.

DON P. WOLF, Division of Reproductive Sciences, Oregon Regional Primate Research Center, Beaverton, Oregon, and Department of Obstetrics/Gynecology, Oregon Health Sciences University, Portland, Oregon, USA.

RICHARD M. WRIGHT, Center for Recombinant Gamete Contraceptive Vaccinogens, Department of Anatomy and Cell Biology, University of Virginia Health Sciences Center, Charlottesville, Virginia, USA.

MARY B. ZELINSKI-WOOTEN, Division of Reproductive Sciences, Oregon Regional Primate Research Center, Beaverton, Oregon, USA.

Part I

IVF in Primates

1

State of the Art and Future Directions in Human IVF

ALAN TROUNSON

The techniques for human *in vitro fertilization* (IVF) by Edwards and Steptoe in the 1970s (1, 2) and Wood and colleagues in the late 1970s and early 1980s (3–5) have been developed into a broad spectrum of treatment strategies that are now known as the new reproductive technologies for assisted human conception. The complex nature of the treatment options that now exist is often not appreciated, and there is frequently a misunderstanding of the interrelationships between the various options and their application. This chapter identifies the treatment options and their relationships to one another.

What is apparent in the operation of infertility clinics is the adherence to certain established protocols for the treatment of patients irrespective of the specific nature of their infertility problems. With the knowledge that we have gained over the last decade and the treatment options available, the approach should be reoriented to the solution of a couple's particular problem rather than to the allocation of patients to a general treatment schedule and the assessment of success based on pregnancy rate per cycle of treatment. It is certain that a substantial quantity of information about the couple's infertility problem is required before commencement of treatment, and a planned strategy aimed at solving the particular problem that limits the possibility of success needs to be devised with some idea of the probability of a successful outcome. Indeed, it is possible to provide couples with the probability of outcome given the options chosen and, in many cases, also provide a guarantee of success. This approach would dramatically alter the way in which infertility services are currently operated and would improve the outcome of treatment.

Superovulation and Natural Cycles

An endocrine evaluation is really necessary before a treatment procedure can be designed for patients. The assessment should include determination of LH, FSH, and prolactin levels in the follicular phase and plasma progesterone and estrogen in the midluteal phase to enable the confirmation of ovulation and diagnosis of such problems as polycystic ovarian syndrome, hyperprolactinaemia, and occult ovarian failure (6). These tests enable some prediction of the likely response to exogenous gonadotropins. Recent records of the patient's menstrual cycle lengths will also allow determination of the most likely days of their midcycle LH surge and ovulation (7), which is useful in planning superovulation regimens and opportunities for recovery of oocytes in natural ovulatory cycles. A number of treatment options are available for patients, and these are discussed in the following sections.

Natural Cycles

Experience varies in the use of natural cycles for IVF and *gamete fallopian tube transfer* (GIFT). Pregnancy rates vary considerably from low rates of 2%–5% per cycle of treatment to high rates of 15%–20% per cycle of treatment (8, 9). There is also great variability in the way follicular growth is monitored, the LH surge detected, and the criteria used for timing the administration of *human chorionic gonadotropin* (hCG). Perhaps the most important considerations are the patient's preference to avoid superovulation, the patient's age, the couple's desire to avoid multiple pregnancy, the certain diagnosis of the cause of infertility, and the need to have a minimal interruption to the couple's lifestyle commitments.

Natural cycle IVF can suit young women with bilateral tubal occlusion, regular menstrual cycles, and partners who have normal semen parameters. Usually, these couples are both working and often wish to continue their careers. Natural cycle GIFT may also suit young couples with unexplained or idiopathic infertility (Table 1.1). Age and regular menstrual cycles are important for the ease of monitoring follicular growth and for high implantation rates because single oocytes or embryos are

TABLE 1.1. Use of natural cycles for GIFT and IVF at the Infertility Medical Centre, Melbourne, 1988–1990.

Factor	Spontaneous LH surge	Timed by hCG injection
Cycles in which follicles aspirated	59	153
Cycles with occytes recovered	40 (68%)	117 (76%)
Number of GIFT cycles	8	16
Pregnancies	1 (13%)	2 (13%)
Number of IVF cycles	32	101
Oocytes fertilized	24 (75%)	65 (64%)
Cycles with embryo transfer	20 (63%)	61 (60%)
Pregnancies	2 (10%)	4 (7%)

Mean age of patients: 38 years.

replaced. It can be argued that natural cycle IVF or GIFT should precede surgery or other interventions in young infertile couples. In the situation of minimal monitoring (single ultrasound or a blood sample for plasma estrogen), patients will return regularly for natural cycle IVF, resulting in high success rates when measured as the resolution of their infertility over a confined period (1–2 years) (9). Couples with unexplained infertility need to be advised about contraception after pregnancy because of the frequent occurrence of fertility following pregnancy after treatment with assisted conception techniques.

Superovulation Using Clomiphene Citrate and hMG

Superovulation using clomiphene citrate (4) or clomiphene citrate and *human menopausal gonadotropin* (hMG) (10) are well-established procedures that rely on the increase of endogenous FSH by clomiphene citrate early in the follicular phase (days 2–8) and supplementation of endogenous FSH with exogenous gonadotropins during the follicular phase to increase the number of growing follicles. Follicle growth and estrogen production need to be monitored in order to anticipate the endogenous LH surge or to administer hCG prior to onset of the endogenous LH surge. It is also customary to assay for LH and progesterone at the time of hCG injection (11) in order to adjust the time of oocyte recovery in patients where the LH surge may have already been initiated.

It is also common to superovulate patients with hMG alone or in combination with more purified preparations of FSH (e.g., Metrodin [Serono]). These protocols provide supplementation to normal endogenous gonadotropin secretion (12). The number of follicles induced to grow is related to the dose of FSH used for superovulation rather than the nature of the preparation, and follicular response varies widely among patients. Increasing the doses of FSH increases the risk of hyperstimulation syndrome, but, generally, also increases the number of oocytes recovered and embryos available for transfer. Recombinant FSH, which is devoid of LH, is now also being trial tested for superovulation. However, it is doubtful whether recombinant FSH will produce better results than the preparations from urinary extracts in the majority of patients. The availability of pure FSH may be of benefit for patients whose follicles prematurely luteinize when given high doses of hMG.

Superovulation Using GnRH Analogs and hMG

Endogenous gonadotropins may frequently result in one or two leading follicles that induce a premature LH surge before the cohort of follicles stimulated by exogenous FSH have reached the appropriate size for production of competent oocytes. GnRH analogs (Leuprolide [Abbot Pharmaceuticals] or Buserelin [Hoechst AG]) can be used to reduce endogenous gonadotropin levels and prevent a precocious LH surge (13). In the method known as *boost*, or *flare*, GnRH analogs are given at the beginning of the menstrual cycle, resulting in an initial boost or flare of endogenous gonadotropin followed by the down-regulation of pituitary gonadotropin secretion. GnRH is commonly started on about day 2 of the menstrual cycle and continued until the day before hCG injection, and hMG or FSH administration begins 1 or 2 days after the start of GnRH and continues until the required number of leading follicles are 16 mm or larger and plasma estrogen is in excess of that expected for the number of follicles at midcycle. Administration of hCG is required to complete follicular and oocyte maturation, and oocytes are recovered 36 h after hCG injection. This superovulation protocol is satisfactory for most IVF and GIFT patients and has a relatively low cancellation rate because of failure of multiple follicular growth.

In order to remove the interference of one or two follicles growing well in advance of the superovulated cohort, a protocol of extended down-regulation of gonadotropin secretion from the pituitary gland may be required (13). This protocol requires a longer period of GnRH administration that begins on about day 21 of the previous menstrual cycle, and hMG administration commences after adequate pituitary and ovarian suppression (estrogen: <180 pmol/L; LH: <2 IU/L; and progesterone: <2 nmol/L). Normally, there are no cysts or follicles larger than 2–5 mm in diameter when hMG is commenced, and a larger dose of hMG and extra time are required to develop a cohort of follicles of the size required using this technique. Generally, this method should be reserved for those patients with inadequate or abnormal endocrine and ovarian responses to boost (14, 15). A simple indicator for unsatisfactory response to boost is the failure to produce a doubling in plasma estrogen levels the day after injection of 1-mg Leuprolide when given as a subcutaneous injection on day 2 of the menstrual cycle (16). The down-regulation method has also been shown to be a satisfactory alternative for those patients who have an inadequate response to superovulation protocols that involve only hMG (17).

Supplementation of Gonadotropins with Growth Hormone

The most common reason for cancellation of an IVF or GIFT cycle is abnormally low (less than the 5th centile) serum estrogen levels in response to gonadotropin stimulation. Patients in this group tend to repeat this problem in subsequent cycles of treatment and have very poor prospects for pregnancy (18). A proportion of these patients have occult ovarian failure (elevated FSH levels and regular menstrual cycles) (6).

Owen et al. (19) have reported that poorly responding patients with polycystic ovaries will respond better when given growth hormone (24 IU every second day) during ovarian stimulation with hMG in the superovulation protocol that involves clomiphene citrate and hMG. Growth hormone stimulates ovarian and hepatic production of IGF-I that in turn may enhance ovarian responses to gonadotropin. Interestingly, the other patients in this category of poor responders showed no improvement in ovarian response over that achieved with gonadotropins. It is probable that the growth hormone will have only a limited use as an aid for superovulation.

Programmed Superovulation Using Oral Contraceptives

Protocols for programmed superovulation have been used for the superovulation of patients to enable the determination of the day of oocyte recovery prior to treatment. These treatments involve the use of either a progestogen (e.g., norethisterone: 10 mg/day) for patients in the second half of their menstrual cycle or combined estrogen and progestogen contraceptive pills for patients in the first half of their menstrual cycles and are normally given for about 17–24 days prior to initiation of superovulation with clomiphene citrate and hMG or hMG alone. The patients are given hCG systematically 9 days after beginning clomiphene citrate and hMG (20). These treatments are designed for reduced cost and efficiency, but do not address individual requirements of patients. They should only be used for patients who have normal endocrine parameters and who respond normally to gonadotropin treatment (Fig. 1.1).

Integration of Treatment Options

It is necessary to begin with a thorough assessment of the patient's ovulatory endocrine profile in order to determine the options best suited for recovery of oocytes (Fig. 1.1). Patients with normal endocrine parameters have the option of superovulation or natural cycles. Those choosing superovulation may simply be treated by boost if they respond to GnRH; a further option is to use clomiphene citrate and hMG if stimulation failure is unsatisfactory with boost. Continual stimulation failure will require down-regulation. Patients with abnormal endocrine profiles require correction of this problem, if possible, or consideration of the problem in the design of superovulation therapy. Patients with occult ovarian failure do not superovulate and face the options of natural cycle IVF or GIFT or oocyte donation (21). Other problem patients may be best treated by down-regulation with GnRH analogs. The polycystic patient may respond better with cotreatment by gonadotropins and growth hormone, clomiphene citrate alone, or with purified preparations of FSH. The options chosen for patients will have a major influence on the outcome of their treatment.

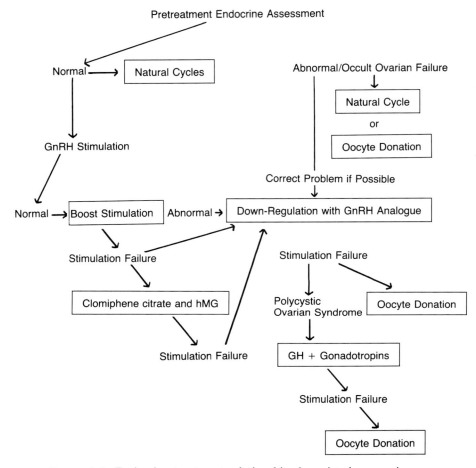

FIGURE 1.1. Endocrine treatment relationships in assisted conception.

Oocyte Maturation In Vitro: An Alternative to Superovulation

Oocyte maturation in vitro (IVM) has been developed as an alternative to superovulation for embryo production in cattle (22). Oocytes may be recovered from the unstimulated ovaries of cows by vaginal ultrasound-guided follicular aspiration several times during the estrous cycle, matured in vitro, fertilized in vitro, and the embryos grown in culture. Cha et al. (23) have shown that 56% of immature oocytes recovered from unstimulated ovaries will mature to metaphase II when cultured in medium containing 50% preovulatory follicular fluid. Of these matured oocytes, 81% fertilized, 64% cleaved normally, and when transferred to

TABLE 1.2. Maturation of human oocytes recovered from patients by laparoscopy and ultrasound-guided follicular aspiration.

Number of attempted recoveries	55
Number of successful recoveries	25 (46%)
Number of oocytes collected	117 (1–20)
Oocytes/attempted recovery	2.13
Oocytes/successful recovery	4.68
Number of oocytes cultured	67
Outcome: GV	5 (8%)
Degenerated	8 (12%)
Metaphase I	10 (15%)
Metaphase II	44 (66%)
Number of oocytes inseminated	28
Outcome: Unfertilized	13 (46%)
1PN	3 (11%)
3PN	1 (4%)
2PN	11 (39%)
Number of 2PN oocytes cultured	11
Cleaved	11 (100%)

one patient with premature ovarian failure, triplets were born following the transfer of 3 embryos.

In our own studies on the maturation of immature oocytes (24), oocytes recovered from ovariectomy specimens and by ultrasound-guided aspiration could be matured and fertilized in vitro (Table 1.2). Present studies are concentrated on the development of follicular aspiration techniques to recover viable oocytes from patients with polycystic ovarian disease in order to avoid the complications associated with attempts to superovulate these patients. A systematic development of this technique would enable the recovery of large numbers of immature oocytes that may be matured, fertilized, and embryos transferred to these patients. Cryopreservation of oocytes for these patients would also be extremely useful.

Oocytes require a short period of time after expulsion of the first polar body to gain full cytoplasmic competence before insemination (5). Insemination of oocytes 1–3 h after extrusion of the polar body results in high rates of fertilization (25) (Table 1.3). The untimely insemination of oocytes can lead to the premature penetration of oocytes by sperm and failure of normal pronuclear development or the failure of fertilization through a nonreversible hardening of the zona that appears to occur when incompletely matured oocytes are exposed to sperm in vitro. Unfortunately, a high proportion of oocytes may be incompletely matured when recovered after superovulation, and this problem can contribute to fertilization failure and abnormalities of embryo development in IVF (Table 1.4).

TABLE 1.3. Fertilization rate and interval between oocyte recovery and insemination.

Interval between oocyte recovery and insemination	Number of oocytes	Fertilization rate (%)	Polypronuclear oocytes (%)
0–1	119	64	4
1–4	227	73	4
4–8	781	80[a]	8
Total	1127	77	7

[a] Significantly higher than other groups.
Source: Adapted from Osborn (25).

TABLE 1.4. Competence of immature oocytes recovered after superovulation.

Stage of maturation before and after culture	Number of oocytes inseminated	Number of oocytes fertilized
Mature (MII)	564	274 (49%)
Metaphase I–MII	65	31 (48%)
GV–MII	61	17 (25%)

Source: Adapted from Osborn (25).

IVF and Embryo Development in Culture

There is a consistently higher success rate for the GIFT procedure where oocytes and sperm are transferred to the fallopian tubes than for the IVF procedure where oocytes are fertilized in vitro and embryos cultured for a period of time before replacement in utero (Fig. 1.2). Analyses of implantation rates for IVF, GIFT, and *tubal embryo stage transfer* (TEST) or ZIFT show a variable pattern. For example, Bollen et al. (26) analyzed the implantation rates for the three procedures involving the replacement of 3 oocytes (GIFT), zygotes (TEST), and embryos (IVF) for a large number of cases (190: GIFT; 161: TEST; and 713: IVF). They found a significant ($P < 0.01$) difference in implantation rates between TEST (18.2%), IVF (13.7%), and GIFT (8.4%). In a smaller study involving a comparison of 34 patients involved in TEST and 44 patients involved in transferred embryos, Toth et al. (27) reported a 9.8% implantation rate in TEST compared to 11.0% in IVF. These results make it difficult to conclude that there is a consistent problem arising from the effects of IVF or embryo culture on embryo viability if implantation rates are similar in GIFT (corrected for fertilization failure), TEST, and IVF. It is also apparent that implantation rates are unaffected by transfer of cleavage-stage embryos to the fallopian tube or uterus (28).

Improved development of human embryos has frequently been reported in new culture media. For example, the medium HTF was claimed

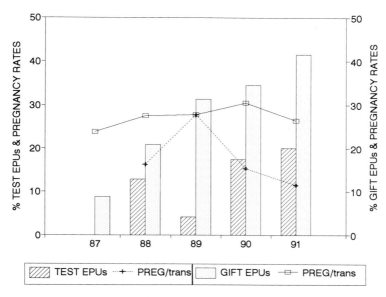

FIGURE 1.2. Proportion of patients treated by GIFT and TEST at the Infertility Medical Centre, Melbourne, 1987–1991, and the pregnancy success rates per cycle of oocyte recovery (EPUS). The remaining patients were treated by IVF (data not shown).

to increase IVF pregnancy rates (29), and human amniotic fluid improved embryo development and IVF success rates (30), but neither showed any significant effect in prospective randomized control trials (31, 32). More recently, feeder cell layers have been reported to improve the development and viability of human embryos. These include human oviduct cells (33) and Vero cells (34). Coculture of embryos with feeder cell layers may benefit embryo viability, but the benefits are probably limited because it is usual to culture human embryos only for about 48 h after insemination. There is no firm evidence to date that early cleavage-stage embryos have any capacity to utilize proteins secreted by cocultured cells. Embryo viability is also likely to be retained by optimization of the ionic, energy, and amino acid components of culture medium and by the prevention of cytotoxic and cytolytic processes, such as the accumulation of superoxide radicals that can damage membranes through lipid peroxidation, ammonium that is a product of amino acid metabolism, and excessive lactate formed as a product of energy metabolism.

Pregnancy in IVF is related to embryo quality and the capacity to select embryos for transfer (Table 1.5). Hence, the establishment of pregnancy involves both a viable embryo(s) and a receptive uterus. It has been calculated that both embryo viability and uterine receptivity are major contributors to successful IVF and that uterine receptivity may be higher

TABLE 1.5. Statistically significant embryo factors associated with pregnancy in IVF.

Factor	Nonpregnant patients	Pregnant patients
Mean number of embryos/collection		
IVF	3.2	5.2
TEST	2.2	5.0
Percentage of embryos transferred	42%	28%
Percentage of embryos frozen		
(of those remaining)	37%	48%
Percentage of embryos discarded	63%	52%
Individual embryos[a]		
Good quality (>7.5)	27%	37%
Poor quality (<3.5)	10%	3%
Average of all embryos recovered[a]		
Good quality (>6.5)	44%	69%
Fair quality (4.5–6.0)	33%	25%
Poor quality (<4.0)	23%	6%

[a] Embryo quality—1 (very poor) to 10 (excellent)—determined by cleavage rate, blastomere symmetry, clarity or granularity of cytoplasm, fragmentation, and general appearance.
Source: Data from the Infertility Medical Centre, Melbourne.

in IVF programs that use hMG rather than clomiphene citrate and hMG (35). Recent calculations based on the incidence of twin ovulations in natural cycles and dizygotic twin births (36) indicate that spontaneous pregnancy is determined solely by embryo viability, with very little if any contribution of uterine receptivity. These observations suggest that any progress toward optimizing IVF needs to include strategies to retain full developmental capacity of oocytes and embryos and superovulation techniques that minimize any adverse effects on uterine receptivity. It has also been frequently reported that human oocytes and embryos derived from superovulation have an unusually high rate of chromosomal abnormalities (in excess of 25% in most reports). We have recently shown (37) that a similar high rate of chromosomal abnormalities occurs in oocytes of spontaneously ovulating women (Table 1.6). This high baseline of chromosomal abnormalities will limit the success of GIFT and IVF, and their identification through embryo biopsy techniques may be required in order to eventually optimize IVF success rates.

Oocyte Donation and Uterine Receptivity with Increasing Maternal Age

It is possible to achieve high success rates following transfer of oocytes or embryos to agonadal women treated with estrogen and progesterone in a sequential regimen to mimic the normal menstrual cycle (38). The

TABLE 1.6. Chromosomal aneuploidy in human oocytes from superovulated and natural cycle donors.

Treatment	Number of oocytes	Number of aneuploids (%)	
Superovulated cycles	68	23 (34%)	NS
Natural cycles	20	4 (20%)	

NS indicates not significantly different.
Source: Data from Gras, McBain, Trounson, and Kola (37).

TABLE 1.7. Human oocyte donation using variable-length estrogen replacement and simplified fixed doses of estrogen and progesterone.

Recipients	Anonymous IVF donor	Known donor	Total
Number of embryo transfers	62	12	74
Number of embryos transferred	114	37	181
Mean number of embryos transferred	2.3	3.1	2.4
Number of pregnancies	14	6	20
Pregnancy rate/transfer (%)	23	50	27

Source: Data from the Infertility Medical Centre, Melbourne, 1988–89. Reprinted with permission from Trounson (39).

technique, known as *oocyte donation*, enables women with ovarian dysgenesis, premature menopause, and absent or nonfunctional ovaries to establish their own pregnancy and maintain the pregnancy to term by continuing ovarian steroid replacement for 6–8 weeks until the time of the luteoplacental shift (39). This technique has been used widely to treat patients with these problems (39–44). The hormonal replacement regimen has been simplified by using a variable-length and constant-dose schedule for follicular phase estrogen and luteal phase progesterone (21, 45), enabling flexibility in the timing of embryo transfer and synchronization of the donor and recipient. The hormonal replacement regimen we presently use is summarized in Figure 1.3, and data comparing original sequential regimen with the constant-dose and variable-follicular phase regimen are shown in Table 1.7.

Using a standardized sequential hormonal replacement regimen in oocyte or embryo recipients, it is possible to direct the effects of superovulation on uterine receptivity and the contributions of the oocyte or embryo and uterine receptivity to reduced fertility and fecundity with increasing age. For example, de Ziegler and Frydman (46) showed that pregnancy rates were much higher in women with artificial menstrual cycles receiving cryopreserved embryos from fertile donors than in patients receiving embryos from regular IVF cycles. This suggests that the fertile donors have better-quality oocytes. In a discussion of these observations, Check (1991) reported that for embryos derived from the same source (regular IVF patients), pregnancy rates were always higher

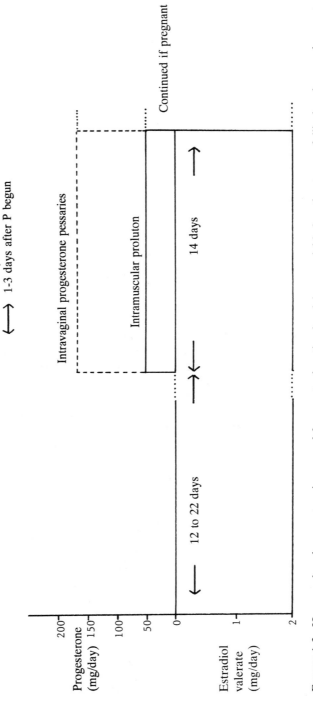

FIGURE 1.3. Hormonal replacement regimen used for oocyte donation involving a variable-length estrogen follicular phase and a constant dose of progesterone and estrogen for the artificial luteal phase. Reprinted with permission from Trounson (39).

when the embryos were transferred to recipients given hormonal replacement therapy than to the original IVF patients (13 of 257 [5%] embryos developed to the third trimester in recipients vs. 5 of 248 [2%] donor embryos transferred; P = 0.02). This data indicates that uterine receptivity is higher in patients with artificial menstrual cycles than with superovulated cycles. The combination of improved oocyte quality from fertile donors and increased uterine receptivity in recipients with artificial cycles would explain the high success rates of oocyte donation (41, 48).

In an interesting study published by Sauer et al. (49), pregnancy rates were similar in patients under and over the age of 40 years who received transferred embryos from younger donors, suggesting that decreased oocyte quality rather than decreased uterine receptivity governs reduced fertility in older women. However, Levran et al. (50) argue from a retrospective analysis of oocyte donation data that pregnancies were more frequent in younger recipients, but that pregnancy rate was not related to the age of the donor, although spontaneous abortion was significantly increased with the older donors. The difference in interpretation of the two sets of data may relate to the age comparisons made. For example, Levran et al. (50) compared patients younger than 33 to those older than 32 years, and Sauer et al. (49) compared patients younger and older than 40 years. It is likely that both oocyte quality and uterine receptivity decline with increasing age, but the relative smaller contribution of decreasing uterine receptivity enables older women to become pregnant at quite high rates when they are given hormonal replacement therapy and transferred embryos derived from oocytes of younger women. This effectively removes the barrier of age for conception, a fact that is beginning to test community attitudes again about the use of the reproductive technologies in patients older than the normal age for menopause.

Male Factor Infertility and Micromanipulation

For patients with suspected male factor infertility, it has been considered that *zygote intrafallopian transfer* (ZIFT) or TEST would be preferable to GIFT or IVF. GIFT results are usually higher than IVF results, but specific problems of fertilization are not identified in GIFT. ZIFT or TEST enables assessment of fertilization, and transfer of the zygotes or embryos to the fallopian tubes may benefit pregnancy success rates. In fact, it is apparent in our own experience that GIFT is always preferable to TEST for the treatment of male infertility.

Calderon et al. (51) reported that a retrospective analysis of the treatment of 371 couples with male factor infertility showed that the pregnancy rate for GIFT was 25% compared with 13% for TEST. Using a randomized prospective control design, Calderon et al. (52) allocated

100 male factor patients to GIFT or TEST. There was no significant difference in the mean age, number of oocytes recovered, or semen parameters between the two groups, but 3.1 oocytes were transferred to the GIFT cases and 2.35 embryos were transferred to the TEST cases. Pregnancy rate per transfer was 25% for GIFT patients and 7% for TEST patients. In addition, fertilization failed completely in 36% of the TEST patients, resulting in no embryos for transfer. These data show that GIFT is the preferred initial option for male factor patients providing they have sufficient motile sperm for transfer with oocytes to the fallopian tubes. Furthermore, it is possible that there may be some advantage for fertilization for male factor patients if their sperm are in the tubal environment with oocytes rather than in vitro. This hypothesis is based on the frequent occurrence of pregnancy in GIFT patients, but failure of sperm to fertilize supernumary oocytes in the same cycle of treatment. Further exploration of this hypothesis is warranted to improve fertilization in vitro for male factor patients whose partners have tubal occlusion.

Insemination of Oocytes in Microvolumes

In severe cases of oligoasthenospermia, there may be insufficient motile sperm that can be harvested for transfer to the fallopian tubes in GIFT or even conventional IVF. Many systems can be used to micronize the IVF procedure, but in our own studies, we use 50-µL drops of medium containing 0.3 to 1.15×10^6 sperm/mL. In a randomized control trial of microdrop insemination (53) and sperm microinjection, involving the microinjection of 5–10 sperm into the perivitelline space of oocytes (54), we showed that microdrop insemination should be used as the first option. Thirteen severely oligoasthenospermic patients were randomly allocated to microdrop insemination, and an equal number allocated to microinjection. In the microdrop insemination 45 of 97 (46%) oocytes were fertilized and another 4% were polypronuclear compared with 32 of 116 (28%) oocytes fertilized in the microinjection patients (another 7% were polypronuclear). These data support the approach that, if possible, microdrop insemination should precede micromanipulation techniques in the severely infertile male.

Micromanipulation Techniques Used to Aid Fertilization

There has been a range of techniques proposed to aid sperm in bypassing the zona pellucida, which is the major barrier preventing sperm access to the oocyte. The zona is a relatively stiff glycoprotein coat that requires hyperactive sperm motility, the acrosome reaction, and release of zona digestins from the acrosomal cap in order for sperm to penetrate. The zona may be perforated—using finely drawn glass needles in a technique known as *partial zona opening* (PZD)—and the oocytes inseminated (55,

56), thus increasing the chance of fertilization. The mechanical method of zona perforation is preferable to acid digestion of a hole using the technique known as *zona drilling* because this latter method appears to increase fragmentation, granulation of cytoplasm, and uneven cleavage (57). One of the major problems with PZD is the high rate of polyspermic fertilization.

Microinjection of sperm into the perivitelline space (58) has also been developed to aid the severely infertile male. The karyotypes of pronuclear oocytes formed after microinjection of sperm from severely infertile men under the zona pellucida were similar to those of pronuclear oocytes formed by IVF using sperm from IVF couples. Pregnancies have been initiated using this technique to inject 1–10 sperm into the perivitelline space (54, 59, 60). These studies have been performed for male factor patients who are repeat failures at IVF or for men with so few motile sperm that microdrop insemination is not possible. Our own clinical data for 1991 (61) is shown in Table 1.8.

Fertilization rates in the microinjection technique are low (20%–30%) despite the injection of multiple sperm and the stimulation of sperm motility with 2-deoxyadenosine and pentoxyfylline. Further research is required to increase the fertilizing capacity of sperm microinjected under the zona pellucida and to limit the occurrence of polyspermic fertilization when multiple sperm are used. A comparison of PZD and microinjection of sperm (1–12) under the zona showed significantly higher fertilization with microinjection (30%) than with PZD (13%) (62). At the present time there is widespread use of the subzonal sperm microinjection technique for the treatment of severe male infertility. Microinjected oocytes can be transferred immediately to the fallopian tubes (MIFT), and this appears to be a highly effective procedure, resulting in pregnancy rates in excess of 25% in our initial clinical trials.

TABLE 1.8. Summary of subzonal sperm microinjection clinical studies, the Infertility Medical Centre, Melbourne, 1991.

Number of cycles of treatment (number of patients)[a]	111 (83)
Number of mature oocytes microinjected	910
Number of monospermic fertilized oocytes	187 (21%)
Number of polyspermic fertilized oocytes	61 (7%)
Total fertilized oocytes	248 (27%)
Number of monospermic fertilized oocytes that cleaved normally	164 (88%)
Number of embryos transferred (number frozen)	142 (10)
Number of cycles resulting in embryo transfer	66 (60%)
Number of preclinical pregnancies	1 (1.5%)
Number of clinical pregnancies	7 (11%)
Total pregnancies	8 (12%)

[a] Patients who had previously failed conventional IVF on at least 2 occasions or those with too few sperm for insemination of oocytes in microdroplets.
Source: Data from Trounson and Sathananthan (61).

It is also possible to microinject sperm (usually frozen-thawed in culture medium) directly into the cytoplasm of oocytes (63). There is an increased risk of damage to oocytes because the plasma membrane has to be perforated by the injection pipette, and one report exists of increased chromosomal abnormalities, including multiple breaks and rearrangements using this technique (64). Pregnancies have been established by direct cytoplasmic injection of oocytes (65). It is important that the loss of oocytes and any increased abnormalities are outweighed by the increased number of fertilized and normally developing embryos to warrant adoption of the technique. Our own studies indicate that this is achievable.

Integration of Treatment Options for Male Factor Infertility

It is important to obtain a complete semen profile from anyone considered to be a male factor patient in order to determine the best options for his treatment (Fig. 1.4). While a range of tests exists for determining the semen profile and possible fertilizing capacity of the patient's sperm, those factors that are crucial include the proportion of normal sperm forms, sperm concentration and number in the ejaculate, and sperm motility. The presence of sperm antibodies, the capacity of sperm to respond to motility stimulants, and the capacity to acrosome-react in vitro can also be useful.

Patients should be treated by GIFT or IVF depending on whether their partners have patent tubes or not. At least 3 cycles of GIFT treatment should be undertaken to enable a reasonable chance of pregnancy. In cases of repeated failure of fertilization with IVF, including supernumary oocytes in GIFT, subzonal microinjection or PZD may be necessary. Patients who have very few motile sperm should be treated by subzonal microinjection of sperm or direct cytoplasmic sperm microinjection. Microdrop insemination would be the first option for patients with low numbers of normal motile sperm, proceeding to subzonal microinjection or PZD if fertilization does not occur in the microdrops. Using this series of options, including cytoplasmic injection for patients with live but poorly motile or immotile sperm, the chance of resolving male factor infertility is high. However, it is essential that the number of oocytes obtained from their partners is maximized to improve the probability of transfer of 3 fertilized oocytes or embryos.

Cryopreservation of Embryos and Oocytes

Fertilized pronuclear oocytes and cleavage-stage embryos can be successfully cryopreserved using slow cooling techniques involving the cryoprotectants 1,2–propanediol and dimethyl sulfoxide (66–68). These

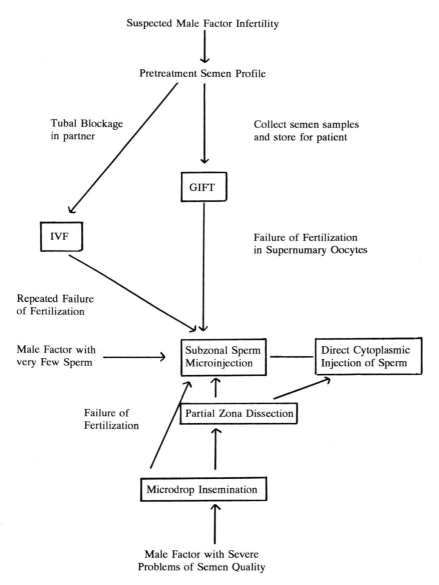

FIGURE 1.4. Treatment options for male factor patients.

techniques are known as *equilibrium freezing* (69), where under slow cooling conditions intracellular water is drawn out of the cell and freezes in the extracellular domain as large, rounded ice crystals. The technique requires sufficient cryoprotectant (typically 1.5 M) to protect the cell from lethal intracellular ice formation and sufficiently slow cooling rates (typically 0.3°C/min) during the phase of ice formation. The solution containing embryos is ice nucleated around −7°C in order to prevent the rapid cooling rebound associated with release of the heat of crystallization that occurs during spontaneous ice formation in supercooled solutions. Slow cooling is continued until the transition temperature is reached (−30°C for propanediol and −36°C for dimethyl sulfoxide), which is when rapid cooling does not cause lethal intracellular ice formation in the dehydrated cells. Under these conditions it can be expected that 70% or more of human embryos will be recovered after thawing without serious cryoinjury, and pregnancy rates for the transfer of 1–3 cryopreserved embryos should be around 20% (68).

More recently, *rapid freezing*, or *vitrification*, techniques have been introduced to cryopreserve embryos (70, 71). These techniques require that the concentration of cryoprotectant solutes be increased to higher levels than those used for slow cooling. If the concentrations exceed 45% (v/v), the solution will vitrify when rapidly cooled, forming a glass rather than ice. Vitrification does not involve any structural rearrangements of the liquid and should avoid cryoinjury by ice crystals and osmotic effects of concentrated ions in the solution. However, at high concentrations, cryoprotectants can be toxic to cells, particularly at room temperature or above. Embryos are usually exposed to the high concentrations of cryoprotectants for brief periods at room temperature or 0°C before rapid cooling by the direct transfer to liquid nitrogen. Using 3.0–4.5 M DMSO as the cryoprotectant and 0.25 M sucrose as an extracellular solute for partial dehydration of cells, very high survival rates can be achieved for ultrarapid freezing of mouse embryos with no significant effect on their viability (72, 73). The concentrated solution effectively dehydrates the cells during the brief exposure (typically 3 min), resulting in high intracellular concentrations of cryoprotectant in the shrunken cells. When rapidly cooled, the solution vitrifies, and when warmed, small innocuous ice crystals will form that do not affect the cells. The application of these techniques to cryopreservation of human embryos has resulted in high survival rates of embryos (74) and pregnancies and births after transfer to patients (67, 75, 76). It is likely that these simple and economical methods will be gradually modified and adopted for clinical use in the future.

Unfertilized oocytes may also be frozen by slow cooling techniques providing embryos are equilibrated with the cryoprotectant (usually dimethyl sulfoxide) at low temperatures (0°C–4°C) (77–80). However, the mature unfertilized oocyte is subject to changes induced by cryoprotectants and cooling. These include the disassembly of the meiotic metaphase II spindle microtubules, dispersal of polar pericentriolar material, hardening of the zona pellucida that reduces the oocyte's capacity to be fertilized, and the formation of micronuclei containing chromatin dispersed from the metaphase plate. These problems and the very low rates of producing viable embryos in the human from cryopreserved oocytes have discouraged clinical use of the technique (81). Further research is required to establish a safe and efficient method for the cryopreservation of unfertilized human oocytes.

Preconception and Preimplantation Embryo Genetic Diagnosis

Micromanipulation techniques can be used to sample the cells of developing preimplantation embryos or the polar body of the mature unfertilized oocyte without seriously affecting their developmental potential (82). It is possible to identify homozygous and heterozygous polar bodies from unfertilized oocytes using DNA primers that recognize a familial genetic disorder and the normal DNA sequence. Diagnosis of the autosomal recessive disease cystic fibrosis has been reported by Verlinsky et al. (83), and the format for the procedure is shown in Figure 1.5 (A is the abnormal gene sequence and B is the normal sequence).

The particular problem with polar body biopsy is that both defective alleles need to be identified in the polar body and both need to be amplified by the *polymerase chain reaction* (PCR) to make a certain diagnosis. The possibility of error may be unacceptably high when there is only the one polar body available for analysis because of amplification failure, contamination, and cell transfer errors (84). If the polar body is diagnosed as homozygous for the normal allele or heterozygous, a further sampling would be required from the embryo that developed in order to confirm that it was not homozygous for the defective allele after fertilization and extrusion of the second polar body (Fig. 1.5). If the polar body contains both defective alleles, the conceptus would be normal or a recessive carrier if fertilized by a sperm with the defective allele. While preconception genetic analysis by polar body biopsy is a very interesting approach, it seems likely to be superseded by the direct biopsy of embryos, where several blastomeres may be aspirated and the results confirmed in replicate samples.

The identification of embryo sex in patients with sex-linked recessive genetic disease has been reported by Handyside et al. (85, 86) and Handyside and Delhanty (87). Blastomeres were biopsied from 4- to 8-cell embryos, and a Y-specific repeat sequence was amplified by PCR to identify male embryos. The authors reported a high rate of implantation of sexed embryos that were transferred, and they confirmed the correct diagnosis of females in 6 of 7 fetuses. Despite the large numbers of repeat sequences targeted for amplification, the identification of sex in individual male cells has a PCR amplification failure of about 15% (87). Similarly, in the studies on polar body biopsy and diagnosis of cystic fibrosis, there was an 18% failure of amplification (83). Co-amplification of an X-repeat

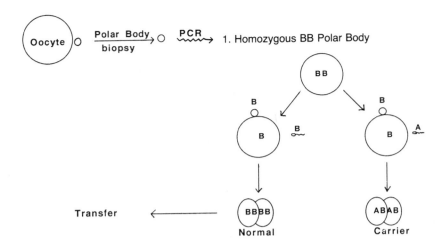
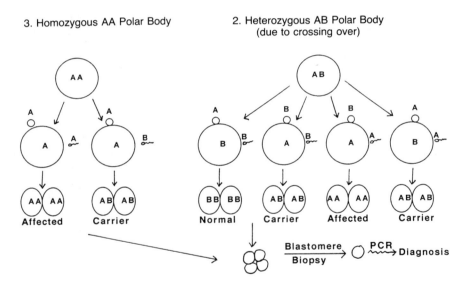

FIGURE 1.5. Preconception genetic diagnosis of the abnormal recessive gene A using polar body biopsy combined when necessary with preimplantation embryo biopsy to ensure that the homozygous AA disorder is avoided in pregnancies at risk for this type of genetic disease.

sequence did not apparently reduce amplification failure, indicating independence of efficiency of the two probes used (87).

Fluorescent in situ hybridization (FISH) is also being explored for the identification of chromosomes. Chromosomes can be identified by FISH in interphase nuclei (88). Using dual FISH to X- and Y-specific probes with different fluorochromes, Griffin et al. (89) reported the correct diagnosis of embryo sex in 89% of metaphase nuclei and 72% of interphase nuclei. Using a minimum of 3–4 cells would give a very accurate diagnosis of sex in preimplantation embryos. This technology is likely to advance rapidly and will provide an option for couples with recessive genetic disease as well as for those at risk of chromosomal abnormalities to the present prenatal diagnostic techniques and consequent therapeutic abortion.

Conclusions

Assisted reproductive technologies will continue to develop and new methodologies will proliferate in order to treat human infertility and identify genetic disease prior to the initiation of pregnancy. The prospects of using the new generation of GnRH antagonists and recombinant gonadotropins may continue the simplification of superovulation procedures and further assist patients who are not responsive to the fertility drugs presently available. While it is too soon to assess the real benefits of growth hormone, it is possible that an improved superovulatory response will be achieved in some patients by the co-administration of growth hormone and gonadotropins. Further research is needed on the recovery of immature oocytes from unstimulated ovaries of patients who need to avoid the use of fertility drugs, but this is likely to be an option for patients in the near future. These developments will provide a wide spectrum of treatment options for the recovery of oocytes for IVF or GIFT.

The capacity to achieve fertilization with severe male factor infertility cases using improved insemination techniques and micromanipulation has been a major advance, and the success rates are likely to improve further in the near future. Research is presently identifying those molecules involved in sperm and oocyte plasma membrane interactions (e.g., the RGD adhesion peptide) (90) and the requirements for decondensation and pronuclear formation of sperm microinjected directly into the ooplasm. It is apparent that embryologists involved in human IVF require micromanipulation skills in order to provide the techniques necessary for present and future IVF. The apparent success of MIFT, where microinjected oocytes are transferred to the fallopian tubes, may further advance our capacity to treat the severely infertile male and our knowledge about the natural requirements for developing the full fertilizing capacity of sperm.

It is apparent that the early cleavage-stage human embryo has very specific nutritional requirements and that embryo viability may be determined by the metabolic capacity of embryos as determined by ultrafluorametric techniques. By protecting the embryos from deleterious substances in culture and by providing the nutritional requirements for optimum metabolism, embryos should be cultured at high rates to advanced preimplantation stages. This will increase the selection advantage for implantation and again reduce the number of embryos required for transfer and the incidence of multiple fetuses and births. It is very likely that the maximum number of embryos transferred will decrease to two.

The capacity to maintain embryo viability in culture will further aid the possibility of detecting chromosomal, genetic, and developmental abnormalities. Analysis of a number of cells from the preimplantation embryo will improve the accuracy of diagnosing genetic and chromosomal errors and, with increasing embryo viability, make these techniques an attractive option to the more conventional methods of prenatal diagnosis. It will also be possible to reduce the incidence of chemical pregnancies that fail to continue development and to reduce the incidence of miscarriage in IVF.

While it may be possible to develop embryonic stem cells in the human, it is unlikely that they will be used to correct developmental or genetic abnormalities. Likewise, the production of genetically identical embryos by microdissection or blastomere fusion into enucleated oocytes, while possible, will not be used in human medicine.

References

1. Edwards RG, Steptoe PC, Purdy JM. Establishing full-term human pregnancies using cleaving embryos grown in vitro. Br J Obstet Gynaecol 1980;87:737–56.
2. Steptoe PC, Edwards RG, Purdy JM. Clinical aspects of pregnancies established with cleaving embryos grown in vitro. Br J Obstet Gynaecol 1980;87:757–68.
3. Lopata A, Johnston IWH, Hoult IJ, Speirs AI. Pregnancy following intrauterine implantation of an embryo obtained by in vitro fertilization of a preovulatory egg. Fertil Steril 1980;33:117–20.
4. Trounson AO, Leeton JF, Wood C, Webb J, Wood J. Pregnancies in humans by fertilization in vitro and embryo transfer in the controlled ovulatory cycle. Science 1981;212:681–2.
5. Trounson AO, Mohr LR, Wood C, Leeton JF. Effect of delayed insemination on in vitro fertilization, culture and transfer of human embryos. J Reprod Fertil 1982;64:285–94.
6. Cameron IT, O'Shea FC, Rolland JM, Hughes EG, deKretser DM, Healy DL. Occult ovarian failure:a syndrome of infertility, regular menses and elevated follicle-stimulating hormone concentrations. J Clin Endocrinol Metab 1988;67(6):1190–4.
7. McIntosh JEA, Matthews CD, Crocker JM, Broom TJ, Cox LW. Predicting the luteinizing hormone surge: relationship between the duration of the follicular and the luteal phases and the length of the human menstrual cycle. Fertil Steril 1980;34:125–30.

8. Foulot H, Ranoux C, Dubuisson J-B, Rambaud D, Aubriot F-X, Poirot C. In vitro fertilization without ovarian stimulation: a simplified protocol applied in 80 cycles. Fertil Steril 1989;52:617–21.
9. Trounson A, Hammarberg K. The benefits of natural cycle IVF. Healthright 1989;8:22–5.
10. McBain JC, Trounson A. Patient management-treatment cycle. In: Wood C, Trounson A, eds. Clinical in vitro fertilization. Berlin: Springer-Verlag, 1984:49–65.
11. Trounson AO, Calabrese R. Changes in plasma progesterone concentrations around the time of the luteinizing hormone surge in women superovulated for in vitro fertilization. J Clin Endocrinol Metab 1984;59:1075–80.
12. Jones HW. In vitro fertilization. In: Behrman SJ, Kistner RW, Patton GW, eds. Progress in infertility. 3rd ed. Boston: Little Brown, 1988:534–61.
13. MacLachlan V, Besanko M, O'Shea F, Wade H, Wood C, Trounson A, Healy DL. A controlled study of luteinizing hormone-releasing hormone agonist (Buserelin) for the induction of folliculogenesis before in vitro fertilization. N Engl J Med 1989;320:1233–7.
14. MacLachlan V, Besanko M, Wade H, Morrow L, O'Shea F, Trounson A, Healy D. Luteinizing-hormone-releasing hormone agonist treatment in patients with previously failed folliculogenesis during in vitro fertilization therapy. In: Jones HW, Schrader C, eds. Fertilization and other assisted reproduction. Ann N Y Acad Sci 1988;541:60–74.
15. Polson DW, MacLachlan V, Krapez JA, Wood C, Healy DL. A controlled study of gonadotrophin-releasing hormone agonist (Buserelin acetate) for folliculargenesis in routine in vitro fertilization patients. Fertil Steril 1991;56:509–14.
16. Winslow KL, Oehninger SC, Toner JP, Acosta AA, Brzyski RG, Muasher SJ. The gonadotrophin-releasing hormone agonist stimulation test—a sensitive predictor of performance in the flare-up in vitro fertilization cycle. Fertil Steril 1991;56:711–7.
17. Dirnfeld M, Gonen Y, Lissak A, et al. A randomised prospective study on the effect of short and long Buserelin treatment in women with repeated unsuccessful in vitro fertilization (IVF) cycles due to inadequate ovarian response. J In Vitro Fertil Embryo Transfer 1991;8:339–43.
18. Pellicer A, Lightman A, Diamond MP, Russell JB, DeCherney AH. Outcome of in vitro fertilization in women with low response to ovarian stimulation. Fertil Steril 1987;47:812–5.
19. Owen EJ, West C, Mason BA, Jacobs HS. Co-treatment with growth hormone of sub-optimal responders in IVF-ET. Hum Reprod 1991;6:524–8.
20. Forman RG, Demouzon J, Feinstein MC, Testart J, Frydman R. Studies on the influence of gonadotrophin levels in the early follicular phase on the ovarian response to stimulation. Hum Reprod 1991;6:113–7.
21. Leeton J, Rogers P, Cameron I, Caro C, Healy D. Pregnancy results following embryo transfer in women receiving low-dosage variable-length estrogen replacement therapy for premature ovarian failure. J In Vitro Fertil Embryo Transfer 1989;6:232–5.
22. Trounson A. The production of ruminant embryos in vitro. Anim Reprod Sci 1992.
23. Cha KY, Koo JJ, Ko JJ, Choi DH, Han SY, Yoon TK. Pregnancy after in vitro fertilization of human follicular oocytes collected from non-stimulated

cycles, their culture in vitro and their transfer in a donor oocyte program. Fertil Steril 1991;55:109–13.
24. Osborn JC, Trounson A. Unpublished data.
25. Osborn JC. Oocyte retrieval and maturation. In:Trounson AO, Gardner DG, eds. Handbook of in vitro fertilization. CRC Press, 1992.
26. Bollen N, Tournaye H, Camus M, Devroey P, Staessen C, Van Steirteghem AC. The incidence of multiple pregnancy after in vitro fertilization and embryo transfer, gamete, or zygote intrafallopian transfer. Fertil Steril 1991;55:314–8.
27. Toth TL, Brzyski RG, Oehninger S, Acosta AA, Toner JP, Muasher SJ. Embryo transfer to the uterus or the fallopian tube after in vitro fertilization yields similar results. Fertil Steril 1992;57:1110–3.
28. Balmaceda JP, Ord T, Alam V, Snell K, Roszjtein D, Asch RH. Embryo implantation rates in oocyte donation: a prospective comparison of tubal versus uterine transfers. Fertil Steril 1992;57:362–5.
29. Quinn P, Kerin JF, Warnes GM. Improved pregnancy rate in human in vitro fertilization with the use of a medium based on the composition of human tubal fluid. Fertil Steril 1985;44:493–8.
30. Gianaroli L, Seracchioli R, Ferraretti AP, Trounson A, Flamigni C, Bovicelli L. The successful use of human amniotic fluid for mouse embryo culture and human in vitro fertilization, embryo culture and embryo transfer. Fertil Steril 1986;46:907–13.
31. Cummins JM, Breen TM, Fuller SM, et al. Comparison of two media in a human in vitro fertilization program: lack of significant difference in pregnancy rate. J In Vitro Fertil Embryo Transfer 1986;3:326–9.
32. Gianaroli L, Trounson A, King C, Farraretti A, Chiappazzo L, Bafaro G. Human amniotic fluid for fertilization and culture of human embryos: results of clinical trials in human IVF programs. J In Vitro Fertil Embryo Transfer 1989;6:213–7.
33. Bongso A, Ng SC, Fong CY, Ratnam SS. Co-cultures: a new lead in embryo quality improvement for assisted reproduction. Fertil Steril 1991;56:179–91.
34. Menezo Y, Guerin J, Czyba J. Improvement of human embryo development in vitro by coculture on monolayers of Vero cells. Biol Reprod 1990;43:301–6.
35. Rogers PAW, Milne B, Trounson AO. A model to show uterine receptivity and embryo viability following ovarian stimulation for in vitro fertilization. J In Vitro Fertil Embryo Transfer 1986;3:93–8.
36. Leeton J, Shaw G, Short RV. Twin ovulations, dizygotic twins and human fecundability. Br J Obstet Gynaecol 1992.
37. Gras L, McBain J, Trounson A, Kola I. The incidence of chromosomal aneuploidies in stimulated and unstimulated (natural) uninseminated human oocytes. Hum Reprod 1992.
38. Lutjen PJ, Trounson A, Leeton J, Findlay J, Wood C, Renou P. The establishment and maintenance of pregnancy using in vitro fertilization and embryo donation with primary ovarian failure. Nature 1984;307:174–5.
39. Trounson A. The development of the technique of oocyte donation and hormonal replacement therapy: is oestrogen really necessary for the establishment and maintenance of pregnancy. Reprod Fertil Dev 1992;4.

40. Navot D, Laufer N, Kopolovic J, et al. Artificially induced endometrial cycles and establishment of pregnancies in the absence of ovaries. N Engl J Med 1986;314:806–18.
41. Rosenwaks Z. Donor eggs: their application in modern reproductive technologies. Fertil Steril 1987;47:895–909.
42. Formigli L, Formigli G, Scelsi R, Roccio C, Belotti G, Noe C. Transfer of preembryos donated and fertilized in vivo to recipients given a steroid replacement therapy. Hum Reprod 1988;3:741–6.
43. Sauer MV, Paulson RJ, Macaso TM, Francis MM, Lobo RA. Oocyte and preembryo donation to women with ovarian failure: an extended clinical trial. Fertil Steril 1991;55:39–43.
44. Navot D, Scott, RT, Droesch K, Veeck LL, Liu H-C, Rosenwaks Z. The window of embryo transfer and the efficiency of human conception in vitro. Fertil Steril 1991;55:114–8.
45. Serhal P, Craft I. Simplified treatment for oocyte donation. Lancet 1987;1:687–8.
46. de Ziegler D, Frydman R. Different implantation rates after transfers of cryopreserved embryos originating from donated oocytes or from regular in vitro fertilization. Fertil Steril 1990;54:682–8.
47. Check JH. Uterine receptivity in subjects with ovarian failure. Letter to the editor. Fertil Steril 1991;55:1208–9.
48. Sauer MV, Paulson RJ, Macaso TM, Francis MM, Lobo RA. Oocyte and preembryo donation to women with ovarian failure: an extended clinical trial. Fertil Steril 1991;55:39–43.
49. Sauer MV, Paulson RJ, Lobo RA. A preliminary report on oocyte donation extending reproductive potential to women over 40. N Engl J Med 1990;323:1157–60.
50. Levran D, Ben-Shlomo I, Dor J, Ben-Rafael Z, Nebel L, Mashiach S. Aging of endometrium and oocytes: observations on conception and abortion rates in an egg donation model. Fertil Steril 1991;56:1091–4.
51. Calderon I, Kovacs G, Cushnahan L, Osborne J. GIFT vs TEST for male infertility [Abstract]. Proc 7th World Congress in IVF and Assisted Procreations, Paris, 1991.
52. Calderon I, Fuscaldo G, Yates C, Azuma K, Osborne J, Wood C. GIFT versus TEST for male infertility, a prospective randomized trial [Abstract]. Proc Am Fertil Soc 48th annu meet, New Orleans, Oct 1992.
53. Sobieszczuk D, Fuscaldo G, Poulos C, McLachlan R, Trounson A. Microdrop insemination vs microinjection/insemination in severe male factor patients [Abstract]. Proc Am Fertil Soc 48th annu meet, New Orleans, Oct 1992.
54. Sakkas D, Lacham O, Gianaroli L, Trounson A. Subzonal sperm microinjection in cases of severe male factor infertility and repeated in vitro fertilization failure. Fertil Steril 1992;57:1279–88.
55. Malter HE, Cohen J. Partial zona dissection of the human oocyte:a nontraumatic method using micromanipulation to assist zona pellucida penetration. Fertil Steril 1989;51:139–48.
56. Cohen J, Malter H, Wright G, Kort H, Massey J, Mitchell D. Partial zona dissection of human oocytes when failure of zona pellucida penetration is anticipated. Hum Reprod 1989;4:435–42.

57. Garrisi GJ, Sapira V, Talansky BE, Navot D, Grunfeld L, Gordon JW. Clinical evaluation of three approaches to micromanipulation-assisted fertilization. Fertil Steril 1990;54:671–7.
58. Laws-King A, Trounson A, Sathananthan AH, Kola I. Fertilization of human oocytes by microinjection of a single spermatozoon under the zona pellucida. Fertil Steril 1987;48:637–42.
59. Ng SC, Bongso A, Sathananthan AH, et al. Pregnancy after transfer of multiple sperm under the zona. Lancet 1988;2:790.
60. Fishel S, Johnson J, Jackson R, Gross S, Antimori S, Versaci C. Subzonal insemination for the alleviation of infertility. Fertil Steril 1990;54:828–35.
61. Trounson A, Sathananthan AH. Fertilization using micromanipulation techniques. In: Trounson AO, Gardner DG, eds. Handbook of in vitro fertilization. CRC Press, 1992.
62. Cohen J, Talansky BE, Malter HM, et al. Microsurgical fertilization and teratozoospermia. Hum Reprod 1991;6:118–23.
63. Lazendorf SE, Slussor MS, Maloney MK, Hodgen GD, Veek LL, Rosenwaks Z. A pre-clinical evaluation of pronuclear formation by microinjection of human spermatozoa into human oocytes. Fertil Steril 1988;49:835–42.
64. Martin RH, Ko E, Rademaker A. Human sperm chromosome complements after microinjection of hamster eggs. J Reprod Fertil 1988;84:179–86.
65. Van Steirtegham A. Unpublished data.
66. Trounson A. Preservation of human eggs and embryos. Fertil Steril 1986;46:1–12.
67. Trounson AO. Cryopreservation. Br Med Bull 1990;46:695–708.
68. Trounson AO, Shaw JM. Embryo cryopreservation. Reprod Med Rev 1992.
69. Mazur P. Equilibrium, quasi-equilibrium, and non-equilibrium freezing of mammalian embryos. Cell Biophys 1990;17:53–92.
70. Rall WF. Factors affecting the survival of mouse embryos cryopreserved by vitrification. Cryobiology 1987;24:387–402.
71. Trounson A, Peura A, Kirby C. Ultrarapid freezing: a new low-cost and effective method of cryopreservation. Fertil Steril 1987;48:822–6.
72. Shaw JM, Trounson AO. Effect of dimethyl sulfoxide and protein concentration on the viability of two-cell mouse embryos frozen with a rapid freezing technique. Cryobiology 1989;26:413–21.
73. Shaw JM, Diotallevi L, Trounson AO. A simple rapid 4.5M dimethyl-sulfoxide freezing technique for the cryopreservation of one-cell to blastocyst stage preimplantation mouse embryos. Reprod Fertil Dev 1991;3:621–6.
74. Trounson AO, Sjoblom P. Cleavage and development of human embryos in vitro after ultrarapid freezing and thawing. Fertil Steril 1988;50:373–6.
75. Gordts S, Roziers P, Campo R, Noto V. Survival and pregnancy outcome after ultrarapid freezing of human embryos. Fertil Steril 1990;53:469–72.
76. Barg PE, Barad DH, Feichtinger W. Ultrarapid freezing (URF) of mouse and human preembryos:a modified approach. J In Vitro Fertil Embryo Transfer 1990;7:355–7.
77. Whittingham DG. Fertilization in vitro and development to term of unfertilized mouse oocytes previously stored at $-196°C$. J Reprod Fertil 1977;49:89–94.
78. Glenister PH, Wood MJ, Kirby C, Whittingham DG. The incidence of chromosome anomalies in first-cleavage mouse embryos obtained from frozen-thawed oocytes fertilized in vitro. Gamete Res 1987;16:205–16.

79. Trounson A, Kirby C. Problems in the cryopreservation of unfertilized eggs by slow cooling in dimethyl sulfoxide. Fertil Steril 1989;52:778–86.
80. Schroeder AC, Champlin AK, Mobrnaten LE, Eppig JJ. Developmental capacity of mouse oocytes cryopreserved before and after maturation in vitro. J Reprod Fertil 1990;89:43–50.
81. Trounson A, Shaw J. The cryopreservation of human eggs and embryos. In: Wallach E, ed. Reproductive medicine and surgery, 1992.
82. Tarin JJ, Handyside AH. Embryo biopsy strategies for preimplantation diagnosis. Fertil Steril 1992.
83. Verlinsky Y, Rechitoky S, Evsikov S, et al. Reconception and preimplantation diagnosis for cystic fibrosis. Prenat Diagn 1992;12:103–10.
84. Navidi W, Arnheim N. Using PCR in preimplantation genetic disease diagnosis. Hum Reprod 1991;6:836–49.
85. Handyside AH, Pattinson JK, Penketh RJA, Delhanty JD, Winston RML, Tuddenham EDG. Biopsy of human preimplantation embryos and sexing by DNA amplification. Lancet 1989:347–9.
86. Handyside AH, Konogianni EH, Hardy K, Winston RML. Pregnancies from biopsied human preimplantation embryos sexed by Y-specific DNA amplification. Nature 1990;344:768–70.
87. Handyside AH, Delhanty DA. Cleavage stage biopsy of human embryos and diagnosis of X-linked recessive disease. In: Edwards RG, ed. Preimplantation diagnosis of human genetic disease. Cambridge, UK: Cambridge University Press, 1992.
88. Grifo JA, Boyle A, Fischer E, et al. Preembryo biopsy and analysis of blastomeres by in situ hybridization. Am J Obstet Gynecol 1990;163:2013–9.
89. Griffin DK, Handyside AH, Penketh RJA, Winston RML, Delhanty JDA. Fluorescent in situ hybridisation to interphase nuclei of human preimplantation embryos with X and Y chromosome specific probes. Hum Reprod 1991;6:101–5.

2

IVF in Nonhuman Primates: Current Status and Future Directions

BARRY D. BAVISTER AND DOROTHY E. BOATMAN

A number of laboratories during the past two decades have contributed to the current technical status of nonhuman primate IVF. A major rationale for interest in nonhuman primate IVF was, and still is, to provide data on early development that could be directly applicable to humans. However, it is ironic that progress in production of human embryos by IVF has always been somewhat more advanced than that in nonhuman primates. Thus, although the feasibility of IVF in monkeys was demonstrated by the early 1970s, the first documented human IVF took place several years earlier (Table 2.1). In the early years of IVF research in humans and in monkeys, it was difficult to demonstrate much progress beyond fertilization itself or cleavage to 2 cells. Yet the birth of the first human IVF baby (1) preceded by several years the first demonstrations of live births in nonhuman primates derived from IVF eggs (2, 3). Nevertheless, the chronology of these events does not mean that nonhuman primate IVF cannot point the way to significant improvements in human IVF technology nor increase understanding of key areas in human reproduction. Rather, the indication is that IVF technology in nonhuman primates may be best employed to examine specific events, such as oocyte maturation, sperm capacitation, or development of new culture media, that are more difficult to study in the context of human clinical IVF. In addition to this role, nonhuman primate IVF can serve as a valuable tool for assisting the propagation of endangered species.

Table 2.1 lists some major milestones in the development of nonhuman primate IVF technology. In entering this area of scientific research, major

---

This chapter is publication number 32-035 of the Wisconsin Regional Primate Research Center.

TABLE 2.1. Chronology of nonhuman primate IVF.

Date	Species	Investigator(s) (ref. no.)	Stages achieved
1969	Human	Edwards, Bavister, Steptoe (44)	IVF only
1972	Squirrel monkey	Cline, Gould, Foley (45)	IVF
1973	Squirrel monkey	Gould, Cline, Williams (46)	2-cells
1978	Human	Steptoe, Edwards (1)	Cleavage; first offspring
1979	Squirrel monkey	Kuehl, Dukelow (47)	Cleavage; no pregnancy
1982	Cynomolgus monkey	Kreitmann, Hodgen (48)	Early morula
1983	Rhesus monkey	Bavister, Boatman, et al. (4)	8-cells/early morula
1983	Chimpanzee	Gould (49)	4-cells
1984	Baboon	Clayton, Kuehl (3)	Offspring
1984	Rhesus monkey	Bavister, Boatman, et al. (2)	Offspring[a]
1984	Cynomolgus monkey	Balmaceda et al. (7)	Cleavage, offspring[b]
1984	Rhesus monkey	Morgan, Boatman, Bavister (50)	IVF to hatched blastocyst
1988	Marmoset monkey	Lopata, Summers, Hearn (51)	Morula/blastocyst, offspring
1989	Rhesus monkey	Wolf et al. (45)	Blastocysts, offspring
1989	Pig-tailed monkey	Cranfield et al. (6)	Early cleavage
1992	Pig-tailed monkey/ lion-tailed monkey	Cranfield et al. (52)	Hybrid offspring
1990	Rhesus monkey	Lanzendorf et al. (43)	Cryopreservation; first twins

Note: List is not exhaustive; key advances are shown here. Dates shown are of the report, not the event.
[a] Genetically confirmed by blood cell antigens.
[b] Includes a cross-species transfer (rhesus monkey recipient).

goals of our laboratory were not only to assist understanding of key regulatory events, but also to develop reproducible protocols that could be used by other laboratories. The IVF protocols devised by us for rhesus monkey gametes (2, 4, 5) were adaptable to other macaques, including cynomolgus (*M. fascicularis*), pig-tailed (*M. nemestrina*), and lion-tailed (*M. silenus*) monkeys (6, 7). Interestingly enough, in one study an offspring was produced following transfer of IVF cynomolgus monkey embryos to a rhesus monkey (7), an outcome that has implications for work with endangered species (see Chapter 20, this volume).

In embarking on nonhuman primate IVF research, we need to be constantly aware that in toto, vastly more human IVF embryos are potentially available for research activities because of the very large number of performance sites (human IVF clinics) around the world. Moreover, the total number of nonhuman primate offspring produced by IVF is very small (<20) compared to >20,000 human IVF babies. In spite of these discrepancies, nonhuman primate research offers some substantial opportunities for increasing understanding about preimplantation embryogenesis. Because the best-quality human IVF embryos are usually transferred back to the patient (donor), embryos made available for experimental studies (*residual embryos*) are often of inferior quality (8, 9). Another constraint on research is that experimental studies on human IVF embryos are restricted by ethical and other considerations (10). Neither of these restrictions applies to nonhuman primate embryos, which can be generated solely for research purposes. In addition, nonhuman primate embryo transfer using surrogate (unstimulated) recipients is routinely performed, separating deleterious in vivo consequences of ovarian superstimulation from the milieu for embryo development.

The following discussion highlights some of the research areas in which nonhuman primate IVF might be most useful in investigating fundamental events of early development, with brief presentations of some key questions. To provide a baseline for this discussion, we first describe the IVF protocol that we devised at the Wisconsin Primate Research Center, which with some modification is being applied in other laboratories and in other macaque species.

Rhesus Monkey IVF Protocol

Oocyte Collection

Ovarian follicles are stimulated with gonadotropins. Initially, we used *pregnant mare serum gonadotropin* (PMSG) with considerable success (2, 4, 11). However, animals developed antibodies after only one treatment cycle (12), and we have changed to a human FSH protocol devised at the Oregon Primate Research Center (13) (see also Chapter 8, this volume). Maturation of oocytes in vivo is induced by injection of hCG (2, 4). About 30 h later, follicular contents are aspirated laparoscopically using a modified Renou device (4). We have also done a small amount of work using unstimulated rhesus monkeys in which a single preovulatory oocyte is aspirated after the LH surge (Morgan, Warikoo, Boatman, Bavister, in preparation). A third approach, which we are using extensively, is to aspirate immature (germinal vesicle stage) oocytes from follicles of ovaries excised from unstimulated monkeys in order to examine the process of oocyte maturation in vitro (14, 15).

Semen Collection and Sperm Treatment

Conscious rhesus monkeys are electroejaculated using penile stimulation (4, 16). We believe that this approach provides better-quality semen than the alternative rectal probe method, but it is impractical for larger primates because of the difficulty of restraining them. Sperm are washed in HEPES-buffered TALP (17), incubated in regular TALP for capacitation for at least 4–6 h but up to 36 h as needed, then stimulated with caffeine/cAMP to induce acrosome reactions and fertilizing ability (5). We found that rhesus monkey sperm would not penetrate zonae pellucidae without this drug treatment, and hyperactivated motility characteristic of capacitated sperm did not occur.

In Vitro Fertilization

Coincubation of spermatozoa and ova (12–18 h) takes place in a modified Tyrode's solution (TALP) originally developed for hamster IVF and embryo culture (17). Inclusion of rhesus serum (2%) was beneficial (4).

Embryo Culture

Originally, TALP containing 4 amino acids was used (4, 17), but this was replaced by modified CMRL medium containing 20% rhesus serum (18). The medium was changed periodically. Some IVF rhesus embryos developed into blastocysts in this medium, and some of these hatched (18).

Embryo Transfer

The nonsurgical route, via cannulation of the uterine cervix, was mostly used, resulting in the birth of 2 IVF offspring (2, 18). Later, we devised a laparoscopic oviductal transfer procedure that led to birth of a third rhesus offspring. This method is preferable because it allows more control over the embryo transfer and because earlier stages of development (2- or 4-cell embryos) can be transferred.

Outcome

The above procedures resulted in about 65%–80% of mature ova being fertilized in vitro (2 polar bodies and/or 2 pronuclei). Of these IVF eggs, almost all (95%) underwent at least one cleavage division; this helps verify the frequency of fertilization because parthenogenetic cleavage rarely occurred in our experiments (4). Approximately 50% of embryos reached the blastocyst stage, with 20% hatching from their zonae pellucidae. Finally, these IVF procedures in our laboratory have produced 3 offspring, all males. One of these ("Petri") was the first genetically confirmed IVF nonhuman primate. These animals have all survived and have sired their own offspring (see "Second-Generation IVF Primates").

Uses of Nonhuman Primate IVF

The establishment of technology for nonhuman primate IVF has been driven by several rationales, the major one being a need for a model for early embryonic development that is more directly applicable to the human situation than are the rodent models.

Evaluation of Oocyte Maturation

Experimental primate IVF is needed as a test system for the competence of oocytes matured in vitro and allows study of factors regulating this process. Application of IVF technology to studies on oocyte maturation has clearly demonstrated the importance of determining the status of cytoplasmic maturation in addition to nuclear maturation (19). Evaluating cytoplasmic maturation involves testing the developmental capability of the egg following fertilization and its eventual ability to produce a viable pregnancy. However, most of the early work on oocyte maturation in primates and rodent species was based solely on nuclear criteria; that is, GV breakdown or achievement of metaphase of the second meiotic division and first polar body extrusion. These criteria are quite inadequate as endpoints for full, normal oocyte maturation (19).

A practical advantage of studying oocyte maturation in primates is the potential for increasing production of embryos, particularly age-matched siblings. If oocytes harvested from immature follicles of a single donor could be induced to undergo maturation in vitro, embryos could be produced without the consequences of in vivo gonadotropin stimulation. If done on a large scale, the contribution of genetically valuable individuals to breeding programs could be greatly enhanced. Recent progress and the current status of oocyte maturation studies in nonhuman primates are discussed in Chapter 7.

Sperm Capacitation

IVF provides the only definitive endpoint for in vitro sperm capacitation; that is, the ability of sperm to penetrate the zona pellucida and fertilize the egg. We found that treating rhesus monkey spermatozoa with a combination of caffeine and cAMP at certain concentrations was required for them to exhibit hyperactivated motility, which was concomitant with acquisition of the ability to penetrate rhesus monkey zonae pellucidae (5). This was one of the earliest demonstrations of tight linkage between hyperactivation and fertilizing ability in mammalian sperm. Neither caffeine nor cAMP alone was effective on rhesus spermatozoa, although Dukelow's laboratory had shown previously that cAMP did enhance the fertilizing ability of squirrel monkey spermatozoa (20). Subsequently, it was found that sperm from other macaque species (*M. fascicularis* and *M. nemestrina*) also required chemical stimulation by the combination of caffeine and cAMP (6, 7). Human spermatozoa do not show a requirement for these drugs. Thus, macaque sperm appear to be different from human sperm in their needs for exogenous chemical stimulation to achieve hyperactivated motility and fertilizing ability. However, this dependency of macaque sperm on chemical stimulators could be exploited to analyze the control of hyperactivation. Another possibility is that through studies on the chemical control of sperm fertilizing ability in monkeys, new treatment protocols might be devised for enhancing IVF using subfertile human semen samples.

The precise mechanisms of sperm capacitation and the acrosome reaction are obscure in all species, but virtually nothing is known about these processes in primates. The value of using macaque sperm to study these events is not generally appreciated. Macaques are useful for two reasons. First, chemical stimulators (caffeine and cAMP) regulate the transition between normal and hyperactivated motility, as already discussed, so that changes in sperm within these two phases and during the transition between them can be examined separately. In contrast, when sperm of some other animals and humans make this transition, they do so in a rather asynchronous manner, and sperm generally cannot be maintained in a capacitated but nonhyperactivated/nonacrosome-reacted condition. This makes it very difficult to specify what changes in sperm take place prior to or after the onset of hyperactivation. The ability to turn on at will the type of motility that is correlated with fertilizing ability is a powerful tool for investigating capacitation, hyperactivation, and acrosome reactions.

A second advantage of using macque sperm is that they show great homogeneity of motion under in vitro conditions, unlike human sperm. For example, under unstimulated conditions (no caffeine/cAMP), nearly all motile rhesus monkey sperm swim with nearly straight trajectories that are easily measurable (5). However, after stimulation with these drugs, the majority of sperm change to the hyperactivated mode of motility. In contrast, human spermatozoa exhibit several different kinds of motility at all times during incubation in vitro (21–23), making it difficult to assess changes in motility patterns; the percentages of sperm showing each type of motility may change with time, but not necessarily with any change in incubation conditions. These motility characteristics make human sperm less suitable for the study of capacitation.

Applications of IVF

There are some valuable applied uses of IVF in nonhuman primates. For example, IVF is used to produce cleavage-stage embryos both for research and for clinical uses, such as micromanipulation, transgenesis, and so forth (24). It is particularly difficult to obtain these stages—that is, 1-cell fertilized eggs through about 8–16 cells—from the oviducts of mated animals because it involves surgical flushing. Another valuable contribution of nonhuman IVF is in the area of conservation of endangered species (see Chapter 20, this volume).

Modified IVF approaches can be very useful for testing sperm fertilizing ability. For example, in the *multiple-sperm penetration assay*, salt-stored zonae pellucidae (their eggs lacking a block to polyspermy) were challenged with capacitated or uncapacitated sperm (25). Each zona pellucida allowed the penetration of many capacitated sperm, eventually filling the perivitelline space, whereas zonae exposed to uncapacitated sperm were unpenetrated. Thus, salt-stored zonae could still discriminate between capacitated and uncapacitated sperm. By counting the sperm that accumulated in the perivitelline space, we obtained a quantitative measure of the capacitation status of the sperm population. This test could be applied to nonhuman primate zonae pellucidae; indeed, we have shown in a qualitative way that it is practical.

Another variation on IVF is the *hemizona assay* (26) in which two problems—scarcity and heterogeneity of primate zona material—are simultaneously alleviated. A zona pellucida from a human (26) or monkey (53) egg is removed microsurgically and cut in halves. Because the two halves are (presumably) identical, they can be exposed to paired treatments (control and experimental) to provide a powerful experimental design. An obvious use is to compare control and suspected infertile sperm samples or control vs. drug-treated sperm. Endpoints are the relative numbers of sperm binding to or penetrating the zona pellucida.

Cell Biology of Fertilization

IVF also provides almost the only practical route to study details of the fertilization process itself. These include zona penetration; sperm-egg membrane fusion; the mechanism of the block to polyspermy; events involved in pronuclear formation (about which little is known even in nonprimate species); the origin of the centrosome, a key structure for the organization of microtubules (27); the biochemical events involved in the completion of the second meiotic division; and the mechanism of cleavage.

Ovarian Stimulation

The protocol used to stimulate development of ovarian follicles is critically important for nonhuman primate IVF for at least two reasons. First, the number of oocytes/ova collected from an animal influences the complexity of experiment that can be done; second, the quality of the eggs affects the outcome of the experiment. A superstimulation protocol may produce many eggs of low developmental capability or a few eggs of high quality (6, 11). If testing sperm zona penetration ability is the objective, rather than embryo development, the former protocol may be desirable. At present, the stimulation protocols developed by VandeVoort et al. (13) are the standard for IVF in macaques. A major problem in nonhuman primate IVF is the unrepeatability of follicular stimulation protocols in the same animal because of the rapid development of an immune response to foreign gonadotropins in most macaque species (12, 24). Different gonadotropin stimulation protocols are discussed in detail in Chapter 8.

Oviductal Influences on Fertilization

The efficiency of sperm capacitation and acrosome reactions apparently is much lower in vitro than in vivo, as evidenced by the fact that very large numbers of sperm are required in order to accomplish fertilization in vitro (this is true in any species, including primates). In contrast, only a few sperm, perhaps as few as one per egg, are sufficient to ensure fertilization of the egg at the site of fertilization in vivo. This difference indicates that conditions in vitro for sperm capacitation/acrosome reactions are quite inferior. This raises the question of oviductal influences on fertilization in mammals, which may be very important for the efficiency of this process. This area has not been given sufficient attention.

The oviductal environment may be important to the efficiency of fertilization in primates. The vast majority of work on IVF has been done in rodents (i.e., mouse, hamster, or rat). Almost inevitably, studies have been done with ovulated eggs that have been exposed to the oviductal environment for some hours before collection and with epididymal spermatozoa. In contrast, IVF in primates, both human and monkey, as well as in cattle, has been done almost exclusively on follicular eggs that have not been exposed to the oviductal environment, together with ejaculated sperm. Spermatozoa may react differently depending whether they are recovered from the epididymis or are ejaculated (28). We might expect follicular eggs to be different in some respects from ovulated eggs; indeed, they appear to be. Not only does the maturity status of eggs differ depending on whether they are collected pre- or postovulation, but exposure to oviductal components may radically change their fertilization characteristics.

In a study using the hamster as a model, the ability of follicular eggs to stimulate acrosome reactions on the zona surface was examined during 0.5-h incubation with capacitated sperm (29). There was a significant increase in the numbers of sperm undergoing acrosome reactions on the zona surface of ovulated eggs compared with follicular eggs, indicating that some peri- or postovulatory change(s) had occurred in the zona pellucida that made it more able to stimulate acrosome reactions. Even more interesting was the demonstration that follicular eggs treated with oviduct fluid, washed, and exposed to capacitated sperm showed a large increase in the percentage of zona-bound sperm that had undergone acrosome reactions, to a level equal to the capacity of ovulated eggs' zonae to stimulate acrosome reactions (29). We may deduce from these experiments that exposure of zonae pellucidae to the oviductal environment is important for increasing the efficiency of sperm:zona interactions leading to fertilization. Absence of this natural conditioning could help explain the low efficiency of IVF with follicular eggs—that is, the requirement for very large numbers of sperm to effect fertilization under in vitro conditions, vastly in excess of the few needed in vivo.

A second study involved development of in vivo fertilized hamster eggs; no comparable data are yet available in primates. Eggs were taken from the oviduct at an estimated time of 3 h post-egg activation (i.e., 3 h after sperm first entered the egg cytoplasm and initiated the sequence of embryonic development [3-h eggs]) or at 6 or 9 h post-egg activation. There was a striking effect of time of exposure to the oviductal environment on the ability of eggs to continue development in vitro to the morula and blastocyst stages (30). The 3-h eggs yielded only about 10% morulae and blastocysts, whereas the 6-h post-activation eggs produced 40% and the 9-h eggs, 70%. We can conclude that oviductal factors, effective during the first few hours after egg activation, are beneficial for hamster embryo development. It would be most interesting to extend these findings to primates.

In the same study (30), the addition of hypotaurine, a major amino acid component of oviductal secretions, to the culture medium substantially increased the developmental capability of in vivo fertilized eggs under in vitro conditions. With the 3-h eggs, development to the 8-cell stage and further was increased significantly by hypotaurine. In the 6-h eggs, there was a significant but smaller effect of hypotaurine, but no effect of hypotaurine in the 9-h eggs. So, very interestingly, hypotaurine seems to be able to compensate in part for a lack of exposure to the oviductal environment; this compensatory effect is reduced with increasing time of exposure of the eggs to the oviduct. These findings strongly indicate a role for hypotaurine on embryo development in vivo that may partly explain the supportive role of the oviduct.

Since in vivo fertilized eggs at a very early stage (3 h after sperm activation) are highly sensitive to the presence of hypotaurine for development, we would expect that unfertilized (ovulated) eggs, which have been in the oviduct even less time, would be even more sensitive to hypotaurine; indeed, they were. In the absence of hypotaurine in the culture environment, there was no development beyond the 4-cell stage in hamster IVF eggs, whereas concentrations of hypotaurine (0.1 mM or greater) supported significant development beyond the 8-cell stage and even to the morula and blastocyst stages (30). If we take the idea of oviductal dependence one step farther, then we might expect that the in vitro fertilizability and/or developmental capacity of follicular eggs, which are the primary source of material used in primate and cattle IVF, would be adversely affected because these eggs have had no contact with the oviductal environment. This hypothesis is currently being tested.

Events in Fertilization

The processes involved in fertilization itself have been very little studied in either human or nonhuman primates. As mentioned earlier, events of zona penetration, membrane fusion, and the like, are particularly amenable to investigation in nonhuman primates because of the absence of constraints on IVF research. One topic that needs mention is the block to polyspermy in primate eggs. Mouse eggs are often used as models for primate in vitro fertilization and development. This may be inappropriate because in some fundamental respects, mouse eggs and embryos are different from many other species.

For example, in the hamster; in human and monkey; and in cow, pig, and sheep, the block to polyspermy seems to be almost entirely at the level of the zona pellucida, while a block at the egg plasma membrane level is either weak or absent. This can be deduced from direct evidence (31) and from the scarcity or absence of perivitelline (supernumerary) sperm. In contrast, there is a strong block to polyspermy in mouse eggs at the egg plasma membrane level, about equivalent to that of the zona pellucida block. (Rabbits are very unusual in that the block is entirely at the egg plasma membrane level.) Now, if mouse eggs are used, for example, to examine sperm microinjection beneath the zona pellucida or for zona drilling (32), the powerful egg plasma membrane block to polyspermy may prevent excess sperm that gain access to the perivitelline space from reaching the egg cytoplasm. These observations should not be directly extrapolated to primate eggs, which appear to have little or no defense against polyspermy once the zona is breached, so polyspermy may result from zona drilling. This is a caveat for the use of micromanipulation as a means to enhance sperm penetration of primate zonae.

Development of IVF Embryos

Finally, we want to mention development of IVF embryos in which there seems to be a rather large number of problems. There is no doubt that present culture media environments used for IVF primate embryos are inadequate since many of the embryos fail to develop to or beyond late-cleavage stages. Several studies have indicated that many or most of these advanced preimplantation embryos produced in vitro may be defective and nonviable (9, 18, 33). In a study of rhesus IVF embryos developing in vitro, we found a linear falloff in developmental capability with time, such that 9% of the embryos ceased development each day of culture (18). This is consistent with several studies on human IVF embryos in culture in which only about 20% reached the blastocyst stage

(34, 35) and with reports that following transfer of IVF embryos at the blastocyst stage, the incidence of pregnancy was lower than expected (36), suggesting poor embryo viability.

One of the major reasons for the inadequacy of culture conditions for primate IVF embryos is the poor state of knowledge about embryo metabolism. However, the energy metabolism of human IVF embryos does seem to be seriously disturbed in culture (37). A major problem noted in rhesus monkey IVF embryos that were cultured to the hatched blastocyst stage—which, admittedly, had ceased cell division—was deterioration of the inner cell mass (33). (This is probably not true in embryos that are recovered from the female reproductive tract at the morula or later stages [38]). There has recently been a resurgence of interest in examining the metabolism and substrate preferences of cultured embryos; hopefully, this will soon lead to formulation of new culture media for primate embryos that better support metabolism and, hence, development.

Second-Generation IVF Primates

Underlying concerns about the long-term consequences of IVF include the fertility of individuals resulting from this procedure and the normality of their offspring. These concerns are allayed because we have numerous offspring representing the next generation of IVF primates at the Wisconsin Primate Center. Our 3 IVF rhesus offspring, all male—Petri (born in 1983), Orwell (1984), and Murphy (1986)—have each sired several offspring (54) for a total of 17 (9 males and 8 females). These second-generation offspring, the first to our knowledge, demonstrate that primates produced by IVF are fertile and that normal offspring can be produced by IVF parents.

Summary: Current Status of Nonhuman Primate IVF

We still do not understand the mechanism of sperm capacitation and acrosome reactions in primates; however, this is true for almost every other group of mammals. Procedures for inducing sperm capacitation in vitro are adequate, but clearly inefficient because of the huge numbers of sperm required to ensure penetration of the egg (zona) in vitro. A recent development from VandeVoort's laboratory at the California Primate Center is the successful cryopreservation of monkey sperm (39), which could have a number of advantages for nonhuman primate IVF. It would be laborsaving and convenient to collect semen and oocytes on different days, and replicates of IVF experiments could become more reproducible using semen from the same ejaculate. Semen for use in IVF could be transported to distant sites, avoiding the problems associated with shipping live animals for this purpose. Moreover, the risk of transmitting specific viruses between animals could be reduced by using banked semen after it has been tested for the organism. In addition, genetically valuable semen could be stored for endangered species IVF work, even after the demise of the donor.

Almost nothing is known about the regulation of oocyte maturation in primates. Our ability to sustain oocyte maturation (GV breakdown through to metaphase of meiosis II and expression of the first polar body) under in vitro conditions is clearly very inadequate at present. The incidence of eggs undergoing maturation in vitro is low, and they are not sufficiently viable (40); however, much remains to be learned about the factors regulating oocyte maturation in primates. There is presently a marked difference between primate oocytes matured in vitro and partially or wholly in vivo-matured oocytes (see Chapter 7, this volume). This situation is in sharp contrast to the technology for in vitro oocyte maturation in cattle in which in vitro development of embryos from IVM/IVF oocytes is readily obtained and numerous offspring have been produced following transfers (41). It is essential to make a distinction between nuclear (morphological) and cytoplasmic (developmental) maturation so that IVF and embryo development must be included as criteria for normality of oocyte maturation. Technology needs to be developed for growing preantral follicles, as in the mouse (42), so that the developmental potential of perhaps hundreds or even thousands of the oocytes contained in these follicles can be realized and exploited for embryo production by IVF.

As the foregoing discussion has illustrated, IVF in primates is used mainly as a quantitative measure of gamete quality or fertilization potential or as a means to produce embryos in substantial numbers. In a qualitative sense—that is, what controls events occurring during fertilization—very little is known in primates. An improved understanding of the mechanisms involved in the process of fertilization (e.g., the block to polyspermy) should assist efforts to develop new IVF protocols. Finally, although viable offspring can be produced from primate IVF embryos, the efficiency of IVF embryo development under in vitro and in vivo conditions is usually low. One exception to this is a study in which IVF rhesus embryos were replaced into the oviduct instead of in the uterus (43). The high success rate in this study may have implications for improving human IVF pregnancy rates. In general, however, there is a strong need to improve embryo development in monkeys and in humans following IVF. This low efficiency is no doubt due both to defective oocytes and to inadequate culture media. In other species, for example, the hamster, the oviduct seems to play a major role during the first few hours after fertilization. If a similar relationship between the early embryo and oviduct occurred in primates and if the activity were characterized, perhaps it could be exploited in IVF protocols to enhance embryo development.

During the last 20 years, major progress has been made in nonhuman primate IVF, beginning with limited morphological signs of fertilization and continuing all the way through to routine production of IVF offspring (Table 2.1). Thus, the technology is now in place to investigate the questions and problems that have been alluded to in this chapter. Once again, however, we need to remind ourselves of a persistent paradox with research into nonhuman primate IVF: Although much of the impetus and support for such studies comes from their perceived relevance to human fertility studies—especially IVF—in key areas advances with human IVF have preceded corresponding progress in nonhuman primates. The remarkable advances made during the past decade or so in human IVF technology have occurred mostly, if not entirely, without any transfer of information or procedures from nonhuman primate IVF. In fact, nonhuman primate IVF has frequently been the beneficiary of progress in human clinical IVF, rather than the reverse. If IVF in monkeys is to establish itself as a useful clinical model, we must emphasize research with nonhuman primates in areas that are particularly difficult to investigate in humans for practical or ethical reasons. As this chapter has shown, there are some distinct parallels between human clinical and nonhuman primate IVF, and the potential for obtaining data on early development in primates from studies on monkey IVF is considerable. We need to keep this in sharp focus during the next decade of discovery.

References

1. Steptoe PC, Edwards RG. Birth after the reimplantation of a human embryo (letter). Lancet 1978;2:366.
2. Bavister BD, Boatman DE, Collins K, Dierschke DJ, Eisele SG. Birth of rhesus monkey infant following in vitro fertilization and non-surgical embryo transfer. Proc Natl Acad Sci USA 1984;81:2218–22.
3. Clayton O, Kuehl TJ. The first successful in vitro fertilization and embryo transfer in a nonhuman primate. Theriogenology 1984;21:228.
4. Bavister BD, Boatman DE, Leibfried ML, Loose M, Vernon MW. Fertilization and cleavage of rhesus monkey oocytes in vitro. Biol Reprod 1983;28:983–99.
5. Boatman DE, Bavister BD. Stimulation of rhesus monkey sperm capacitation by cyclic nucleotide mediators. J Reprod Fertil 1984;77:357–66.
6. Cranfield MR, Schaffer N, Bavister BD, et al. Assessment of oocytes retrieved from stimulated and unstimulated ovaries of pig-tailed macaques (*Macaca nemestrina*) as a model to enhance the genetic diversity of the captive lion-tailed macaque (*Macaca silenus*). Zoo Biol 1989;suppl 1:33–46.
7. Balmaceda JP, Pool TB, Arana JB, Heitman TS, Asch RH. Successful in vitro fertilization and embryo transfer in cynomolgus monkeys. Fertil Steril 1984;42:791–5.

8. Menezo YJR, Guerin JF, Czyba JC. Improvement of human early embryo development in vitro by coculture on monolayers of Vero cells. Biol Reprod 1990;42:301–6.
9. Winston NJ, Braude PR, Pickering SJ, et al. The incidence of abnormal morphology and nucleocytoplasmic ratios in 2-, 3- and 5-day human preembryos. Hum Reprod 1991;6:17–24.
10. Austin CR. Dilemmas in human IVF practice. In: Bavister BD, Cummins J, Roldan ERS, eds. Fertilization in mammals. Norwell, MA: Serono, Symposia, USA, 1990:373–9.
11. Boatman DE, Morgan PM, Bavister BD. Variables affecting the yield and developmental potential of embryos following superstimulation and in vitro fertilization in rhesus monkeys. Gamete Res 1986;13:327–38.
12. Bavister BD, Dees HC, Schultz RD. Refractoriness of rhesus monkeys to repeated gonadotropin stimulation is due to formation of non-precipitating antibodies. Am J Reprod Immunol Microbiol 1986;11:11–6.
13. VandeVoort CA, Baughman WL, Stouffer RL. Comparison of different regimens of human gonadotropins for superovulation of rhesus monkeys: ovulatory response and subsequent luteal function. J In Vitro Fertil Embryo Transfer 1989;6:85–91.
14. Morgan PM, Warikoo PK, Bavister BD. In vitro maturation of ovarian oocytes from unstimulated rhesus monkeys: assessment of cytoplasmic maturity by embryonic development after in vitro fertilization. Biol Reprod 1991;45:89–93.
15. Schramm RD, Tennier MT, Boatman DE, Bavister BD. Chromatin configurations and meiotic competence of oocytes are related to follicular maturation in non gonadotropin-stimulated rhesus monkeys. Biol Reprod (in press).
16. Mastroianni L, Manson WA. Collection of monkey semen by electroejaculation. Proc Soc Exp Biol Med 1963;112:1025–7.
17. Bavister BD, Leibfried ML, Lieberman G. Development of preimplantation embryos of the golden hamster in a defined culture medium. Biol Reprod 1983;28:235–47.
18. Boatman DE. In vitro growth of nonhuman primate pre- and peri-implantation embryos. In: Bavister BD, ed. The mammalian preimplantation embryo. New York: Plenum Press, 1987:273–308.
19. Bavister BD. Oocyte maturation and in vitro fertilization in the rhesus monkey. In: Stouffer RL, ed. The primate ovary. New York: Plenum Press, 1987:119–37.
20. Chan PJ, Hutz RJ, Dukelow WR. Nonhuman primate in vitro fertilization: seasonality, cumulus cells, cyclic nucleotides, ribonucleic acid, and viability assays. Fertil Steril 1982;38:609–15.
21. Burkman LJ. Characterization of hyperactivated motility by human spermatozoa during capacitation: comparison of fertile and oligozoospermic sperm populations. Arch Androl 1984;13:153–65.
22. Burkman LJ. Temporal pattern of hyperactivation-like motility in human spermatozoa [Abstract]. Biol Reprod 1986;34(suppl 1):226.
23. Robertson L, Wolf DP, Tash JS. Temporal changes in motility parameters related to acrosomal status: identification and characterization of populations of hyperactivated human sperm. Biol Reprod 1988;39:797–805.

24. Wolf DP, Thomson JA, Zelinski-Wooten MB, Stouffer RL. In vitro fertilization-embryo transfer in nonhuman primates: the technique and its applications. Mol Reprod Dev 1990;27:261-80.
25. Boatman DE, Andrews JC, Bavister BD. A quantitative assay for capacitation: evaluation of multiple sperm penetration through the zona pellucida of salt-stored hamster eggs. Gamete Res 1988;19:19-29.
26. Burkman LJ, Coddington CC, Franken DR, Kruger TF, Rosenwaks Z, Hodgen GD. The hemizona assay (HZA): development of a diagnostic test for the binding of human spermatozoa to the human hemizona pellucida to predict fertilization potential. Fertil Steril 1988;49:688-97.
27. Schatten G, Simerly C, Schatten H. Maternal inheritance of centrosomes in mammals? Studies on parthenogenesis and polyspermy in mice. Proc Natl Acad Sci USA 1991;88:6785-9.
28. Florman HM, Tombes RM, First NL, Babcock DF. An adhesion-associated agonist from the zona pellucida activates G protein-promoted elevations of internal Ca^{2+} and pH that mediate mammalian sperm acrosomal exocytosis. Dev Biol 1989;135:133-46.
29. Boatman DE. Oviductal modulators of sperm fertilizing ability. In: Bavister BD, Cummins J, Roldan ERS, eds. Fertilization in mammals. Norwell, MA: Serono Symposia, USA, 1990:223-38.
30. Barnett DK, Bavister BD. Hypotaurine requirement for in vitro development of golden hamster one-cell embryos into morulae and blastocysts, and production of term offspring from in vitro fertilized ova. Biol Reprod 1992.
31. Stewart-Savage J, Bavister BD. A cell surface block to polyspermy occurs in golden hamster eggs. Dev Biol 1988;128:150-7.
32. Gordon JW. Zona drilling: a new approach to male infertility. J In Vitro Fertil Embryo Transfer 1990;7:223-8.
33. Enders AC, Schlafke S, Boatman DE, Morgan PM, Bavister BD. Differentiation of blastocysts derived from in vitro fertilized rhesus monkey oocytes. Biol Reprod 1989;41:715-27.
34. Bolton VN, Hawes SM, Taylor CT, Parsons JH. Development of spare human preimplantation embryos in vitro: an analysis of the correlations among gross morphology, cleavage rates, and development to the blastocyst. J In Vitro Fertil Embryo Transfer 1989;6:30-5.
35. Lopata A, Hay DL. The potential of early human embryos to form blastocysts, hatch from their zona and secrete HCG in culture. Hum Reprod 1989;4:87-94.
36. Bolton VN, Wren ME, Parsons JH. Pregnancies after in vitro fertilization and transfer of human blastocysts. Fertil Steril 1991;55:830-2.
37. Gott AL, Hardy K, Winston RML, Leese HJ. Non-invasive measurement of pyruvate and glucose uptake and lactate production by single human preimplantation embryos. Hum Reprod 1990;5:104-8.
38. Seshagiri PB, Hearn JP. Protein-free culture media that support in vitro development of rhesus monkey blastocysts. ARTA 1992.
39. Tollner TL, VandeVoort CA, Overstreet JW, Drobnis EZ. Cryopreservation of spermatozoa from cynomolgus monkeys (*Macaca fascicularis*). J Reprod Fertil 1990:90:347-52.
40. Morgan PM, Warikoo PK, Bavister BD. In vitro maturation of ovarian oocytes from unstimulated rhesus monkeys: assessment of cytoplasmic

maturity by embryonic development after in vitro fertilization. Biol Reprod 1991;45:89–93.
41. Parrish JJ, First NL. Bovine in vitro fertilization. In: Dunbar BS, O'Rand MG, eds. A comparative overview of mammalian fertilization. New York: Plenum Press, 1991:351–62.
42. Eppig JJ, Schroeder AC. Capacity of mouse oocytes from preantral follicles to undergo embryogenesis and development to live young after growth, maturation, and fertilization in vitro. Biol Reprod 1989;41:268–76.
43. Lanzendorf SE, Zelinski-Wooten MB, Stouffer RL, Wolf DP. Maturity at collection and the developmental potential of rhesus monkey oocytes. Biol Reprod 1990;42:703–11.
44. Edwards RG, Bavister BD, Steptoe PC. Early stages of fertilization in vitro of human oocytes matured in vitro. Nature (London) 1969;221:632–5.
45. Cline EM, Gould KG, Foley CW. Regulation of ovulation, recovery of mature ova and fertilization in vitro of mature ova of the squirrel monkey (*Saimiri sciureus*) [Abstract]. Fed Am Soc Exp Biol 1972;31:360.
46. Gould KG, Cline EM, Williams WL. Observations on the induction of ovulation and fertilization in vitro in the squirrel monkey (*Saimiri sciureus*). Fertil Steril 1973;24:260–8.
47. Kuehl TJ, Dukelow WR. Maturation and in vitro fertilization of follicular oocytes of the squirrel monkey (*Saimiri sciureus*). Biol Reprod 1979;21:545–56.
48. Kreitman O, Lynch A, Nixon WE, Hodgen GD. Ovum collection, induced luteal dysfunction, in vitro fertilization, embryo development and low tubal ovum transfer in primates. In: Hafez ESE, Semm K, eds. In vitro fertilization and embryo transfer. Lancaster, UK: MTP Press, 1982:303–24.
49. Gould KG. Ovum recovery and in vitro fertilization in the chimpanzee. Fertil Steril 1983;40:378–83.
50. Morgan PM, Boatman DE, Collins K, Bavister BD. Complete preimplantation development in culture of in vitro fertilized rhesus monkey oocytes [Abstract]. Biol Reprod 1984;30(suppl 1):96.
51. Lopata A, Summers PM, Hearn JP. Births following the transfer of cultured embryos obtained by in vitro and in vivo fertilization in the marmoset monkey (*Callithrix jacchus*). Fertil Steril 1988;50:503–9.
52. Cranfield MR, Berger NG, Kempske S, Bavister BD, Boatman DE, Ialeggio DM. Macaque monkey birth following transfer of in vitro fertilized, frozen-thawed embryos to a surrogate mother [Abstract]. Theriogenology 1992;37:197.
53. Oehninger S, Scott RT, Coddington CC, Franken DR, Acosta AA, Hodgen GD. Validation of the hemizona assay in a monkey model: influence of oocyte maturational stages. Fertil Steril 1989;51:881–5.
54. Bavister BD, Boatman DE. "Test-tube" primates: the next generation. Hum Reprod 1992;7:1035.

3

Assisted Reproduction in the Great Apes

KENNETH G. GOULD AND JEREMY F. DAHL

Assisted reproduction with great apes will be considered here as *facilitated* reproduction in order to include the experimental and clinical approaches related to fertility evaluation (including correction of infertility) and optimization of natural fertilization as well as the more invasive methods of reproduction enhancement associated with *artificial insemination* (AI), *in vitro fertilization* (IVF), gamete storage, and surgical intervention. Those techniques are only now becoming effectively applied to the great apes. In addition, the existing data pertaining to early termination of pregnancy, obstetrical intervention, and neonatal and infant mortality can provide valuable guidance regarding the development of strategies for assisted reproduction. Such strategies are, therefore, to be directed to initiation of pregnancy, maintenance of pregnancy, reduction of neonatal loss, provision of means for "normal" postnatal development, and arrangement for social conditions conducive to subsequent natural breeding and rearing of offspring.

Significance

The reasons for investigating assisted reproduction include, as a primary focus, maintenance of the captive population of all species, with subsequent potential support for that species in the wild. *Pan troglodytes* (common chimpanzee) and *Homo sapiens* (human) resemble each other in many ways, reflected in demonstrated similarities in physiology (1), endocrinology (both normal and abnormal) (2–4), immunology (5–8), and reproduction (2, 5, 9). As a result of these phylogenetic similarities and the larger size of the captive population relative to that of the other species, the chimpanzee has great potential as a highly appropriate model for biomedical research.

However, responsible use of the chimpanzee in this manner, or use of any other species, must be predicated upon the existence of a genetically stable, self-sustaining (or growing), captive population. With regard to the great apes, this criterion is further mandated by the increasing pressure on the wild population for survival as a result, in large part, of decreasing habitat. Each of the species, in the wild, is listed in the CITES as endangered. That pressure on the one hand effectively prohibits access to the wild population for research subjects, but also increases the urgency of development for effective methods for assisted breeding of the captive population.

The demographic and genetic status of the captive populations of each of the great ape species is monitored within the zoological community by the *Species Survival Plan/Program* (SSP) for each species. While the common chimpanzee population is sufficiently large that the biomedical population is monitored and managed separately from the zoological population, the orangutan and gorilla populations are treated as single units; the bonobo is managed on an international scale as a single population.

While medical interventions are increasing, in certain of the wild populations, especially that of the mountain gorilla (*Gorilla beringei*) in Rwanda, it is not yet known whether techniques developed for assisted reproduction of the captive population can be used directly and effectively in the wild. Techniques for semen collection and preservation have the most immediate potential to provide additional genetic material from the wild without the need to reduce the wild population (10, 11).

Historical Perspective

The effective mounting of deliberate and successful efforts for assisted reproduction in the great apes presumes an understanding of the normal reproductive process adequate to permit identification and diagnosis of infertility and those conditions—physical, physiological, and social—that impair reproduction. Such information began to be collected, with specific reference to the chimpanzee, with the work of Yerkes, Coolidge, Carpenter, Elder, Nissen, and others in the 1920s and 1930s (12–15). The studies of Yerkes and Elder, together with those of their colleagues, showed that the duration of the menstrual cycle of the chimpanzee female was slightly longer than that of the woman (15). Extension of such observations to characterization of the cycle was facilitated by the presence of a significant perineal swelling in the female that is a sensitive external indicator of the endocrine changes in the ovarian cycle (15). Those studies demonstrated similarities in reproductive behavior and physiology between the chimpanzee and human. In retrospect, they provided information regarding the time of ovulation, as inferred from the incidence of pregnancy after timed matings (13, 15), and associated physical, behavioral, and endocrine changes in the menstrual cycle that preceded publication of similar data in the human. Review and analysis of early work was presented in 1929 (16).

In the decades between 1940 and 1970 and with increased reporting in the 1970s, advancing technology and techniques were applied to investigation of the endocrine changes in the chimpanzee (2, 17–20). However, in large part, the information obtained was not applied to the captive ape population as a whole (21), probably because of the difficulty of reproducing the assays used (18) and the logistical problems associated with collection of body fluids from apes in the zoological facilities of the time. In the last 20 years, there has been an enormous increase in the accessibility of hormone assays, increased interest in their application to all great apes (22–28), improvement of physical facilities for urine collection, and increased availability of safe, predictable pharmaceuticals for chemical immobilization of the animals. It is probably true to say that current practices associated with assisted fertility in the apes would have been severely curtailed without the availability of ketamine HCl (e.g., Ketaset, Aveco, Fort Dodge, IA) and (more recently) tiletamine zolezapam (Telazol, A.H. Robins, Richmond, VA).

There has been a significant increase in the documentation of methods for assisted fertility in the great apes in the last decade. A number of national and international conferences have been held that provided information relevant to this subject, and semiannual meetings are conducted by each of the SSPs related to the management of the captive population.

Problems with Assisted Breeding of Great Apes

The endangered, or threatened, status of the species is itself a restriction on the development of assisted reproduction techniques as the potential interference with, and delay of, natural reproduction associated with such studies can significantly influence the productivity of the captive population. This is clearly evident with regard to the bonobo, where the total captive population is less than 90 individuals.

It is also evident that the effect of efforts toward assisted reproduction can be minimized by the naturally long interbirth interval of the apes. This promotes an immediate dilemma with regard to the use of certain husbandry techniques otherwise suitable for promoting a more rapid numerical increase in the population; for example, shortening of the interbirth interval by removal of the neonate from the mother. Concern for the use of such husbandry strategies is of greater importance as we appreciate the complexity and potential significance (still inadequately described, let alone quantified) of social influences after birth on the future development of the neonate as measured by adequate reproductive and maternal function (29–32).

In addition, there is a scientific and statistical difficulty in the application of novel techniques for assisted reproduction that is related to the restricted population size. Diagnosis and treatment of infertility and provision of assistance in reproduction must usually be conducted on an individual basis. While such individuality of treatment is also true for the human, the techniques used for human treatment have been evaluated in a large human population prior to their invocation for the clinical treatment of any individual. In this regard, the similarities between the human and great apes can be valuable insofar as it is has been shown that treatments effective in the human can often have predictable results in the ape (33–35). However, this is not always true, as the response of the chimpanzee—and to some extent the gorilla—to ovarian stimulation with menotropins has not been as great as anticipated (Table 3.1) (35).

The size of the total captive population also influences utilization of such procedures as cryopreservation, where it is not possible, as a standard procedure, to use a competitive basis to select those males "suitable" to be involved in a semen banking program (34, 36, 37). Competitive selection is used in, for example, the cattle industry, where less than 1 in 10 bulls is selected to be used as a repeated donor for a cryopreservation program. This means that there is an increased significance related to analysis of semen parameters associated with the successful freezing of ape semen, oocytes, and embryos.

Male Evaluation

Current fertility evaluation of the male ape is primarily dependent on the measurement of physical parameters and secondarily dependent on the evaluation of endocrine parameters. Information on the "normal" genitalia (19, 38–41) and semen characteristics of the chimpanzee is available and although still limited in scope, provides information on the normal semen characteristics (9, 42–44), alteration of the semen parameters with season (45), and the development of normal characteristics

TABLE 3.1. Comparison of Metrodin and Pergonal for follicle development in the chimpanzee.

	Follicles developed		Oocytes from large follicle (%)	
	Large	Small	Mature at recovery	Total matured
75 IU Metrodin day 4–8	1.50	1.50	20	80
day 3–9	2.30	3.30	45	60
75 IU Pergonal day 4–8	1.00	1.60	0	40
75 IU Pergonal day 3–9	1.50	3.00	33	57
225 IU Pergonal day 3–7	1.00	3.30	33	44
225 IU Metrodin day 3–7	2.00	3.00	50	87
150 IU Metrodin (PUMP)	1.00	2.25	50	50
350 IU Pergonal (PUMP)	1.25	2.50	40	60

N = 4 for each treatment; numbers are the mean.

TABLE 3.2. Semen parameters for great apes.

Method	Volume	Count	Viability (%)	Motility (%)	N	Reference
Pan troglodytes						
AV	2.40	620.00	70.35	NA	15	Gould et al., 1982
AV	3.40	2561.00	80.00	76	25	Gould et al., 1992
Mast	3.40	784.00	68.00	55	68[a]	Marson et al., 1989
Mast	3.78	540.00	65.00	NA	27	Gould et al., 1982
RPE	1.00	280.00	84.40	NA	9	Gould et al., 1982
RPE	1.90	743.00	52.70	NA	11	Bader, 1983
RPE	2.50	1000.00	89.00	85	6	Gould, unpubl.
Pan paniscus						
RVE	1.50		Good quality		1	Matern, 1983
RPE	1.52	423.00	81.00	81	4[b]	Gould, unpubl.
AV	1.80	1303.00	40.00	36	14[b]	Gould, unpubl.
Gorilla gorilla						
RPE					24	Seager et al., 1982
RPE	0.31	15.20		11	2	Platz et al., 1980
RPE	0.38	41.00	54.00	32	9	Gould and Kling, 1982
RPE	0.62	121.00	62.00	51	12	Gould and Martin, unpubl.
Pongo pygmaeus						
RPE	1.10	61.00	59.00	47	5	Warner et al., 1974
Mast	6.50	164.00	NA	60	25[b]	VandeVoort et al., 1992

AV = artificial vagina; Mast = masturbation; RPE = rectal probe electroejaculation. Figures in the "Count" column are $\times 10^{-6}$/mL.
[a] Multiple, frequent ejaculates.
[b] Single male.

during puberty (44). However, information in these areas for the other apes remains scant (19, 40, 46–53) (Table 3.2).

It has long been recognized that an evaluation of sperm viability and motion characteristics could aid in the identification of normal, and thus potentially fertile, males (54); and it is undoubtedly true that "semen analysis remains the cornerstone of male fertility evaluation. . . ." (55). Evaluation of semen motility has traditionally been a subjective exercise, based on initial work in the rabbit and man (54), with the observer assigning a numerical value to the sample. The measurement of semen quality using conventional techniques has been standardized by the World Health Organization (56). Frequently scored on a 5-point scale (0–4), such assignment has been subject to variable interpretation. Thus, does a sample with 75% of the sperm swimming in a progressive manner rate a higher (better) score than one with 50% of the sperm swimming, but at a faster rate and in a more active manner than the first? Such questions have been addressed by modification and extension of the rating scale (57).

Early reports of quantitation of sperm motion were based on painstaking measurement and analysis of individual frames of a film or videotape record or used computer-assisted plotting of sperm motion on a

TABLE 3.3. Motion analysis parameters.

Frame rate (frames/sec)	30
Duration of data capture (frames)	90
Minimum path length (frames)	30
Minimum motile speed (μm/sec)	10
Maximum burst speed (μm/sec)	350
Distance scale factor (μm/pixel)	1.3880
Camera aspect ratio	1.0590
ALH path smoothing factor (frames)[a]	7
Cent. X search neighborhood (pixels)[b]	4
Cent. Y search neighborhood (pixels)[b]	2
Cent. cell size minimum (pixels)[c]	2
Cent. cell size maximum (pixels)[c]	10
Path max. interpolation (frames)[d]	5
Path prediction percentage (%)[e]	0
Depth of sample (μm)	10
Video processor model	VP110

[a] Factor used to reduce the effect of random movement on calculation of lateral head displacement (ALH).
[b] Area of the frame (in pixels) to be checked as part of the area into which the subject could move from one frame to the next.
[c] Definition of the subject size to be considered as a sperm head.
[d] Number of frames to be used to account for missing data due to sperm swimming "out of focus."
[e] This value accounts for the predictability of the sperm path. As sperm do not travel in a predictable direction, this value is low; here, it is set at zero.

frame-by-frame basis (58, 59). Methods are now available for sophisticated analysis of sperm viability, motility, fertilizing capacity, and membrane composition (37, 60–63). *Computer-assisted motion analysis* (CAMA) of sperm motion characteristics is practical using videotape recordings, as they are based upon digital images derived from a *charge-coupled device* (CCD) with functional dimensions measured in pixel units. Current CAMA makes it practical to analyze a sufficient number of samples/individual sperm to be of use in the determination of motility parameters that may be correlated to potential fertilizing capacity (61, 64, 65). There are several commercially available CAMA systems, and the data presented here are derived from Motion Analysis Systems (Santa Rosa, CA) equipment and programs using a VP110 processor (37).

It must be emphasized, however, that the reproducibility by CAMA of sperm motion parameters is dependent on the baseline parameters provided by the investigator with regard to recorder calibration, field sampling size and scale, and limiting values for such parameters as velocity and path linearity (60, 61, 66); thus, it is important when translating information from one species to another and from one laboratory to another to be aware of the measurement parameters used. The potential for fertility evaluation of currently available methods has been critically evaluated (55). Information from CAMA is recovered by analysis of video recordings of sperm motility made immediately (<30 min) after collection, after liquefaction of the semen sample (2–4 h after collection) and, when used for *sperm penetration assay* (SPA) analysis, after overnight incubation at 4°C (67). Our analysis of great ape semen is made using CAMA according to the parameters shown in Table 3.3. Current CAMA makes it practical to analyze a sufficient number of samples/individual sperm to be of use in the determination of motility

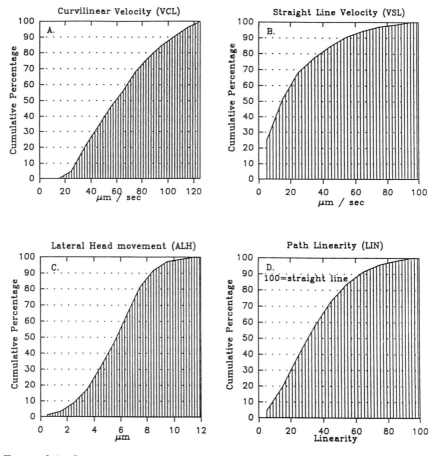

FIGURE 3.1. Cumulative histograms of motion parameters of chimpanzee sperm. Data were derived from analysis of at least 4500 sperm from 6 adult males using the parameters described in the text.

parameters that may be correlated to the potential fertilizing capacity of chimpanzee sperm (Fig. 3.1). Such analysis has not yet been routinely applied to the other apes, although preliminary data suggest that CAMA can provide useful correlative information (Fig. 3.2) (68).

There is relatively limited information available on the normal endocrine profiles of developing and adult apes (Fig. 3.3) and on the response of males to endocrine manipulation designed to test the integrity of the hypothalamo-pituitary axis; most available information relates to the chimpanzee (22, 24, 69–75) (Figs. 3.4 and 3.5). In general, the

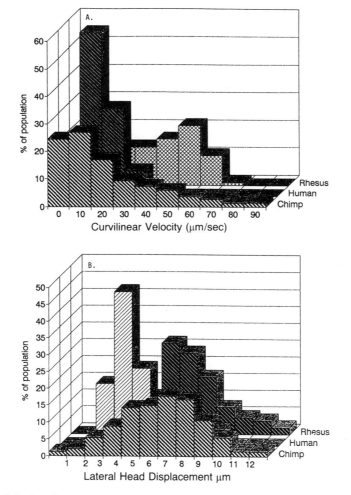

FIGURE 3.2. Species comparison for two parameters of sperm motility. Data were derived from analysis of at least 2500 sperm for each species.

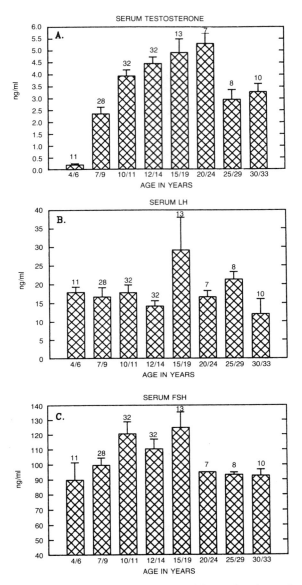

FIGURE 3.3. Changes in serum hormone levels with age for the male chimpanzee. Numbers above each bar are the sample size. Data are presented as mean ± SEM.

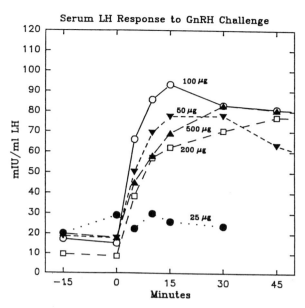

FIGURE 3.4. Serum LH response in the adult male chimpanzee to IV challenge with GnRH.

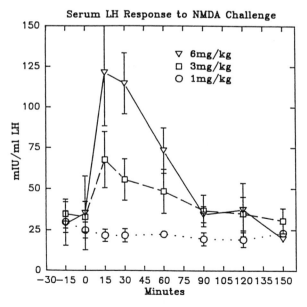

FIGURE 3.5. Serum LH response in the adult male chimpanzee to IV challenge with NMDA.

normal levels of circulating hormones and the normal levels for pituitary response to challenge for the great apes are within the range considered normal for the human male (76). There is an indication that altered pituitary hormone levels are associated with infertility in the gorilla (23); this has suggested the use of clomiphene citrate (Clomid) in a manner analogous to that used in the human male (77–79) as a method for stimulation of pituitary function in order to improve semen quality. Infertility of other male apes has not been demonstrated as a result of a spontaneous/pathological alteration in endocrine profile.

At this time, there is little evidence of male infertility in the apes as a result of infection (35). It has been shown that the male chimpanzee is susceptible to experimental infection with *Mycoplasma genitalium* with the induction of an obvious genital tract infection and shedding of the organism (80, 81). While spontaneous infection with ureaplasmas and mycoplasmas has been demonstrated by recovery of the organisms from the genital tract of approximately one-third of males evaluated, it has not been possible, as in humans, to correlate those isolations with infertility (80, 81). The possible role of *Chlamydia spp.* has not been evaluated. *Herpes hominis* type II (human genital herpes) has been identified in the chimpanzee (82), but again, in a manner similar to that in humans, despite the presence of clear and characteristic lesions, the infection has not been correlated with infertility. Earlier efforts to associate an apparent lowered fertility of the captive male gorilla population with mumps infection were inconclusive. There has been a single report of arteriosclerosis of the spermatic arteries of the chimpanzee that is possibly associated with infertility despite continued spermatogenesis. That association, however, is circumstantial (83).

At this time, therefore, it seems more likely that social effects, rather than infectious agents, are involved in reduction of male productivity in the apes. This analysis is supported by the successful breeding of more than 25% of gorilla males previously considered infertile following introduction to novel partners and situations according to the recommendations of the Gorilla SSP. In light of the above observations, it is not surprising that literature on the correction of male ape infertility is meager.

Female Evaluation

With regard to the female, especially of the chimpanzee, the situation is somewhat different. There is documentation of the physical and endocrinological parameters of the normal menstrual cycle for the chimpanzee (1, 2, 19, 26, 69, 84–87), gorilla (21, 25, 88), and, to a lesser extent, the orangutan and gibbon (27, 28, 89–91). Attempts, or failed attempts, at natural and/or artificial breeding of the species have prompted physical and endocrinological evaluations (88, 92, 93). There is a growing documentation of normal and abnormal parameters for the menstrual cycle and for pregnancy (2, 94–98).

In studies related to analysis of luteal phase adequacy of the chimpanzee, the ano-genital swelling of 9 common chimps has been closely monitored (96) using a refined methodology developed with *Pan paniscus* as subjects (99). This noninvasive procedure of visual documentation clearly involves the minimum amount of manipulation of the subject, but has the potential for recognizing subtle differences in the progress of the ovarian cycle. Use of this monitoring method permitted the detection of cycles in which there was an *SLL* pattern of swelling—that is, a relatively *short* preswollen phase, followed by a *long* swollen phase, and a *long* postswollen phase. Dahl et al. (96) found that SLL cycles had relatively higher levels of *pregnanediol* (PdG) in the luteal phase (3000–3500 ng/mg Cr, days 6–9) (Fig. 3.6) and that ovulation occurred on either day 15 or 16 of the swollen phase so that ovulation could be predicted and AI timed accordingly. Conversely, it would appear that the *long* preswollen phase in *LSS* cycles might be predictive of relatively low progesterone concentrations in the luteal phase; recognition of this pattern could be used as an indicator that the post-insemination luteal phase might require progesterone supplementation as in the gorilla (88).

At this time there is limited information on the successful correction of endocrine infertility other than the termination of such otherwise "natural" phenomena as lactational amenorrhea. Suckling in humans is well documented to have a contraceptive effect as it produces a prolonged postpartum amenorrhea (100, 101). The effectiveness of suckling as a contraceptive in women may be limited to the first 6 weeks postpartum and to the frequency of the suckling behavior. Although menstrual cyclicity may recommence in the first 6 months postpartum, pregnancy rates still remain low.

Monitoring the Yerkes colony of chimpanzees for subjects suitable for artificial insemination—with a focus on those with a history of poor fertility but a valuable genome—involves identification of postpartum individuals with a *prolonged postpartum amenorrhea* (pPA) (102). This has been identified as being longer than 60 days and associated with stimulation of the nipple in the absence of an infant; stimulation can be either by the individual or by a cagemate (102). The amenorrhea appears to be caused by an increase in prolactin concentration consequent to the *nipple stimulation behavior* (NSB); an increase in prolactin is found in humans on self-nipple stimulation (103). Treatment with a dopamine receptor agonist (pergolide) disrupts the prolactin release, and subjects return to cyclicity (102). Experience with the chimpanzee colony at Yerkes reveals that many subjects who have returned to cyclicity postpartum or after treatment with pergolide during a pPA fail to become

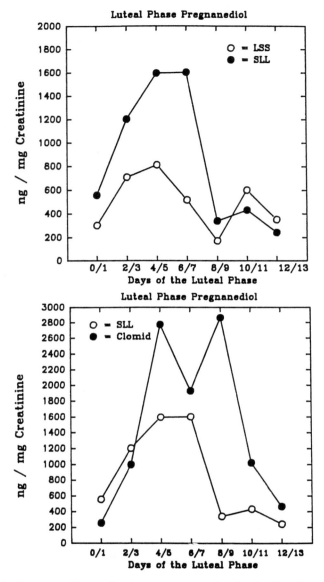

FIGURE 3.6. Concentrations of urinary pregnanediol in the female chimpanzee related to the perineal swelling pattern (top) or supplementation with clomiphene citrate (50 mg/day) during the follicular phase of the cycle (bottom).

pregnant even though exposed to a fully functional and fertile male or repeated AI procedures.

The swelling patterns of these subjects were scrutinized by comparison with the two distinctive types previously identified (96). The relatively infertile subjects were found to exhibit LSS cycles and cycles with erratic swelling patterns unlike LSS or SLL patterns. The behavior of subjects with LSS and erratic patterns was observed in order to determine whether nipple stimulation was occurring and with what intensity. All subjects examined to date with erratic cycle patterns exhibited NSBs, albeit with some variability, and prolactin was again implicated as a proximate cause. Above some critical level it was possible that prolactin might be disrupting either the follicular or the luteal phase. It was hypothesized that the latter was more likely due to the association between prolactin levels and progesterone secretion from the corpus luteum (101, 104).

An attempt to explore this hypothesis was made by monitoring NSBs during treatment with pergolide (the *dopamine receptor agonist* [DRAg]) using subjects exhibiting either pPA or erratic cycles. The DRAg did not extinguish the behavior in most cases, and NSBs rose to high levels in some pPA subjects during the swelling cycle occurring consequent to the DRAg treatment. Erratic cyclicity continued during the treatment, and it was concluded that (i) a factor other than prolactin levels had to be implicated in the disruption of the ovarian cycle in these subjects, and (ii) prolactin was not part of the positive feedback mechanism that was reinforcing the NS behavior. An alternative potential candidate for this effect is *oxytocin* (OT), as this is secreted consequent to suckling and is presumably released when the nipple is self-suckled or manipulated during NSBs. Thus, it appears that breeding of chimpanzees exhibiting contraceptive NSBs may be assisted by appropriate treatment with an OT antagonist; OT appears to be the releaser for prolactin (105, 106) and appears to have a role in the function of the corpus luteum (107, 108).

There are sporadic reports of abnormal endocrine development (4), and such pathological situations as the presence of extensive pelvic adhesions, oviductal occlusion (35, 98), and membranous dysmenorrhoea (endometrial casting) have also been recorded (109, 110). Although it has been suggested that endometrial casting is associated with the loss of an extrauterine pregnancy (111), this does not appear to be the case in those instances studied in the chimpanzee. The etiology of this event thus remains unknown. However, although the fertility/parity of the females described from the TNO colony was not given (109), females in the Yerkes colony that have demonstrated endometrial casting have subsequently conceived and delivered normal infants (110). Endometriosis has not been identified as contributing to infertility in the apes, although documentation in other nonhuman primates and in the woman provides no reason to believe that it cannot occur.

Abortion in the gorilla has been associated with infection by *Shigella flexneri*, but this has not been demonstrated for the chimpanzee (112). It has been shown that the female chimpanzee is susceptible to experimental infection with *Mycoplasma genitalium* with the induction of an obvious genital tract infection and shedding of the organism (81). While spontaneous infection with ureaplasmas and mycoplasmas has been demonstrated by recovery of the organisms from the genital tract of approximately two-thirds of females evaluated, as in women, it has not been possible to correlate those isolations with reduced fertility (80, 81). The possible role of *Chlamydia spp.* and other organisms has not been evaluated.

Assisted Reproduction

Assisted reproduction in the apes, then, comprises the use of techniques directed to conservation of the species as well as to alleviation of infertility per se. With regard to the male, this is achieved by collection and storage of semen, using collection methods and storage methods derived from domestic animals and the human. Semen can be collected from apes by a number of methods. These include recovery after deposition within or near the female tract (50), by collection at time of automasturbation (48), by collection into an artificial vagina (113), or by collection at the time of electrical stimulation of the pelvic organs of an anesthetized male (114, 115). The quality of the collected semen varies according to the collection method used (116, 117), as well as with individual male characteristics. Thus, certain males will provide their best samples after masturbation, but others will provide poorer samples by this method as a consequence of certain behaviors associated with the collection; for example, placing the ejaculate in the mouth prior to donation to the collector. In addition, while spontaneous masturbation often results in the ejaculation of high-quality samples, by the time of collection of such samples, the fraction of the sample that was liquid at the time of release may have been removed or depleted, a maneuver that can significantly reduce the total sperm count of the sample.

There have been some documented differences in semen parameters and quality associated with method and frequency of collection (42, 117), but the correlation of motility parameters with fertilizing capacity has not been adequately demonstrated. In the chimpanzee there is a suggestive, though as yet insignificant, association of alteration in swimming speed with season and birth rate subsequent to natural mating, with lower speed being associated with reduced birth rate (118) (Fig. 3.7). Further, there is a significant correlation of higher linearity of sperm motility with successful artificial insemination (41.7 ± 2.7 vs. 30.1 ± 3.6; $P = \langle 0.02 \rangle$) (43). The techniques of semen collection, evaluation, concentration, and freezing will be of great assistance in those cases where semen quality is a limiting incompatibility. In addition, should a genetically valuable pair of individuals not mate, or if they are separated geographically, artificial insemination can be invoked.

It is important, however, to realize that the present success rate for AI is only around 22% for the chimpanzee (49, 119), less for the gorilla (~15%),

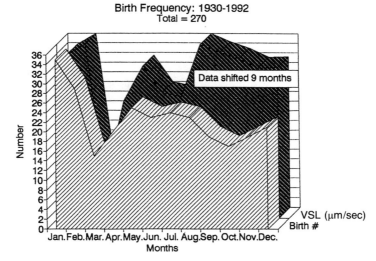

FIGURE 3.7. Relationship of birth frequency to sperm straight-line velocity (VSL). Data have been shifted by 255 days to account for the gestation period. The correlation approaches significance (P = 0.058).

and has yet to be demonstrated in the bonobo and orangutan. During the past five years, the control rate for AI using our standard method (49, 120) was 27% (11/40). However, review of the history using this method suggests that when insemination was performed within a hormonally synchronized cycle *and* was performed on day 15 or 16 of the swollen phase, as monitored by a refined scaling system for documentation of changes in perineal swelling (96, 99), the success rate after a single intrauterine insemination of $>15 \times 10^6$ motile sperm was 71% (5/7). However, it is not yet possible to predict when that precise timing will occur, suggesting a need for further refinement of our methods for monitoring follicle development. Such refinement is constrained by the necessity for anesthesia of the female prior to ultrasound monitoring of follicle development or blood collection for assay of serum hormone levels, which emphasizes the value of detailed and controlled inspection of the ano-genital swelling.

With regard to the female, assisted reproduction is achieved by careful documentation of the menstrual cycle (96), reinforcement of the natural cycle (49), monitoring of the natural time of ovulation and/or induction of ovulation with hCG (49, 96, 120), attempted stimulation of multiple-follicle development and collection of many oocytes for in vitro fertilization (35), and progesterone supplementation of the luteal phase to increase the chances of successful implantation and pregnancy (88).

Induction of multiple-follicle development in the chimpanzee (Table 3.1) and gorilla (reviewed in 35) has not usually been as successful as in the woman; however, this situation may be overcome by the use of a more aggressive induction regimen using Metrodin administration (6 ampules/day) after pituitary suppression by administration of GnRH agonist (35). This requirement may reflect the source of the hormones used. It is not usually practical to use PMSG or other antigenic agents in the apes as a result of the endangered status of the species and the potential risk of overall impairment of fertility associated with antibody production (121, 122). However, a combination of techniques used for induction of follicle development in women has recently been shown to be effective in the gorilla, with subsequent recovery of oocytes using transrectal ultrasound-guided follicle aspiration (35). IVF has been demonstrated in the chimpanzee and gorilla (98, 123, 124) using methods directly derived from those reported for the woman, but term pregnancy has not resulted from embryo transfer (35, 98, 123, 124), probably as a result of the few embryos available for transfer. This situation can be addressed either by improvement of methods for induction of multiple-follicle development or by the cryopreservation of IVF embryos over several cycles (125–128). The logistical feasibility of such an approach is, however, doubtful.

Luteal phase inadequacy has been implicated as a factor for reduced fertility in the gorilla with continued pregnancy resulting from progesterone supplementation of the luteal phase (88). In chimpanzees the detection of cycles in which there is an LSS pattern of swelling (i.e., a relatively long preswollen phase followed by a short swollen and short postswollen phase prior to menstruation) is associated with relatively low levels of PdG during the luteal phase (1500–1900 ng/mg Cr). Ovulation cannot be predicted adequately for AI (Fig. 3.6), and a clomid/hCG regimen for synchronization of ovulation time (49, 129) is indicated.

Suggested Protocols for Assisted Breeding

While it is evident that each situation must be considered alone, there are certain stages to the application of assisted breeding for the great apes that can be used in all cases. First, it is important to evaluate the social situation: Is the "infertile" pair socially compatible; has breeding activity been observed and if so, has it occurred with any regular pattern that could relate to the female cycle? Second, when the pair is compatible and breeding has been observed, it is necessary to consider the ages of both individuals: Are both of an age considered potentially fertile—that is, approximately 7–9 for males and 9–11 for females? Third, is there evidence of cyclicity in the female as shown by regular perineal swelling and overt menses in the chimpanzee spp. and possible evidence of menses in the gorilla? In the orangutan and gorilla, it may be necessary to arrange for collection of urine on as close to a daily basis as possible to monitor the urine for blood using Hemastix. It may be possible to train the females for genital inspection to permit monitoring of changes in labial tumescence (25, 130). It is evident that these evaluations will, themselves, be time consuming and involve a number of months.

Should the above evaluations fail to provide reasons for the lack of pregnancy, then it is logical to proceed with more specific evaluation. For the male, this would comprise immobilization, collection of blood for analysis of serum levels of steroid and protein hormones, collection of semen sample(s) by use of *rectal probe electroejaculation* (RPE) if the male would not provide samples by other means, and analysis of semen parameters using conventional (56) or CAMA (43, 55) analysis. If practical, functional analysis using SPA (67, 131, 132) would also be appropriate. If the results of the first assay show semen quality to be poor, this would prompt a second evaluation.

It is particularly evident in the gorilla that results of semen analysis are difficult to use in a predictive fashion. Any sample that contains motile sperm must be considered indicative of potential fertility (23, 133). Only after repeated failure to obtain sperm-rich semen samples should more invasive diagnostic procedures, such as testis biopsy, be undertaken. Also, in the gorilla as well as other species, recovery of poor samples should prompt the use of more complex semen collection techniques to ensure that the poor results obtained are not the result of retrograde ejaculation at the time of electrostimulation (133). Although not yet documented, it is possible that retrograde ejaculation could account for apparent male infertility at the time of natural mating.

If the female shows no evidence of menstrual cyclicity or irregular cyclicity, then cycle synchronization can be undertaken using administration of progestational agents with the intent of synchronizing the cycle subsequent to progesterone withdrawal (134) or by suppression of the endogenous cycle with GnRH agonist as suggested for synchronization of the human cycle (33, 135). In addition, when practical, the perineal swelling should be observed on a daily or semidaily basis in order to determine the type of cycle (SLL or LSS) being exhibited in order to time an artificial insemination attempt or to predict the need for use of progesterone supplementation in the luteal phase. It is pertinent to note that for some chimpanzee females at the Yerkes Center, a series of several LSS cycles were infertile, while the first SLL cycle was associated with successful induction of pregnancy after natural mating.

Should there remain no evidence of a potentially fertile cycle pattern, then synchronization of the cycle to permit use of AI should be undertaken (49). The overnight preincubation of semen at wet ice temperature (67) results in a significant increase in the percentage penetration of oocytes in the SPA. While that observation has yet to be correlated with improved results of AI in the chimpanzee, it does suggest a possible advantage to the use of a combined fresh and incubated semen sample for AI in the great apes.

When these approaches have been shown to be unsuccessful, it is appropriate to proceed with techniques to verify normal female function. Such assays involve use of ultrasound to document the appropriate development of ovarian follicles in the menstrual cycle. Lack of such development would be expected to have been detected by alteration in the endocrine pattern of the menstrual cycle as monitored by urinary hormone assay (86, 88, 136). In the situation where the duration and endocrinology of the menstrual cycle appear normal, but there is no pregnancy subsequent to repeated exposure to an apparently fertile male, it is appropriate to proceed to evaluation of the female using more invasive techniques—such as hysterosalpingography (124, 137) and laparoscopy for identification of pelvic pathology (138, 139)—and to consideration of ovarian stimulation for purposes of oocyte recovery and IVF (35, 140).

It is evident that the more invasive methods for fertility assistance in the great apes still require development. Thus, although demonstrated to be feasible, AI cannot be viewed as a panacea, as the realistic success rate is less than 25%, with the implication that more than 4 cycles of intervention must be undertaken to achieve a single pregnancy. Such technological advances as gamete and embryo micromanipulation, though demonstrated in other species both human and nonprimate, have yet to be developed for use in the great apes. However, procedures demonstrated in the human for microinjection of spermatozoa with alleviation of infertility associated with low sperm number and high rate of sperm abnormality hold promise for application in such species as the gorilla when "normal" semen parameters are poor. Such potential uses have recently been enumerated (98).

It is evident that a priority for assisted fertilization in the great apes must be the diagnosis of the reason for nonproductivity, a diagnosis that precedes that of infertility. The significant role of social and environmental factors must be carefully considered prior to initiation of more aggressive techniques for diagnosis and correction of infertility and establishment of pregnancy. The availability of sophisticated methods for ovulation induction, oocyte recovery, IVF, and gamete cryopreservation will have an increasing role in the conservation of endangered species, both as methods for assisted reproduction and, potentially, for reproductive amplification by use of related species as surrogate mothers in a manner analogous to that demonstrated in nonprimate species.

Acknowledgments. This research was supported, in part, by NIH Grant RR-00165 from the National Center for Research Resources to the Yerkes Regional Primate Research Center and by Grants RR-03587,

HD-26076, and RR-05994 to Kenneth G. Gould. The Yerkes Center is fully accredited by the American Association for Accreditation of Laboratory Animal Care.

References

1. Graham CE. The chimpanzee: a unique model for human reproduction. In: Antikatzides Th, Erichsen S, Spiegel A, eds. The laboratory animal in the study of reproduction: 6th ICLA symposium, Thessaloniki, 1975. Stuttgart, New York: G.F. Verlag, 1976:29–38.
2. Graham CE, Gould KG, Collins DC, Preedy JRK. Regulation of gonadotropin release by luteinizing hormone-releasing hormone and estrogen in chimpanzees. Endocrinology 1979;105:269–75.
3. Winter JSD, Faiman C, Hobson WC, Prasad AV, Reyes FI. Pituitary-gonadal relations in infancy, I. Patterns of serum gonadotropin concentrations from birth to four years of age in man and chimpanzee. J Clin Endocrinol Metab 1975;40:545–51.
4. Winterer J, Merriam GR, Gross E, et al. Idiopathic precocious puberty in the chimpanzee: a case report. J Med Primatol 1984;13:73–9.
5. Socha WW. Blood groups as genetic markers in chimpanzees: their importance for the national chimpanzee breeding program. Am J Primatol 1981;1:3–13.
6. Murayama Y, Fukao K, Noguchi A, Takenaka O. Epitope expression on primate lymphocyte surface antigens. J Med Primatol 1986;15:215–66.
7. Haas GG, Nahhas F. Failure to identify HLA, ABC and Dr antigens on human sperm. Am J Reprod Immunol Microbiol 1986;10:39–46.
8. Bontrop RE, Broos LAM, Pham K, Bakas RM, Otting N, Jonker M. The chimpanzee major histocompatibility complex class II *DR* subregion contains an unexpectedly high number of beta-chain genes. Immunogenetics 1990;32:272–80.
9. Martin DE, Gould KG. The male ape genital tract and its secretions. In: Graham CE, ed. Reproductive biology of the great apes: comparative and biomedical perspectives. New York: Academic Press, 1981:127–62.
10. Durrant B. Semen collection, evaluation and cryopreservation in exotic animal species: maximizing reproductive potential. ILAR News 1990;32:2.
11. Ballou JD. Potential contribution of cryopreserved germ plasm to the preservation of genetic diversity and conservation of endangered species in captivity. Cryobiology 1992;29:19–25.
12. Riesen A. Introduction. In: Bourne GH, ed. Progress in ape research. New York: Academic Press, 1977:1–4.
13. Yerkes RM, Elder JH. Oestrus, receptivity and mating in the chimpanzee. Comp Psychol Monogr 1936;13:1–39.
14. Young WC, Yerkes RM. Factors influencing the reproductive cycle in the chimpanzee: the period of adolescent sterility and related problems. Endocrinology 1943;33:121–54.
15. Elder JH, Yerkes RM. The sexual cycle of the chimpanzee. Anat Rec 1936;76:119–43.
16. Yerkes RM, Yerkes AW. The great apes. New Haven: Yale University Press, 1929.

17. Clark G, Birch HG. Observations on the sex skin and sex cycle in the chimpanzee. Endocrinology 1948;43:218–31.
18. Graham CE. A survey of advances in chimpanzee reproduction. In: Bourne GH, ed. Progress in ape research. New York: Academic Press, 1977:177–90.
19. Hobson W, Fuller GB. LH-RH induced gonadotropin release in chimpanzees. Biol Reprod 1977;17:294–7.
20. Notelovitz M. Using devices that measure bone mass. Technology 1985:61–7.
21. Nadler RD, Graham CE, Collins DC, Gould KG. Plasma gonadotropins, prolactin, gonadal steroids, and genital swelling during the menstrual cycle of lowland gorillas. Endocrinology 1979;105:290–6.
22. Nadler RD, Wallis J, Roth-Meyer C, Cooper RW, Baulieu EE. Hormones and behavior of prepubertal and peripubertal chimpanzees. Horm Behav 1987;21:118–31.
23. Gould KG, Kling OR. Fertility in the male gorilla (*Gorilla gorilla*): relationship to semen parameters and serum hormones. Am J Primatol 1982;2:311–6.
24. Martin DE, Swenson RB, Collins DC. Correlation of serum testosterone levels with age in male chimpanzees. Steroids 1977;29:471–81.
25. Nadler RD, Collins DC, Miller LC, Graham CE. Menstrual cycle patterns of hormones and sexual behavior in gorillas. Horm Behav 1983;17:1–17.
26. Reyes FI, Winter JSD, Faiman C, Hobson WC. Serial serum levels of gonadotropins, prolactin and sex steroids in the non-pregnant and pregnant chimpanzee. Endocrinology 1975;96:1447–53.
27. Nadler RD. Reproductive behavior and endocrinology of orang utans. In: de Boer LEM, ed. The orang utan: its biology and conservation. The Hague, Holland: Dr. W. Junk, 1982:231–48.
28. Inaba T, Imori T, Saburi T. Urinary estrogen levels during the menstrual cycle of the orangutan. Jpn J Vet Sci 1983;45:857–9.
29. Nicholson NA. Maternal behavior in human nonhuman primates. In: Loy JD, Peters CB, eds. Understanding behavior: what primate studies tell us about human behavior. New York: Oxford University Press, 1991.
30. Hannah AC, Brotman B. Procedures for improving maternal behavior in captive chimpanzees. Zoo Biol 1990;9:233–40.
31. Fritz J, Fritz P. The hand-rearing unit: management decisions that may affect chimpanzee development. In: Clinical management of great apes. New York: A.R. Liss, 1985:1–34.
32. Bard K. Maternal competence in chimpanzees. Early Dev Parenting 1992.
33. Patton PE, Eaton D, Burry KA, Wolf DP. The use of gonadotropin-releasing hormone agonist to regulate oocyte retrieval time. Fertil Steril 1990;54:652–5.
34. Gould KG. Techniques and significance of gamete collection and storage in the great apes. J Med Primatol 1990;19:537–51.
35. Hatasaka HH, Schaffer NE, Chenette PE, et al. Strategies for ovulation induction and oocyte retrieval in the lowland gorilla. Theriogenology 1992.
36. Harcourt AH, Fossey D, Stewart KJ, Watts DP. Reproduction in wild gorillas and some comparisons with chimpanzees. J Reprod Fertil 1980;suppl 28:59–70.

37. Gould KG, Styperek RP. Improved methods for freeze preservation of chimpanzee sperm. Am J Primatol 1989;18:275–84.
38. Dahl JF, Nadler RD. The male external genitalia of the extant Hominidae. Am J Phys Anthropol 1990;81:211–2.
39. Crisp E. On the os penis of the chimpanzee (*Troglodytes niger*) and of the orang (*Simia satyrus*). Proc Zoolog Soc Lond 1865:48–9.
40. Mayer C. Zur antomie des orang-utang und des chimpanse. Archiv fur Naturgeschichte, Berlin 1856;22:281–304.
41. Hill WCO. Note on the male external genitalia of the chimpanzee. Proc Zoolog Soc Lond 1946;116:129–32.
42. Marson J, Gervais D, Meuris S, Cooper RW, Jouannet P. Influence of ejaculation frequency on semen characteristics in chimpanzees (*Pan troglodytes*). J Reprod Fertil 1989;85:43–56.
43. Gould KG, Young LG, Smithwick E, Phythyon S. Semen characteristics of the adult male chimpanzee (*Pan troglodytes*). Am J Primatol 1992.
44. Marson J, Meuris S, Cooper RW, Jouannet P. Puberty in the male chimpanzee: progressive maturation of semen characteristics. Biol Reprod 1991;44:448–55.
45. Vogel K, Gould KG. Seasonal change in semen parameters in the chimpanzee associated with fertilizing capacity [Abstract]. Fertility in the Great Apes 1989:46.
46. Dahl JF. The external genitalia of the orang utan. In: Schwartz JH, ed. The biology of the orang utan. New York: Oxford University Press.
47. Dixson AF, Knight J, Moore HDM, Carman M. Observations on sexual development in male orang-utans (*Pongo pygmaeus*). Int Zoo Yearbook 1982;22:222–7.
48. VandeVoort CA, Neville LE, Tollner TL, Field LP. Non-invasive semen collection from adult orang-utans. Zoo Biol 1992.
49. Gould KG, Martin DE. Artificial insemination of nonhuman primates. In: Benirschke K, ed. Primates, the road to self-sustaining populations. New York: Springer Verlag, 1986:425–43.
50. Bader H. Electroejaculation in chimpanzees and gorillas and artificial insemination in chimpanzees. Zoo Biol 1983;2:307–14.
51. Seuanez HN, Carothers AD, Martin DE, Short RV. Morphological abnormalities in spermatozoa of man and great apes. Nature 1977;270:345–7.
52. Matern B. Problems, experiences in performing artificial insemination in bonobos (*Pan paniscus*). Zoo Biol 1983;2:303–6.
53. Dahl JF, Gould KG, Nadler RD. Testicle size of orang-utans in relation to body size. Am J Phys Anthropol 1992.
54. Emmens CW. The motility and viability of rabbit spermatozoa at different hydrogen ion concentrations. J Physiol 1947;106:471–81.
55. Comhaire FH, Huysse S, Hinting A, Vermeulen L, Schoonjans F. Objective semen analysis: has the target been reached? Hum Reprod 1992;7:237–41.
56. World Health Organization. Laboratory manual for the examination of human semen and semen-cervical mucus interaction. Cambridge, UK: Cambridge University Press, 1987:1–67.
57. Harrison RM. Normal sperm parameters in *Macaca mulatta*. Lab Primate Newslett 1975;14:10–3.

58. Holt WV, Moore HDM, Hillier SG. Computer assisted measurement of sperm swimming speed in human semen; correlation of results with in vitro fertilization assays. Fertil Steril 1985;44:112–9.
59. Katz DF. Sperm motility assessment by videomicrography. Fertil Steril 1981;35:188–93.
60. Mack SO, Wolf DP, Tash JS. Quantitation of specific parameters of motility in large numbers of human sperm by digital image processing. Biol Reprod 1988;38:270–81.
61. Knuth UA, Nieschlag E. Comparison of computerized semen analysis with the conventional procedure in 322 patients. Fertil Steril 1988;49:881–5.
62. Gould KG, Young LG. Acquisition of fertilizing capacity by chimpanzee sperm. Folia Primatol 1990;54:105–8.
63. Young LG, Gould KG. Surface components of ejaculated chimpanzee sperm. Arch Androl 1982;8(1):15–20.
64. Vantman D, Banks SM, Koukoulis G, Dennison L, Sherins RJ. Assessment of sperm motion characteristics from fertile and infertile men using a fully automated computer-assisted semen analyzer. Fertil Steril 1989;51(1):156–61.
65. Chan SYW, Wang C, Ng M, et al. Evaluation of computerized analysis of sperm movement characteristics and differential sperm tail swelling patterns in predicting human sperm in vitro fertilizing capacity. J Androl 1989;10:133–8.
66. Mack SO, Tash JS, Wolf DP. Effect of measurement conditions on quantification of hyperactivated human sperm subpopulations by digital image analysis. Biol Reprod 1989;40:1162–9.
67. Bolanos JR, Overstreet JW, Katz DF. Human sperm penetration of zona-free hamster eggs after storage of the semen for 48 hours at 2°C to 5°C. Fertil Steril 1983;39:536–41.
68. Holt WV. Collection, assessment, and storage of sperm. In: Benirschke K, ed. Primates: the road to self-sustaining populations. New York: Springer-Verlag, 1986:413–24.
69. Hobson WC, Fuller GB, Winter JSD, Faiman C, Reyes FI. Reproductive and endocrine development in the great apes. In: Graham CE, ed. Reproductive biology of the great apes. New York: Academic Press, 1981:83–103.
70. Doering CH, McGinnis PR, Kraemer HC, Hamburg DA. Hormonal and behavioral response of male chimpanzees to a long-acting analogue of gonadotropin-releasing hormone. Arch Sex Behav 1980;9:441–50.
71. Young LG, Gould KG, Smithwick EB. Selected endocrine parameters of the adult male chimpanzee. Am J Primatol 1992.
72. Marson J, Meuris S, Cooper RW, Jouannet P. Puberty in the male chimpanzee: time-related variations in luteinizing hormone, follicle-stimulating hormone, and testosterone. Biol Reprod 1991;44:456–60.
73. Copeland KC, Eichberg JW, Parker CR, Bartke A. Puberty in the chimpanzee: somatomedin-C and its relationship to somatic growth and steroid hormone concentrations. J Clin Endocrinol Metab 1985;60(6):1154–60.
74. Fuller GB, Faiman C, Winter JSD, Reyes FI, Hobson WC. Sex-dependent gonadotropin concentrations in infant chimpanzees and rhesus monkeys. Proc Soc Exp Biol Med 1982;169:494–500.

75. Nadler RD, Roth-Meyer C, Wallis J, et al. Hormonal and compartmental correlates during adrenarche in the chimpanzee. C R Acad Sci [III] 1984;298:409–13.
76. Young LG, Gould KG, Smithwick E. Endocrine parameters of the adult male chimapanzee (*Pan troglodytes*). Am J Primatol 1992.
77. Fishel SB, Webster J, Faratian B, Jackson P, Shelton K, Johnson J. Establishing pregnancies after follicular stimulation for IVF with clomiphene citrate and human menopausal gonadotrophin only. Hum Reprod 1991;6:106–12.
78. Check JH, Nowroozi K, Lee M, Adelson H, Katsoff D. Evaluation and treatment of a male factor component to unexplained infertility. Arch Androl 1991;25:199–211.
79. Marshburn PB, Stovall DW, Hammond MG, Talbert LM, Shabanowitz RB. Fertility rates in men with normal semen characteristics: spermatozoal testing by induction of the acrosome reaction and Wright-Giemsa staining for subtle abnormal forms. Obstet Gynecol 1991;77:961–2.
80. Taylor-Robinson D, Barile MF, Furr PM, Graham CE. Ureaplasmas and mycoplasmas in chimpanzees of various breeding capacities. J Reprod Fertil 1987;81:169–73.
81. Tully JG, Taylor-Robinson D, Rose DL, Furr PM, Graham CE, Barile MF. Urogenital challenge of primate species with *Mycoplasma* genitalium and characteristics of infection induced in chimpanzees. J Infect Dis 1986;153:1046–54.
82. McClure HM, Swenson RB, Kalter SS, Lester TL. Natural genital herpesvirus hominis infection in chimpanzees (*Pan troglodytes* and *Pan paniscus*). Lab Anim Sci 1980;30:895–901.
83. Pearson GR, Slinger WB. Arteriosclerosis of the spermatic arteries of a chimpanzee (*Pan troglodytes*). Vet Pathol 1982;19:710–2.
84. Graham CE. Menstrual cycle of the great apes. In: Graham CE, ed. Reproductive biology of the great apes. New York: Academic Press, 1981:1–43.
85. Graham CE, Gould KG, Wright K, Collins DC. Luteal estrogen secretion and decidualization in the chimpanzee. In: Chivers DJ, Ford EHR, eds. Recent Adv Primatol 1978:209–11.
86. Graham CE, Collins DC, Robinson H, Preedy JRK. Urinary levels of estrogens and pregnanediol and plasma levels of progesterone during the menstrual cycle of the chimpanzee: relationship to the sexual swelling. Endocrinology 1972;91:13–24.
87. Young WC, Orbison WD. Changes in selected features of behavior in pairs of oppositely-sexed chimpanzees during the sexual cycle after ovariectomy. J Comp Psychol 1944;37:107–43.
88. Czekala NM, Reichard T, Lasley BL. Assessment of luteal phase competency by urinary hormone evaluation in the captive female gorilla. Am J Primatol 1991;24:283–8.
89. Carpenter CR. The menstrual cycle and body temperature in two gibbons (*Hylobates lar*). Anat Rec 1941;79:291–6.
90. Nadler RD. Sexual and reproductive behavior. In: Schwartz JH, ed. Orangutan biology. New York: Oxford University Press, 1988.
91. Dahl JF, Nadler RD. Genital swelling in females of the monogamous gibbon, *Hylobates (H.) lar*. Am J Phys Anthropol 1992.

92. Kraemer HC, Horvat JR, Doering C, McGinnis PR. Male chimpanzee development focusing on adolescence: integration of behavioral with physiological changes. Primates 1982;23:393–405.
93. Coe CL, Connolly AC, Kraemer HC, Levine S. Reproductive development and behavior of captive female chimpanzees. Primates 1979;20:571–82.
94. Nadler RD, Dahl JF, Collins DC, Gould KG, Wilson DC. Effects of oral contraceptives on chimpanzees: a preliminary report. In: Eley RM, ed. Comparative reproduction in mammals and man. Proc Conf National Centre for Research in Reproduction, Nairobi, 1987:30–3.
95. Pache TD, Hop WC, Wladimiroff JW, Fauser BCJM, De Jong FH. Growth patterns of nondominant ovarian follicles during the normal menstrual cycle. Fertil Steril 1990;54:638–42.
96. Dahl JF, Nadler RD, Collins DC. Monitoring the ovarian cycles of *Pan troglodytes* and *P. paniscus*: a comparative approach. Am J Primatol 1991;24:195–209.
97. Graham CE, Struthers EJ, Hobson WC, Faiman C. Prolonged postpartum amenorrhea in chimpanzees: treatment and etiology. In: Eley RM, ed. Comparative reproduction in mammals and man. National Museums of Kenya, Nairobi, 1989:8–11.
98. Loskutoff NM, Kraemer DC, Raphael BL, Huntress SL, Wildt DE. Advances in reproduction in captive, female great apes: value of biotechniques. Am J Primatol 1991;24:151–66.
99. Dahl JF. Cyclic perineal swelling during the intermenstrual intervals of captive female pygmy chimpanzees (*Pan paniscus*). J Hum Evol 1986;15:369–85.
100. Schams D. Hormonal control of lactation. In: Elliot K, Fitzsimons DW, eds. Breast-feeding and the mother. Amsterdam: North Holland, 1976: 27–43.
101. Tyson JE. Neuroendocrine control of lactational infertility. J Biosoc Sci 1977;4:23–9.
102. Graham CE, Struthers EJ, Hobson WC, et al. Postpartum infertility in common chimpanzees. Am J Primatol 1991;24:245–55.
103. Kolodny RC, Jacobs LS, Darhaday WH. Mammary stimulation causes prolactin secretion in non-lactating women. Nature 1972;238:284–5.
104. McNatty KP, Sawyers RS. Relationship between the endocrine environment within the graffian follicle and the subsequent rate of progesterone secretion by human granulosa cells in vitro. J Endocrinol 1975;66:391–400.
105. Samson WK, Lumpkin MD, McCann SM. Evidence for a physiological role for oxytocin in the control of prolactin secretion. Endocrinology 1986;119:554–60.
106. Pettibone DJ, Clineschmidt BV, Anderson PS, et al. A structurally unique, potent and selective oxytocin antagonist derived from *streptomyces silvensis*. Endocrinology 1989;125:217–22.
107. Luck MR. A function for ovarian oxytocin. J Endocrinol 1989;121:203–4.
108. Bennegard B, Hahlin M, Dennefors B. Antigonadotropic effect of oxytocin on the isolated human corpus luteum. Fertil Steril 1987;47:431–5.
109. Solleveld HA, van Zwieten MJ. Membranous dysmenorrhea in the chimpanzee (*Pan troglodytes*): a report of four cases. J Med Primatol 1978;7:19–25.

110. Gould KG, Martin DE. The female ape genital tract and its secretions. In: Graham CE, ed. Reproductive biology of the great apes: comparative and biomedical perspectives. Academic Press, 1981:105–25.
111. Novak ER, Woodruff JD. Novak's gynecologic and obstetric pathology. Philadelphia: Saunders, 1974.
112. Swenson RB, McClure HM. Septic abortion in a gorilla due to *Shigella flexneri*. Ann Proc Am Assoc Zoo Vet 1974:195–6.
113. Fussell EN, Franklin LW, Franta RC. Collection of chimpanzee sperm with an artificial vagina. Lab Anim Sci 1973;23:252–5.
114. Roussel JD, Austin CR. Improved electroejaculation of primates. J Inst Anim Tech 1968;19:22–32.
115. Gould KG, Warner H, Martin DE. Rectal probe electroejaculation of primates. J Med Primatol 1978;7:213–22.
116. Check JH, Bollendorf AM, Press MA, Breen EM. Noninvasive techniques for improving fertility potential of retrograde ejaculates. Arch Androl 1990;25:271–6.
117. Gould KG, Mann DR. Comparison of electrostimulation methods for semen recovery in the rhesus monkey (*Macaca mulatta*). J Med Primatol 1988;17:95–103.
118. Gould KG, Dahl JF. Reproduction in chimpanzees: with reference to problems of fertility and infertility. Wien Klin Wochenschr 1992.
119. Hardin CJ, Liebherr G, Fairchild O. Artificial insemination in chimpanzees. Int Zoo Yearbook 1975;15:132.
120. Gould KG. Ovulation detection and artificial insemination. Am J Primatol 1982;1:15–25.
121. Wolf DP, Thomson JA, Zelinski-Wooten MB, Stouffer RL. In vitro fertilization-embryo transfer in nonhuman primates: the technique and its applications. Mol Reprod Dev 1990;27:261–80.
122. Lanzendorf SE, Zelinski-Wooten MB, Stouffer RL, Wolf DP. Maturity at collection and the development potential of rhesus monkey oocytes. Biol Reprod 1990;42:703–11.
123. Gould KG. Ovum recovery and in vitro fertilization in the chimpanzee. Fertil Steril 1983;40:378–83.
124. Huntress SL, Loskutoff NM, Raphael BL, et al. Pronucleus formation following in-vitro fertilization of oocytes recovered from a gorilla (*Gorilla gorilla*) with unilateral endometrioid adenocarcinoma of the ovary. Am J Primatol 1989;18:259–66.
125. Troup SA, Matson PL, Critchlow JD, Morroll DR, Lieberman BA, Burslem RW. Cryopreservation of human embryos at the pronucleate, early cleavage, or expanded blastocyst stages. Eur J Obstet Gynecol Reprod Biol 1991;38:133–9.
126. Toner JP, Brzyski RG, Oehninger S, Veeck LL, Simonetti S, Muasher SJ. Combined impact of the number of pre-ovulatory oocytes and cryopreservation on IVF outcome. Hum Reprod 1991;6:284–9.
127. Fehilly CB, Cohen J, Simons RF, Fishel SB, Edwards RG. Cryopreservation of cleaving embryos and expanded blastocysts in the human: a comparative study. Fertil Steril 1985;44:638–44.
128. De Ziegler D, Frydman R. Different implantation rates after transfers of cryopreserved embryos originating from donated oocytes or from regular in vitro fertilization. Fertil Steril 1990;54:682–8.

129. Gould KG, Martin DE, Warner H. Improved method for artificial insemination in the great apes. Am J Primatol 1985;8:61–5.
130. Nadler RD. Cyclicity in tumescence of the perineal labia of female lowland gorillas. Anat Rec 1975;181:791.
131. Yanagimachi R, Yanagimachi H, Rogers BJ. The use of zona-free animal ova as a test system for the assessment of the fertilizing capacity of human spermatozoa. Biol Reprod 1976;15:471–6.
132. Ansari AH, Gould KG. Contraception and the cervix. Adv Contracept 1986;2:101–15.
133. Schaffer NE, Jeyendran RS, Beehler B. Improved sperm collection from the lowland gorilla: recovery of sperm from bladder and urethra following electroejaculation. Am J Primatol 1991;24:265–72.
134. Pope NS, Gould KG. Synchronization of the menstrual cycle through the use of an oral progestin in the rhesus monkey. Theriogenology 1990:34(1).
135. Norman RJ, Warnes GM, Wang X, Kirby CA, Matthews CD. Differential effects of gonadotrophin-releasing hormone agonists administered as desensitizing or flare protocols on hormonal function in the luteal phase of hyperstimulated cycles. Hum Reprod 1991;6:206–13.
136. Czekala NM, Mitchell WR, Lasley BL. Direct measurements of urinary estrone conjugates during the normal menstrual cycle of the gorilla (*Gorilla gorilla*). Am J Primatol 1987;12:223–30.
137. Ramsay E, Tytle T, Carey J. Hysterosalpingography in gorilla. J Zoo Anim Med 1985;16:85–9.
138. Cambre RC, Wildt DE. Laparoscopy of female gorillas as an initial aid in reproductive evaluation. Ann Proc Am Assoc Zoo Vet 1980:42–3.
139. Wildt DE, Chakraborty P, Cambre RC, Howard J, Bush M. Laparoscopic evaluation of the reproductive organs and abdominal cavity content of the lowland gorilla. Am J Primatol 1982;2:29–42.
140. Yee B, Loskutoff NM, Cambre RC, et al. A preliminary study on the use of a long-acting GnRH agonist prior to ovarian stimulation for oocyte retrieval in the western lowland gorilla. Theriogenology 1990;33:358.

4

Assisted Reproduction in New World Primates

W. RICHARD DUKELOW

The first half of this century gave us increased knowledge of the basic physiology of a number of nonhuman primate species. This usually was oriented to increased breeding efficiencies of captive colonies and, occasionally, information that would eventually have application to what we now refer to as the new reproductive biotechnology or, in the human reproductive field, *assisted reproduction*. Such findings related to the natural time of ovulation, methods to induce ovulation, optimal times for mating, methods of semen collection, artificial insemination, and embryo transfer. Probably the first successful *in vitro fertilization* (IVF) in nonhuman primates was by Kraemer with the baboon in 1972.

In the 20 years since, successful IVF has been achieved in 6 nonhuman primate species, with live births resulting from IVF in 3 species. IVF has been achieved in 2 of the New World species: the squirrel monkey (*Saimiri sciureus*) and the common marmoset (*Callithrix jacchus*). IVF of the squirrel monkey was first reported by two groups in 1972, and this model has been extensively used to study the normality of embryonic development, both biochemically and chromosomally.

A variety of follicular induction techniques can be used with the squirrel monkey both to superovulate or to control ovulation to single or double ovulations. Generally, purified FSH has been used for ovulation induction since first reported in this species in 1970 and, combined with *human chorionic gonadotropin* (hCG), for IVF studies since 1972. In the squirrel monkey, repeated use of a moderate regimen of FSH and hCG does not cause refractoriness. *Pregnant mare serum* (PMS) can be used as well, but because of the high antigenicity of the preparation, it has not been used extensively in New World species. Because most embryos produced by IVF in the squirrel monkey were used in metabolic and chromosomal studies, few efforts were made to transfer IVF embryos for pregnancy, and no offspring have been produced by this method from this species. Approximately 2000 squirrel monkey oocytes have been in vitro fertilized and used in metabolic studies.

Studies on the common marmoset have been primarily oriented to preimplantation development and maternal recognition of pregnancy. In this species a prostaglandin analog and hCG were utilized in inducing follicular growth, and one birth following IVF has been recorded. Developmental studies of the embryos, both with in vivo and in vitro fertilization, are significant as they provide comparative data and also demonstrate the usefulness of the animals as models for human reproductive events.

Some preliminary studies with other new biotechnologies have been completed. These include successful xenogenous fertilization, freezing of the mature oocyte with subsequent in vitro fertilization, and embryo splitting of the early squirrel monkey oocyte with subsequent development.

Several aspects of the basic physiology of the New World species influence such studies. New World species are generally high-stress animals, and this can adversely affect hormonal patterns, implantation, and normal gestation. Their lack of menstruation and often unusual luteal tissue arrangements can be challenging. Finally, the high frequency of twinning in some of the Callithricidae offer unique challenges for researchers.

Historical Background

The successful artificial insemination of dogs was accomplished over 200 years ago, and by the 1840s scientists had described the ovum and suggested its role in the reproductive process. During the subsequent 110 years, embryo transfer of in vivo fertilized oocytes was successful, and there were innumerable attempts of the obvious experiment of placing oocytes and sperm in a culture dish to achieve IVF. Despite occasional reports of success, there was no consistent and repeatable procedure developed to carry out IVF and to successfully transfer the embryo (*embryo transfer* [ET]) to establish a pregnancy in any species.

A major discovery in the field occurred in 1951 when sperm capacitation was described, and this led rapidly to the successful IVF and, subsequently, the ET with production of living young (1, 2). The subsequent 18 years allowed scientists to establish basic techniques and examine IVF in a wide variety of laboratory and domestic species. In 1969 Edwards et al. (3) reported the first successful IVF of human oocytes and, 9 years later reported the first birth of an IVF/ET human baby (4). Since that time, over 30,000 children have been born by the IVF/ET procedure. The basic technique has not only led to production of offspring from many species, but has led to alternative procedures termed assisted reproduction with wide-ranging applications.

IVF/ET in Nonhuman Primates

Following efforts in laboratory and domestic animals, as well as the human primate, several laboratories attempted IVF in the nonhuman primate species. Probably the first successful IVF was with the baboon and reported by Kraemer in 1972 at the International Congress of Animal Reproduction and Artificial Insemination meeting in Munich. These studies were published in a much later report (5). The first report of a primate birth following IVF/ET was by Clayton and Kuehl (6) and was also in the baboon. Complete reviews of IVF/ET of Old World and great ape species are included elsewhere in these proceedings.

Among the New World species, only two have been extensively studied. These are the squirrel monkey (*Saimiri sciureus*) and the common marmoset (*Callithrix jacchus*). The reproductive biology in New World species has been reviewed (7). While these studies emanate from only a few laboratories, the work is built upon early basic studies that established procedures for the collection and handling of viable gametes. In fact, such procedures were better developed in the late 1960s for the New World species than for the traditional rhesus monkey and led to the use of such procedures in the squirrel monkey initially. In 1967 Bennett published techniques for sperm collection, ovulation induction, artificial insemination, and embryo recovery that provided important procedures in this species (8–10).

Seasonality

The squirrel monkey experiences a distinct seasonality in the wild that, upon capture and transfer to the northern hemisphere, begins a gradual shift (11) of the birth season from January–March to June–August. This seasonality is manifest not only in the natural mating of the animal but in its response to ovulation induction regimens as well (12, 13). Furthermore, the seasonality can be observed in the animal's behavior by its response to exogenous hormones (14). Despite this seasonality ovulation can be induced in the animal throughout the year, and the oocytes collected are able to be fertilized in vitro.

After adaptation to captivity squirrel monkeys will have a natural conception rate of 50%–70%, but this is coupled with a relatively high incidence of stillbirths or abortions (16.8%) and neonatal death loss (34.3%), probably reflecting a relatively high maternal weight: infant birth weight ratio (7:1) (15). The normal estrual cycle of the squirrel monkey is 8–10 days. There is no menstruation, and vaginal cytology is not as regular an indicator of cycle stage as is seen with other laboratory species.

The common marmoset does not show the extreme seasonality in captivity as does the squirrel monkey. This species, which also does not menstruate, exhibits a cycle of 28.6 days. In the wild this species demonstrates an unusual suppression of fertility where only one female in a social group breeds. Under laboratory conditions one can test this unusual phenomenon, but the mechanism by which it is controlled is not well understood (16).

Follicular Development and Induced Ovulation

A wide variety of regimens can be used to induce follicular development and ovulation in the New World primates. In actual practice, for IVF, follicular growth is stimulated and the oocytes recovered under laparoscopic or ultrasonic guidance prior to the time of actual ovulation. This results in the recovery of oocytes at varying stages of maturation, and in vitro maturation of oocytes is required for IVF of all recovered oocytes.

The original regimen used with the squirrel monkey (9) involved the use of 9 days of 200 IU of PMS with 250–500 IU of hCG given on the last 4 days. This results in superovulation, and the timing of ovulation is difficult to control. Additionally, PMS is extremely antigenic, and repeated use of this regimen renders the animal refractory. In our laboratory the luteal phase of the normal cycle was mimicked with 5 days of treatment with progesterone (5 mg/day) prior to the start of an ovulation induction regimen consisting of domestic animal source pituitary FSH (1 mg/day for 4 days) followed by a single injection of 100–250 IU of hCG on day 4 (17, 18). This results in single or double ovulations.

In contrast to experience with other animals, this ovulation induction scheme can be repeated with squirrel monkeys over several years with little or no refractoriness to the gonadotropins. Squirrel monkeys have been subjected to this regimen 20 or more times (at intervals of 3 weeks or more) with no significant effect on the percentage of animals ovulating or the number of ovulations per animal (19). *Human menopausal gonadotropin* (hMG) has also been used as an FSH source with the squirrel monkey, but resulted in cystic conditions within the ovary and was thus discontinued.

During the anovulatory season of the squirrel monkey (July–September in northern climates), the effectiveness of the FSH-hCG regimen decreases due to a decreased sensitivity of the monkey to the level of FSH. Increasing the daily dose of FSH to 2 mg or giving 1 mg/day for 5 rather than 4 days enhances ovulation (Kuehl). This suggests that the normal seasonal level of FSH (equivalent to 1 mg/day) is near the minimum threshold dose required for follicular stimulation.

In the marmoset the prostaglandin F2α analog (cloprostenol, Estrumate®, ICI, Macclesfield, UK) is given at the rate of 0.5 µg 13–33 days after the previous ovulation and is followed 7.5–8 days later by an injection of 75-IU hCG (20, 21). This results in an average of 2.2 ovulations per animal with IVF rates of 70%–79%.

Sperm Collection

Sperm collection from New World monkeys has almost universally employed the technique of electroejaculation (20, 22, 23) under anesthesia. This involves the use of a small rectal probe with the animal in a restraining device. The semen coagulates soon after ejaculation, but will liquify after incubation in collection medium for a few minutes. Treatment of sperm to enhance motility and capacitation is discussed in the following IVF section. Epididymal sperm have also been utilized for IVF in the marmoset (21).

IVF

The first successful IVF in squirrel monkeys was reported in 1972 by two laboratories and subsequently published (24–28). While a variety of experimental techniques have been developed, the optimal procedure appears to be collection of oocytes by laparoscopy and incubation for 21 h prior to addition of the sperm fraction. Over this period of time, there is a gradual increase in the number of oocytes matured. The tissue culture medium 199 with 20% agamma fetal calf serum has been used with the addition of 72-µg/mL pyruvate and 10-µM dbcAMP to stimulate sperm capacitation and motility (29). The effect of dbcAMP appeared to affect only the sperm and had no effect on the oocyte or on subsequent embryo cleavage in vitro. Under these conditions IVF rates of 70%–90% were achieved, a rate similar to that obtained with other primates including humans.

In the marmoset oocytes are recovered surgically 24 h after administration of hCG and incubated in minimum essential medium (α-MEM, Flow Laboratories, Irvine, Scotland). This basic medium is supplemented with either 10% heat-inactivated human cord serum (20) or marmoset serum (21). Incubation of oocytes for 2–5, 9–11, and 21–29 h before insemination resulted in fertilization rates of 50%, 86%, and 90%, respectively.

Culture media and conditions have varied among laboratories with no evident superiority of one system over another. Readers are referred to publications from individual laboratories in this regard.

Preimplantation Embryo Development

Basically three types of criteria have been utilized to measure the normality of in vivo fertilized embryos recovered from the reproductive tract and cultured and in vitro fertilized oocytes cultured in vitro. These are developmental, chromosomal, and biochemical normality. Table 4.1 illustrates comparison of both in vivo and in vitro fertilized embryos of several primate species relative to normal development. Since differences between the two methods of fertilization were not significant, the values are pooled, and the relative uniformity of development between species is evident. Despite minor differences the similarity of development of the primate species to the human is much closer than the developmental rates of rodents or other species. This emphasizes the usefulness of nonhuman primate models in general and precludes the concept of a single nonhuman primate species as the only acceptable model for human reproduction.

Extensive chromosomal or biochemical analysis of fertilized embryos is not ethically possible with human oocytes and offers one of the major advantages of the application of IVF and other biotechnologies to nonhuman primates. One of the technical problems of determining chromosomal normality in multicelled embryos is the tendency for mixing of chromosomes after disruption of the blastomeric membranes due to the fixing process. In 1976 a technique was developed for the chromosomal study of Chinese hamster embryos that avoided this problem (30). This technique was subsequently modified (31) for studies of New World primate embryos before and after IVF.

Thus, studies were carried out on follicular oocytes recovered after squirrel monkey ovulation induction and on embryos following IVF. By the metaphase II stage, the incidence of abnormalities observed were 7.4%–14%, a value comparable to that found with other laboratory species and with normal oocytes. In similar studies with IVF embryos, of

TABLE 4.1. Comparative rates of nonhuman primate preimplantation development.

Species	Two polar bodies	Hours after fertilization					
		2-cell	4-cell	8-cell	16-cell	Morula	Blastocyst
Squirrel monkey	6–22	16–40	46–52	57–72			96
Marmoset monkey		24–48	48–72		96–144	120–192	168–240
Rhesus monkey		24–36	36–48	48–72	72–96	88–111	144–192
Cynomolgus monkey		16–36	37–48	49–72	61–108	97–120	
Baboon	6–24	24	48	48–72	96–120	120–148	96–144
Human	12	30–38	38–52	51–72	85–96	96–135	123–147

Data from Kuehl and Dukelow (26); Asakawa, Chan, and Dukelow (33); and (40–54).

877 oocytes recovered the incidence of abnormalities was 9%–16%, with the common abnormalities being missing or extra chromosomes (32, 33). These levels are comparable with those of other species that have been studied for IVF, and the levels are comparable to those found with in vivo (natural) fertilization. There was no indication of an effect of IVF on increasing the level of chromosomal abnormalities. Interestingly, using the chromosomal analysis technique, the incidence of triploidy in in vitro fertilized squirrel monkeys was 16.7%. Triploidy is common in all laboratory species using IVF and has been reported with human IVF (34). To the best knowledge of this author, no chromosomal studies have been carried out following IVF on other New World primate species.

The biochemical or metabolic aspects of development have received little attention following IVF of New World species. There have been squirrel monkey studies on protein synthesis, uptake of steroid hormones, oxygen consumption, and overall viability of IVF embryos (35). Incorporation of labeled leucine as an indicator of protein synthesis declined with oocyte maturation and remained constant after IVF as assessed by autoradiography. There was a nonsignificant increase at first cleavage. Uptake of estradiol and progesterone increased after IVF in these embryos, but there were no further changes in uptake of either steroid at first cleavage.

Uridine incorporation and uptake, as a measure of RNA synthesis, decreased in oocytes recovered from squirrel monkeys 36 h after hCG administration compared to oocytes recovered at 16 h. An approximate doubling of uridine incorporation occurred after fertilization, with further increase as development progressed after the first cleavage division, an event similar to that observed with mouse embryos. Uridine incorporation has also been studied on a comparative basis for nonfertilized oocytes in squirrel monkeys and humans (36). There was a significant decline in both the uptake and incorporation of uridine in squirrel monkey oocytes 36 h after hCG administration, and, similarly, RNA synthesis diminished in human oocytes collected 35 h after hCG compared with 12 h after hCG.

Embryo Transfer

Few offspring of New World species have been produced by IVF/ET, reflecting not a difficulty of the procedure (although generally the size of the animals may present an obstacle), but rather an emphasis by past researchers on examining the development of the embryo itself rather than utilizing embryos for transfer. No uterine transfers have been made with IVF-produced squirrel monkey embryos.

In the marmoset 3 recipients received embryos produced by IVF. In 2 of the recipients, a 4-cell and a 6-cell embryo were placed in the uterus after 3 days of in vitro culture. The third recipient received a single 6-cell embryo. Of the first 2 animals (receiving 2 embryos each), one became pregnant and delivered twins. The animal receiving the single embryo became pregnant and, after a gestation period of 144 days, gave birth to a male infant (20).

Related Biotechnologies

Other related biotechnologies have been employed with New World primates on a limited scale that have applications to future studies (37). *Xenogenous fertilization* involves the collection of sperm and ova from a single species with their subsequent placement for fertilization into the oviduct of a foreign species, normally a pseudopregnant rabbit. This procedure resulted in a 35% fertilization rate for the squirrel monkey (38). Embryo transfer of these embryos has not been attempted, but similar studies with domestic animals have resulted in pregnancies and live deliveries. Although the procedure may appear unusual, the use of the pseudopregnant rabbit oviduct for embryo culture has a long and illustrious history dating back to 1951, and early studies involved the transport of embryos over long distances in live rabbits. The procedure is also a forerunner of the *gamete intrafollicular transfer* (GIFT) procedure used with humans, the only difference being that the oviductal fertilization occurs in a laboratory species. Squirrel monkey oocytes have also been successfully frozen, thawed, and xenogenous or in vitro fertilized (39).

In the early 1980s technology became available for the splitting or bisecting of embryos in culture. This technology has been extensively utilized to produce identical twins and triplets with domestic animals. In New World primates the procedure has been used in a limited fashion to biopsy cells from early developing blastocysts (39), and the embryos continued to develop through at least another cleavage stage. More research is needed on this important aspect of embryo manipulation.

Transgenic Primates

With the development of advanced techniques for genetic engineering, the possibility of producing transgenic primates for research becomes important. No such animals have been produced from any nonhuman primate species, but certainly the technology exists for the basic handling of sperm, oocytes, embryos, and blastocysts. The technology exists to remove or inject single cells or chemical substances directly into the cells of the embryo or into the blastocoele. Oocytes, embryos, blastocysts, or blastomeres can be frozen and held indefinitely.

Conclusions

The clinical importance of human IVF/ET has led the way in pioneering advances in the field of reproductive biotechnology. Studies with non-human primates of several species have supplemented these advances with contributions on the basic biology of the IVF embryos, contributions that cannot be made from an ethical standpoint with human embryos. The New World primates, because of early studies on gamete handling techniques, have provided a substantial number of such basic contributions to a better overall understanding of human preimplantation biology. These species offer continuing opportunities to enhance our knowledge of the new biotechnologies in the future.

Acknowledgments. Some of the studies reported here were supported by NIH Grants HD-07534 and ES-04911, the Health Science Research Foundation, and the Population Medicine Center of Michigan State University. The author thanks Mrs. L.M. Cleeves for assistance in preparing the manuscript and Dr. J.P. Hearn of the Wisconsin Regional Primate Research Center for important discussions on the manuscript itself.

References

1. Chang MC. Fertilizing capacity of spermatozoa deposited into the fallopian tubes. Nature 1951;168:697–8.
2. Chang MC. Development of fertilizing capacity of rabbit spermatozoa in the uterus. Nature 1955;175:1036.
3. Edwards RG, Bavister BD, Steptoe PC. Early stages of fertilization in vitro of human oocytes matured in vitro. Nature 1969;221:632–5.
4. Steptoe PC, Edwards RG. Birth after the reimplantation of the human embryo. Lancet 1978;2:366.
5. Kraemer DC, Flow BL, Schriver MD, Kinney GM, Pennycook JW. Embryo transfer in the nonhuman primate, feline and canine. Theriogenology 1979;11:51–62.
6. Clayton O, Kuehl TJ. The first successful in vitro fertilization and embryo transfer in a nonhuman primate. Theriogenology 1984;21:228.
7. Hearn JP. Reproduction in New World primates. Lancaster, UK: MTP Press, 1983.
8. Bennett JP. Semen collection in the squirrel monkey. J Reprod Fertil 1967;13:353–5.
9. Bennett JP. The induction of ovulation in the squirrel monkey (*Saimiri sciureus*) with pregnant mares serum (PMS) and human chorionic gonadotrophin (HCG). J Reprod Fertil 1967;13:357–9.
10. Bennett JP. Artificial insemination of the squirrel monkey. J Endocrinol 1967;37:473–4.

11. DuMond FV. The squirrel monkey in a seminatural environment. In: Rosenblum LA, Cooper RW, eds. The squirrel monkey. New York: Academic Press, 1968:88–145.
12. Harrison RM, Dukelow WR. Seasonal adaption of laboratory-maintained squirrel monkeys (*Saimiri sciureus*). J Med Primatol 1973;2:227–83.
13. Kuehl TJ, Dukelow WR. Ovulation induction during the anovulatory season in *Saimiri sciureus*. J Med Primatol 1975;4:23–31.
14. Jarosz SJ, Kuehl TJ, Dukelow WR. Vaginal cytology, induced ovulation and gestation in the squirrel monkey (*Saimiri sciureus*). Biol Reprod 1977;16:97–103.
15. Dukelow WR. The squirrel monkey (*Saimiri sciureus*). In: Hearn JP, ed. Reproduction in New World primates. Lancaster, UK: MTP Press, 1983:149–80.
16. Hearn JP. The common marmoset (*Callithrix jacchus*). In: Hearn JP, ed. Reproduction in New World primates. Lancaster UK: MTP Press, 1983:182–215.
17. Dukelow WR. Induction and timing of single and double ovulations in the squirrel monkey (*Saimiri sciureus*). J Reprod Fertil 1970;22:303–9.
18. Dukelow WR. Human chorionic gonadotropin: induction of ovulation in the squirrel monkey. Science 1979;206:234–5.
19. Dukelow WR, Theodoran CG, Howe-Baughman J, Magee WT. Ovulatory patterns in the squirrel monkey (*Saimiri sciureus*). Anim Reprod Sci 1981;4:55–63.
20. Lopata A, Summers PM, Hearn JP. Births following the transfer of cultured embryos obtained by in vitro and in vivo fertilization in the marmoset monkey (*Callithrix jacchus*). Fertil Steril 1988;50:503–9.
21. Wilton LJ, Marshall VS, Piercy E. In vitro fertilization and embryonic development in the marmoset monkey (*Callithrix jacchus*). Proc 2nd Internatl Conf on Adv in Reproductive Res in Man and Animals, Natl Museums of Kenya, Nairobi, 1992:11.
22. Kuehl TJ, Dukelow WR. A restraint device for electroejaculation of squirrel monkeys (*Saimiri sciureus*). Lab Anim Sci 1974;24:364–6.
23. Cui K-H, Flaherty SP, Newble CD, Guerin MV, Napier AJ, Matthews CD. Collection and analysis of semen from the common marmoset (*Callithrix jacchus*). J Androl 1991;12:214–20.
24. Johnson MJ, Harrison RM, Dukelow WR. Studies on oviductal fluid and in vitro fertilization in rabbits and nonhuman primates. Fed Proc 1972;31:278.
25. Cline EM, Gould KG, Foley CW. Regulation of ovulation, recovery of mature ova and fertilization in vitro of mature ova of the squirrel monkey (*Saimiri sciureus*). Fed Proc 1972;31:277.
26. Kuehl TJ, Dukelow WR. Fertilization in vitro of *Saimiri sciureus* follicular oocytes. J Med Primatol 1975;4:209–16.
27. Dukelow WR, Kuehl TJ. In vitro fertilization of nonhuman primates. La Fecondation, Collaque de la Societe Nationale pour L'Etude de la Sterilite et de la Fecondite, Masson, Paris, 1975:67–80.
28. Gould KG, Cline EM, Williams WL. Observations of the induction of ovulation and fertilization in vitro in the squirrel monkey (*Saimiri sciureus*). Fertil Steril 1973;24:260–8.

29. Chan PJ, Hutz RJ, Dukelow WR. Nonhuman primate in vitro fertilization: seasonality, cumulus cells, cyclic nucleotides, ribonucleic acid and viability assays. Fertil Steril 1982;38:609–15.
30. Kamiguchi Y, Funaki K, Mikamo K. A new technique for chromosome study of murine oocytes. Proc Jpn Acad 1976;52:316–9.
31. Mizoguchi H, Dukelow WR. Gradual fixation method for chromosomal studies of squirrel monkey oocytes after gonadotropin treatment. J Med Primatol 1981;10:180–6.
32. Asakawa T, Dukelow WR. Chromosomal analyses after in vitro fertilization of squirrel monkey (*Saimiri sciureus*) oocytes. Biol Reprod 1982;26:579–83.
33. Asakawa T, Chan PJ, Dukelow WR. Time sequence of in vitro and chromosomal normality in metaphase I and metaphase II of the squirrel monkey (*Saimiri sciureus*) oocyte. Biol Reprod 1982;27:118–24.
34. Lopata A, Brown JB, Leaton JF, McTalbot J, Wood C. In vitro fertilization of preovulatory oocytes and embryo transfer in infertile patients. Fertil Steril 1978;30:27–35.
35. Hutz RJ, Phan PJ, Dukelow WR. Nonhuman primate in vitro fertilization: biochemical changes associated with embryonic development. Fertil Steril 1983;40:521–4.
36. Hutz RJ, Holzman GV, Dukelow WR. Synthesis of ribonucleic acid in oocytes collected from squirrel monkeys and humans following chorionic gonadotropin administration. Am J Primatol 1983;5:267–70.
37. Dukelow WR. Modern trends in embryo research. Proc 2nd Internatl Conf on Adv in Reproductive Res in Man and Animals, Natl Museums of Kenya, Nairobi, 1992.
38. DeMayo FJ, Mizoguchi H, Dukelow WR. Fertilization of squirrel monkey and hamster ova in the rabbit oviduct (xenogenous fertilization). Science 1980;208:1468–9.
39. DeMayo FJ, Rawlins RG, Dukelow WR. Xenogenous and in vitro fertilization of frozen/thawed primate oocytes and blastomere separation of embryos. Fertil Steril 1985;43:295–300.
40. Ariga S, Dukelow WR. Recovery of preimplantation blastocysts in the squirrel monkey by a laparoscopic technique. Fertil Steril 1977;28:577–80.
41. Avendano S, Croxatto HD, Pereda J, Croxatto HB. A seven-cell human egg recovered from the oviduct. Fertil Steril 1975;26:1167–72.
42. Bavister BD, Boatman DE, Liebfried L, Loose M, Vernon MW. Fertilization and cleavage of rhesus monkey oocytes in vitro. Biol Reprod 1983;28:983–99.
43. Boatman DE. In vitro growth of nonhuman primate and pre- and peri-implantation embryos. In: Bavister BD, ed. The mammalian preimplantation embryo. New York: Plenum, 1987:273–308.
44. Croxatto HB, Diaz S, Fuentealga B, Croxatto HD, Carrillo D, Fabres C. Studies on the duration of egg transport in the human oviduct, I. The time interval between ovulation and egg recovery from the uterus in normal woman. Fertil Steril 1972;23:458–77.
45. Edwards RG, Steptoe PC. Physiological aspects of human embryo transfer. In: Behrman SJ, Kistner RW, eds. Progress in infertility. Boston: Little Brown, 1975:377–409.

46. Fujisaki M, Suzuki M, Kohno M, Cho F, Honjo S. Early embryonal culture of the cynomolgus monkey (*Macaca fascicularis*). Am J Primatol 1989;18:303–13.
47. Hearn JP. The embryo-maternal dialogue during early pregnancy in primates. J Reprod Fertil 1986;76:809–19.
48. Hertig AT, Rock J, Adams EC, Mulligan WJ. On the preimplantation stages of the human ovum: a description of four normal and four abnormal specimens ranging from the second to the fifth day of development. Contrib Embryol, Carnegie Inst 1954;35:199–230.
49. Kraemer DC, Hendrickx AG. Description of stage I, II, and III. In: Hendrickx AG, ed. Embryology of the baboon. University of Chicago Press, 1971:45–52.
50. Kreitmann O, Lynch A, Nixon WE, Hodgen GD. Ovum collection, induced luteal dysfunction, in vitro fertilization, embryo development and low tubal ovum transfer in primates. In: Hafez ESE, Semm K, eds. In vitro fertilization and embryo transfer. Lancaster, UK: MTP Press, 1982:303–24.
51. Kuehl TJ, Dukelow WR. Maturation and in vitro fertilization of follicular oocytes of the squirrel monkey (*Saimiri sciureus*). Biol Reprod 1979;21:545–56.
52. Lewis WH, Hartman CG. Tubal ova of the rhesus monkey. Contrib Embryol Carnegie Inst 1941;29:9–14.
53. Pope CE, Pope VZ, Beck LR. Development of baboon preimplantation embryos to post-implantation stages in vitro. Biol Reprod 1982;27:915–23.
54. Wolf DP, VandeVoort CA, Meyer-Haas GR, et al. In vitro fertilization and embryo transfer in the rhesus monkey. Biol Reprod 1989;41:335–46.

5
IVF-ET in Old World Monkeys

Don P. Wolf and Richard L. Stouffer

IVF-ET is an emerging technology in *nonhuman primates* (NHP) since it is based on relatively limited experience in only a few species. Thus, while successful protocols are available for rhesus and cynomolgus monkeys, consistent or wide-spread application is not yet the order of the day. Undoubtedly, substantial species-dependent differences will be discovered, such as the extended menstrual cycle length recently described by us in *Macaca nigra* (1). This picture is in marked contrast with the extensive experience base now available in the human.

An IVF-ET program using rhesus macaques was initiated by the authors at the *Oregon Regional Primate Research Center* (ORPRC) in 1986 with the expressed purpose of catalyzing studies of ovarian function, oogenesis, fertilization, embryogenesis, and implantation in this species. It was and continues to be our objective to enhance the possibility of employing the assisted reproductive technologies to preserve endangered nonhuman primates, to propagate models for studying human diseases, and to use in the management of captive colonies—some via contraception, others through assisted propagation—as well as in a number of basic and applied research applications (2). The close phylogenetic relationship between humans and the rhesus monkey supports the notion that our studies may advance clinical activities in such areas as preimplanation genetic diagnosis, the improvement of culture conditions for IVF-produced embryos, and the use of ooplasmic sperm injection in the treatment of male infertility.

History

Following the discovery of mammalian sperm capacitation in 1951 (3, 4), semen collection (5) and characterization was described in NHP along with artificial insemination in squirrel (6) and rhesus monkeys (7). Successful IVF is, of course, dependent on both a reliable sperm source and the recovery of high-quality, mature oocytes or oocytes that can be matured readily in vitro. Ovulatory patterns (8, 9) and ovarian hyperstimulation (10, 11) in rhesus macaques were studied in the 1970s, while the successful IVF of oocytes collected from superovulated rhesus or cynomolgus macaques occurred in the decade of the 1980s (12–14). The first IVF success in rhesus macaques was reported in 1984 (13), and the total number of reported infants up through 1991 for rhesus and cynomolgus monkeys combined is only 14 (2, 12). Currently, we are unaware of any other NHP program that is routinely conducting IVF.

Problems that continue to limit IVF activities in Old World macaques include (i) animal availability: In the United States these macaques are restricted despite being the most common species; (ii) expense: as an example at ORPRC, the lease fee for in-house investigators is $625 with per diem of $4.25; (iii) the fact that single-use females can be superovulated only once because of immunologic sequelae to the human gonadotropic hormones administered; and (iv) semen collection: it requires electroejaculation, and washed sperm capacitation is unusually difficult.

Semen Collection, Processing, and Sperm Capacitation

Electroejaculation is the accepted method of semen collection from NHP and is typically performed using direct penile or rectal probe stimulation. With rhesus macaques we prefer a penile band electroejaculation method with controlled stimulus current, a technique that provides consistent, successful, and humane sample collection (15). This approach requires animal conditioning/acceptance, and preliminary attempts have not been successful with *Macaca nigra*. Factors that influence semen quality in addition to the collection technique include (i) abstinence time: Animals masturbate frequently, and it is best to collect early in the day; late samples contain fewer sperm with an increased number of immature forms; (ii) animal handlers: Sample quality varies markedly with the individuals conducting the procedure; some cannot obtain a usable specimen, while others are always successful; and (iii) urine contamination: The sample collected by electroejaculation is susceptible to problems secondary to urine contamination.

Rhesus monkey semen is characterized by the presence of a large coagulum. The sample should liquify at room temperature for approximately 5 min before the liquid portion is removed by aspiration. The resultant sperm population is remarkably uniform in structure and morphology (16). In 6 males collected by the penile band technique, we reported a mean percent motility of 82.6% for the sperm in 80 ejaculates with liquid volumes ranging from 0.22 to 0.56 mL and a mean number of sperm per sample of 328 million (15).

Sperm processing usually involves semen dilution with 3 or more volumes of medium containing a room-air stable buffer (e.g., TALP-

HEPES), followed by centrifugation (360 × g for 7 min) and supernatant removal. This process may be repeated before the final sperm pellet is resuspended in bicarbonate-buffered medium at 20 million sperm/mL and held at 37°C until use. We have found the hard way that adequate semen dilution is required to remove putative *decapacitation factors* (DF). In our original procedure we used a fixed volume of diluent (3 mL) when processing semen and observed a highly variable IVF outcome with a mean fertilization level of 16.7% (37/221). This mean fertilization level jumped to 70% (58/83) when much larger dilution volumes (30×) were used, and we speculated that incomplete removal of DF from larger-volume semen samples could account for the lower than expected fertilization levels; that is, fertilization success was inversely related to semen volume.

In order to corroborate this suspicion, an add-back experiment was conducted in which the supernatant from the first semen wash (TALP-bicarbonate was used instead of TALP-HEPES) was added back to sperm processed by the modified protocol. The resultant fertilization levels were markedly reduced (7%, 1/14) even if this supernatant fraction had been boiled prior to add back (0%, 0/6). These preliminary observations are consistent with the presence of a heat-stable DF activity in rhesus monkey semen.

In nonhuman primates capacitation is induced by washed sperm exposure to activating agents that increase intracellular cyclic nucleotide levels; for instance, *dibutyryl cyclic adenosine monophosphate* (dbcAMP) and caffeine. An association has been made between this exposure, hyperactivated motility, and fertility (14, 17) in rhesus and cynomolgus macaques. The ability to bind to the zona pellucida and undergo an acrosome reaction are other sperm capabilities prerequisite to penetration of the egg. In free-swimming populations of washed sperm, hyperactivation and acrosomal status can be monitored, the former by computer-assisted video image analysis systems (CASA or CASMA) (18, 19) and the latter by indirect immunofluorescence with monoclonal antibodies or lectins.

We have employed a CASMA system (19) to quantify the individual parameters of motility in rhesus monkey sperm. This system supports both head- (Fig. 5.1A) and flagellar-based analyses (Fig. 5.1B). The trajectories of several hundred stimulated and nonstimulated sperm were analyzed in an effort to define a unique subpopulation of hyperactivated cells. In addition to curvilinear velocity ($>180\,\mu$/sec), several additional track parameters that monitor linearity were used to define hyperactivation. A unique subpopulation of sperm was characterized after exposure to activating concentrations of dbcAMP and caffeine (Fig. 5.1C). The extent of hyperactivation was studied in 4 different males. All showed highly significant increases in the number of hyperactivated sperm within minutes of activator exposure. Significant interanimal differences were also noted in the level of hyperactivation (range: 30%–80%). We are

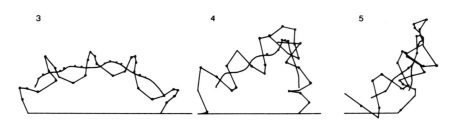

Track	Vc	Va	Vn	L Abs	L Alg	LI	PRc	PR
1	209.2	122.6	118.5	3.0	1.7	0.97	0.59	0.57
2	145.0	93.6	77.5	14.1	1.8	0.83	0.65	0.53
3	175.2	97.1	65.2	12.7	5.9	0.67	0.55	0.37
4	234.6	95.6	48.7	30.3	9.7	0.51	0.41	0.21
5	217.5	95.6	26.2	22.0	10.1	0.27	0.44	0.12

A

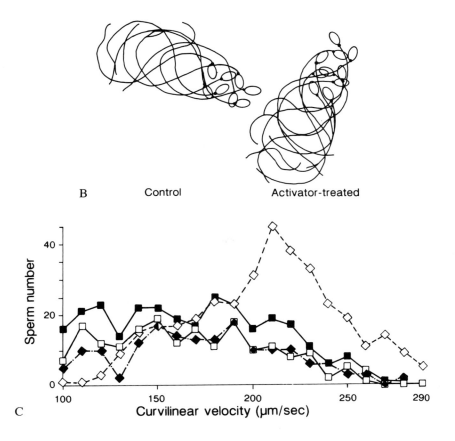

B Control Activator-treated

C

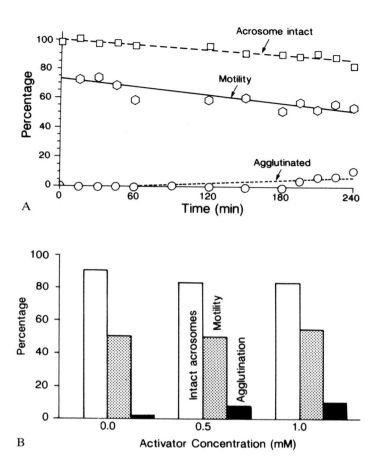

FIGURE 5.2. *A:* The effect of rhesus monkey sperm exposure to activator. Washed sperm (from 2 different males) were preincubated for 3 h in TALP medium containing 3% albumin, as described in reference 20, before exposure to 1 mM each dbcAMP and caffeine. At timed intervals after activator addition, sperm were removed for quantitation of motility, acrosomal status, and agglutination. *B:* The effect of activator concentration on acrosomal status, percent motility, and agglutination of rhesus monkey sperm (2 males). Sperm were preincubated for 3 h in TALP plus 3% albumin before activator exposure. Results represent sperm aliquots characterized after 60 min of activator exposure.

◄─────────────────────────────────────

FIGURE 5.1. *A:* Rhesus monkey sperm head trajectories and quantitative descriptions of each as determined by CASMA. For details, see reference 19. *B:* Computer-generated images of capacitated and hyperactivated rhesus monkey sperm. *C:* Percentage distribution of curvilinear velocities for nonlinear rhesus monkey sperm trajectories. Control and activator-treated cells were measured at ambient temperature. (Open diamonds = activator-treated, nonlinear; closed diamonds = activator-treated, linear; open boxes = control, nonlinear; closed boxes = control, linear.)

interested in correlating the extent of hyperactivation with fertility in the context of an IVF assay. This approach will require that a sperm concentration dependency be established for each animal in the IVF of sibling oocytes.

In an effort to address the question of why macaque sperm require exposure to dbcAMP and caffeine before acquiring IVF capacity, we have also monitored changes in the acrosomal status of rhesus monkey sperm incubated under capacitating conditions in vitro (Fig. 5.2A) (unpublished observations). Over a 6-h time course at 37°C, a modest decline in the percentage of motile cells was associated with an increase in the percentage of acrosome-reacted sperm. After an hour of incubation, sperm agglutination also became apparent. In an activator concentration series (Fig. 5.2B), an activator-dependent increase in sperm agglutination and, perhaps, also in motility was noted along with a small increase in the percentage of acrosome-reacted cells. These small but consistent increases in acrosome-reacted cells—if significant and reflective of a fertile sperm subpopulation with intact but "sensitized" acrosomes—could provide a mechanism, in addition to motility-related effects, for dbcAMP-caffeine activation of sperm fertility.

The *sperm penetration assay* (SPA) using zona-free hamster eggs as surrogates has been used to evaluate the fertility potential of macaque sperm. Technical problems associated with this application are that sperm bind to eggs in massive numbers, rendering scoring difficult, if not impossible. Nevertheless, successful use of this assay that confirms the need for sperm exposure to dbcAMP-caffeine has been reported (21).

Sperm Cryopreservation

Although full-term pregnancies were recently reported following artificial insemination with frozen-thawed semen in the cynomolgus macaque (22), the cryopreservation of macaque sperm has met with limited success. Using modifications of the standard methods for human semen cryopreservation, we have conducted a systematic approach to improve the postthaw quality of rhesus monkey sperm (Thomson, Wolf, unpublished observations). Controlled rates of cooling were compared with the relatively rapid and uncontrolled rates obtained with a liquid nitrogen vapor procedure. The use of a controlled-rate freezer did not improve outcome. The optimum concentration of cryoprotectant was evaluated in TES-Tris and commercially available egg yolk buffer. With glycerol as cryoprotectant (2%, 4%, and 6% concentrations were equivalent), motility fell off dramatically when thawed sperm were washed and incubated in TALP-bicarbonate at room temperature for several hours (Fig. 5.3, top), and an even faster decline was noted at 37°C (Fig. 5.3, bottom). At the higher concentration of glycerol (8%), sperm survival

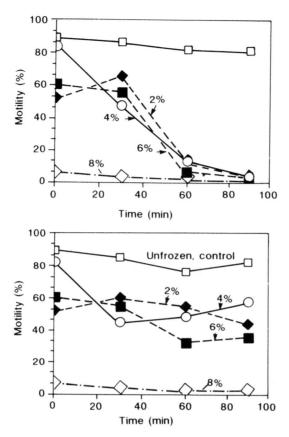

FIGURE 5.3. Postthaw motility of rhesus macaque sperm frozen in 2%, 4%, 6%, and 8% glycerol in the presence of TES-Tris egg yolk buffer, measured by direct microscopic observation after cryoprotectant removal and culture in TALP-bicarbonate. *Top:* Incubated at ambient temperature. *Bottom:* Incubated at 37°C.

was minimal. Based on high-resolution morphological analysis, we have concluded that rhesus monkey sperm are more susceptible to acrosomal damage than are human sperm treated in the same manner. It should be obvious from these remarks that improvements in macaque sperm freezing protocols are needed.

Oocyte Collection

Follicle stimulation protocols have been developed for the recruitment of large follicle numbers in support of egg pickup procedures for IVF (23). We monitor the follicular response by ultrasonography and peripheral

TABLE 5.1. Egg collection and maturity data for the 1991–92 IVF season at the Oregon Regional Primate Research Center.

Protocol	Animals	Atr	Gv	M1	M2	Ovs	Total
1	23	36	70	234	121	1	462
2	4	13	29	96	32	2	172
3	4	10	33	89	24	0	156
4	6	12	50	97	9	0	168
5	6	4	33	75	9	2	123
6	4	2	13	57	31	0	103
Totals	47	77	228	648	226	5	1184
Mean/animal		1.6 (6.5)	4.9 (19.3)	13.8 (54.7)	4.8 (19.1)	0.1 (0.5)	25.2

Numbers in parentheses are percentages. Atr = atretic; Gv = germinal vesicle intact; M1 = metaphase I; M2 = metaphase II; Ovs = ovulation sites.

levels of estradiol and use a laparoscopic approach in follicular aspiration. To illustrate the extent of activity of the ORPRC core, in the 1991–92 IVF season, 54 cycles were initiated using either a standard plan A + hCG protocol (23) or an experimental variation thereof (Table 5.1), culminating in 47 cycles to surgery and the recovery of 1184 eggs. This represents an average of 25 eggs/animal, 74% of which were mature (metaphase II) or maturing (metaphase I) at collection. Less than 5% of these oocytes were characterized as atretic at pickup, while 19% were *germinal vesicle* (GV) intact, presumably arising from the inadvertent aspiration of small follicles.

Oocyte Maturation

The production of live young from *in vitro-matured* (IVM), GV-intact oocytes would allow the rescue of genetic material from important females who die before reproducing, circumvent the need for ovarian hyperstimulation, provide a much larger potential source of oocytes, and lead to knowledge that may improve human IVF success. We have begun studies of oocyte maturation using fully grown, GV-intact oocytes (>50 µm diameter, enclosed with >2 layers of tightly condensed cumulus) collected from ovaries at necropsy, ovariectomy (both from gonadotropin-stimulated and nonstimulated animals), and from stimulated animals following follicular aspiration. Based on results with over 500 oocytes, 24%–30% of immature oocytes matured in vitro to metaphase II within 36–40 h of collection. These levels were obtained in the presence or absence of gonadotropins in the culture medium, although the rate of maturation may be faster in the presence of gonadotropins (24). In our preliminary studies, 55%–62% of these IVM oocytes underwent fertilization and early development. The viability of embryos produced by this approach is currently being tested by ET to synchronized rhesus monkey recipients.

Zona *hardening* during in vitro maturation may provide an explanation for the failure of some IVM oocytes to fertilize. This possibility can be evaluated with micromanipulation procedures where individual sperm can be placed in the perivitelline space. Current technology does not support the routine cryopreservation of primate oocytes despite the obvious advantages that would arise.

IVF

Fertilizability of mature eggs can be as high as 95% (25), but more typically is in the 50%–75% range using motile sperm concentrations of 50,000–100,000/mL. Zona hardening, as mentioned above, may contribute to IVF failure in some oocytes. Oocyte quality may also be a limiting factor based on a recent report by Johnson and coworkers (26). A triple fluorochrome staining technique for chromatin, microtubules, and filamentous actin was applied to the oocytes collected from hyperstimulated rhesus macaques. Based on these parameters a number of anomalies were described, and the authors concluded that follicle size, cumulus morphology, and peripheral estradiol levels were of little predictive value vis-à-vis oocyte quality.

Preliminary results are also available suggesting that nuclear maturation can occur in the absence of ooplasmic maturation (27), an event that may lead to an apparently normal, mature oocyte that fails to fertilize. These observations probably apply to primates in general since fertilization levels in women may vary from 65% of all inseminated oocytes to 85%–95% of pregraded or selected mature oocytes. Interestingly, we have also found in natural cycle IVF in women that the egg recovered from the dominant follicle in a nonstimulated cycle has only about a 70% probability of fertilizing.

The cryopreservation of IVF-produced macaque embryos is based on extensive experience with human embryos using propanediol as the cryoprotectant. The ORPRC program routinely banks embryos at the pronuclear or early cleavage stage, and our inventory has now reached the point that experiments requiring 50–100 embryos can realistically be considered.

In Vitro Culture of Rhesus Monkey Embryos

Success in culturing rhesus (28) and cynomolgus (17, 29) monkey embryos in media supplemented with serum has been reported. The developmental efficiency of IVF-produced embryos is high and apparently medium independent through the first 2–3 cell divisions, but drops during subsequent developmental stages. This probably reflects the existence of a developmental block at the 8- to 16-cell stage at about the time activation of the embryonic genome occurs. The timing of rhesus monkey development to hatched blastocyst has been monitored (30); however, in vitro-produced blastocysts appear different than their in vivo counterparts, characterized by substantial amounts of syncytial trophoblast but inadequate development of the inner cell mass constituents (31).

In preliminary attempts at ORPRC (VandeVoort, Stouffer, Wolf, unpublished observations), nonfrozen embryos were cultured in TALP fortified with 20% heat-inactivated midcycle rhesus monkey serum. Five embryos of the 11 cultured developed to hatched blastocysts in 9–10 days. These hatched blastocysts displayed the behavior described by Boatman (30) of floating in culture following expansion and separation from the zona pellucida. After several additional days in culture, embryo shrinkage was accompanied by swelling and substrate attachment. By high-resolution electron microscopic analysis, trophoblastic differentiation was evident, but the inner cell mass was not as extensive or as well defined as that described by Enders and Schlafke (32) for in vivo-produced embryos. Subsequent culture attempts with cryopreserved embryos, have produced similar results in the 6 additional blastocysts examined.

While it is entirely possible that the developmental potential of these embryos is normal (an ET series is planned), these observations could reflect autocrine or paracrine regulation of cell growth that is subnormal under our existing in vitro culture conditions. Thus, improvements in blastocyst quality might be obtained by somatic cell coculture approaches (33, 34) or by the addition of exogenous growth factors (35). Somatic cell coculture of in vivo-produced embryos from rhesus macaques has been described (36). In our preliminary experimentation (Thomson and Wolf, unpublished observations), TALP containing 20% *fetal calf serum* (FCS) and the same medium with Vero cells were used to culture matched sibling groups of thawed IVF-derived embryos (n = 25 and 25). Similar development occurred to the morula stage (80% vs. 84%). None of the control embryos hatched, while 24% of the embryos developed to the blastocyst stage during coculture. Currently, we are extending the somatic cell culture technique to include fresh bovine oviduct and cumulus cells and buffalo rat liver cells in CMRL-1066 medium with 10% FCS (37).

ET

The success of an IVF-ET program is ultimately demonstrated by live births following the ET of IVF-produced embryos. Using an oviductal transfer approach (Fig. 5.4), we have now reported the birth of 7 infants

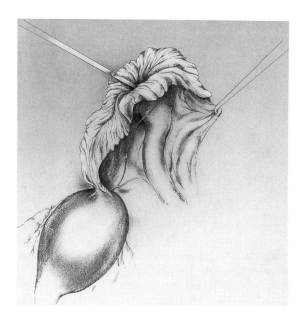

FIGURE 5.4. Illustration of the tubal ET procedure employed by the ORPRC team to transfer IVF-produced embryos.

upon transfer of 14 oviductal-stage embryos to 7 synchronized recipients (2). Based on these results we concluded that the oviductal-stage embryos produced by IVF and subsequently stored frozen are normal. ET trials currently ongoing or planned will allow evaluation of the normalcy of in vitro-matured oocytes, subsequently fertilized in vitro, as well as uterine-stage embryos grown in the presence and absence of somatic cells. A summary of implantation and pregnancy rates for macaque embryos produced by IVF is presented in Table 5.2. With an overall implantation rate of 14% (15/111) and a pregnancy rate of 20% (11/54), these values compare favorably with human IVF outcome despite the very limited experience.

Summary

In considering the IVF-ET technology in Old World macaques, it seems apparent that the protocols available for sperm collection, processing, and capacitation are adequate. Clearly, unlike the situation with IVF-derived embryos, improved cryopreservation capabilities are desirable for both sperm and oocytes. The IVF of IVM oocytes is promising, although the efficiency of IVM at only 25%–30% must be improved. Our ability to fertilize mature rhesus monkey oocytes is comparable to the human

TABLE 5.2. Implantation and pregnancy rates in rhesus and cynomolgus macaques following the transfer of IVF-produced embryos.

Species	Stage frozen	Stage transferred	Implantation rate (%)	Pregnancy rate (%)	Reference
Cynomolgus	NA	2- to 4-cell	2/24 (8.3)	1/10 (10.0)	12
	2- to 8-cell	2- to 8-cell	3/25 (12.0)	3/9 (33.0)	38
Rhesus	NA	3-cell to early blastocyst	3/48 (6.3)	3/28 (10.7)	30
	3- to 6-cell	3- to 6-cell	1/6 (17.0)	1/3 (33.0)	39
	1- to 6-cell	2- to 6-cell	6/8 (75.0)	3/4 (75.0)[a]	25

The "Implantation rate" is the number of infants/number of embryos transferred. The "Pregnancy rate" is the number of live births or clinical pregnancies/number of animals receiving embryos. NA = not applicable.
[a] All twin pregnancies.

despite the fact that the experience base is minuscule in comparison. Finally, improved protocols for the long-term culture of IVF-produced embryos using somatic cell cocultures seem at hand. This accomplishment might well bring the first dividend from the NHP experience to the clinical practice in humans.

Acknowledgments. The authors express their appreciation to all current and past members of the IVF-ET program at ORPRC for their contributions to these studies; to Drs. Godke and Goodeaux and coworkers for contributions to our somatic cell coculture efforts; and to Dr. Hoskins and collaborators for CASMA measurements of sperm motility. A special thanks to Dr. Vaughn Critchlow, Director of ORPRC, for his vision and support in developing a nonhuman primate model for IVF-related research, to Dr. David Hess and his associates in the Hormone Assay Laboratory, and to Ms. Patsy Kimzey for her outstanding editorial and secretarial assistance. We also thank Dr. Jim Hutchison and Serono Laboratories, Inc., for their generous supply of hormone preparations. This work was supported by NIH Grants RR-00163, HD-18185, and HD-20869 (R.L.S.), HD-22408 (R.L.S.), HD-28484 (D.P.W.), and Serono Laboratories, Inc. This is publication number 1850 of the Oregon Regional Primate Research Center.

References

1. Thomson JA, Hess DL, Dahl KD, Iliff-Sizemore SA, Stouffer RL, Wolf DP. The Sulawesi crested black macaque (*Macaca nigra*) menstrual cycle: changes in perineal tumescence and serum estradiol, progesterone, follicle-stimulating hormone, and luteinizing hormone levels. Biol Reprod 1992;46:879–84.

2. Wolf DP, Thomson JA, Zelinski-Wooten MB, Stouffer RL. In vitro fertilization-embryo transfer in nonhuman primates: the technique and its applications. Mol Reprod Dev 1990;27:261–80.
3. Austin CR. Observation on the penetration of the sperm into the mammalian egg. Aust J Sci Res [B] 1951;4:581–96.
4. Chang MC. Fertilizing capacity of spermatozoa deposited into fallopian tubes. Nature 1951;168:697–8.
5. Mastroianni L Jr, Manson WA Jr. Collection of monkey semen by electroejaculation. Proc Soc Exp Biol Med 1963;112:1025–7.
6. Bennett JP. Artificial insemination of the squirrel monkey. J Endocrinol 1967;37:473–4.
7. Settlage DSF, Swan S, Hendrickx AG. Comparison of artificial insemination with natural mating technique in rhesus monkeys, *Macaca mulatta*. J Reprod Fertil 1973;32:129–32.
8. Riesen JW, Meyer RK, Wolf RC. The effect of season on occurrence of ovulation in the rhesus monkey. Biol Reprod 1971;5:111–4.
9. Weick RF, Dierschke DJ, Karsch FJ, Butler WR, Hotchkiss J, Knobil E. Periovulatory time courses of circulating gonadotropic and ovarian hormones in the rhesus monkey. Endocrinology 1973;93:1140–7.
10. Simpson ME, van Wagenen G. Induction of ovulation with human urinary gonadotrophins in the monkey. Fertil Steril 1962;13:140–52.
11. Batta SK, Stark RA, Brackett BG. Ovulation induction by gonadotropin and prostaglandin treatments of rhesus monkeys and observations of the ova. Biol Reprod 1978;18:264–78.
12. Balmaceda JP, Pool TB, Arana JB, Heitman TS, Asch RH. Successful in vitro fertilization and embryo transfer in cynomolgus monkeys. Fertil Steril 1984;42:791–5.
13. Bavister BD, Boatman DE, Collins K, Dierschke DJ, Eisele SG. Birth of rhesus monkey infant after in vitro fertilization and nonsurgical embryo transfer. Proc Natl Acad Sci USA 1984;81:2218–22.
14. Boatman DE, Bavister BD. Stimulation of rhesus monkey sperm capacitation by cyclic nucleotide mediators. J Reprod Fertil 1984;71:357–66.
15. Lanzendorf SE, Gliessman PM, Archibong AE, Alexander M, Wolf DP. Collection and quality of rhesus monkey semen. Mol Reprod Dev 1990;25:61–6.
16. Harrison RM. Semen parameters in *Macaca mulatta*: ejaculates from random and selected monkeys. J Med Primatol 1980;9:265–73.
17. Fujisaki M, Suzuki M, Kohno M, Cho F, Honjo S. Early embryonal culture of the cynomolgus monkey (*Macaca fascicularis*). Am J Primatol 1989;18:303–13.
18. Mack SO, Wolf DP, Tash JS. Quantitation of specific parameters of motility in large numbers of human sperm by digital image processing. Biol Reprod 1988;38:270–81.
19. Stephens DT, Hickman R, Hoskins DD. Description, validation, and performance characteristics of a new computer-automated sperm motility analysis system. Biol Reprod 1988;38:577–86.
20. Bavister BD, Boatman DE, Leibfried L, Loose M, Vernon MW. Fertilization and cleavage of rhesus monkey oocytes in vitro. Biol Reprod 1983;28:983–99.

21. Binor Z, Rawlins RG, Van der Ven H, Dmowski WP. Rhesus monkey sperm penetration into zona-free hamster ova: comparison of preparation and culture conditions. Gamete Res 1988;19:91-9.
22. Tollner TL, VandeVoort CA, Overstreet JW, Drobnis EZ. Cryopreservation of spermatozoa from cynomolgus monkeys (*Macaca fascicularis*). J Reprod Fertil 1990;90:347-52.
23. Stouffer RL, Zelinski-Wooten MB, Aladin Chandrasekher Y, Wolf DP. Stimulation of follicle and oocyte development in macaques for IVF procedures. In: Wolf DP, Stouffer RL, Brenner RM, eds. In vitro fertilization and embryo transfer in primates. New York: Springer-Verlag, 1993. *(See Chapter 8, this volume.)*
24. Alak BM, Wolf DP. Rhesus monkey oocyte maturation in vitro: kinetics and effects of gonadotropin supplementation [Abstract 1]. Serono Symposia Program and Abstracts: In Vitro Fertilization and Embryo Transfer in Primates, Beaverton, OR, May 28-31, 1992:27.
25. Lanzendorf SE, Zelinski-Wooten MB, Stouffer RL, Wolf DP. Maturity at collection and the developmental potential of rhesus monkey oocytes. Biol Reprod 1990;42:703-11.
26. Johnson LD, Mattson BA, Albertini DF, et al. Quality of oocytes from superovulated rhesus monkeys. Hum Reprod 1991;6:623-31.
27. Flood JT, Chillik CF, van Uem JFHM, Iritani A, Hodgen GD. Ooplasmic transfusion: prophase germinal vesicle oocytes made developmentally competent by microinjection of metaphase II egg cytoplasm. Fertil Steril 1990;53:1049-54.
28. Morgan PM, Warikoo PK, Bavister BD. In vitro maturation and fertilization of rhesus monkey oocytes [Abstract I-11]. Serono Symposia Program and Abstracts: The Primate Ovary, Beaverton, OR, May 16-17, 1987:28.
29. Kreitmann O, Lynch A, Nixon WE, Hodgen GD. Ovum collection, induced luteal dysfunction, in vitro fertilization, embryo development and low tubal ovum transfer in primates. In: Hafez ESE, Semm K, eds. In vitro fertilization and embryo transfer. New York: Alan R Liss, 1982:303-24.
30. Boatman DE. In vitro growth of non-human primate pre- and peri-implantation embryos. In: Bavister BD, ed. The mammalian preimplantation embryo. New York: Plenum Press, 1987:273-308.
31. Enders AC, Boatman D, Morgan P, Bavister BD. Differentiation of blastocysts derived from in vitro-fertilized rhesus monkey ova. Biol Reprod 1989;41:715-27.
32. Enders AC, Schlafke S. Differentiation of the blastocyst of the rhesus monkey. Am J Anat 1981;162:1-21.
33. Menezo YJR, Guerin JF, Czyba JC. Improvement of human early embryo development in vitro by coculture on monolayers of *Vero* cells. Biol Reprod 1990;42:301-6.
34. Godke RA, Blakewood EG, Thibodeaux JK. In vitro co-culture of mammalian embryos. In: Seibel MM, Richards C, Kiessling A, eds. Technical advances in infertility. Boston: Springer-Verlag, 1992.
35. Paria BC, Dey SK. Preimplantation embryo development in vitro: cooperative interactions among embryos and role of growth factors. Proc Natl Acad Sci USA 1990;87:4756-60.
36. Goodeaux LL, Thibodeaux JK, Voelkel SA, et al. Collection, co-culture and transfer of rhesus preimplantation embryos. ARTA 1990;1:370-9.

37. Zhang L, Wolf DP, Thomson JA, Goodeaux LL, Thibodeaux JK, Godke RA. Developing frozen-thawed IVF-derived rhesus monkey embryos to the hatched blastocyst stage using somatic cell co-culture [Abstract 23]. Serono Symposia Program and Abstracts: In Vitro Fertilization and Embryo Transfer in Primates, Beaverton, OR, May 28–31, 1992:38.
38. Balmaceda JP, Heitman TO, Garcia MR, Pauerstein CJ, Pool TB. Embryo cryopreservation in cynomolgus monkeys. Fertil Steril 1986;45:403–6.
39. Wolf DP, VandeVoort CA, Meyer-Haas GR, et al. In vitro fertilization and embryo transfer in the rhesus monkey. Biol Reprod 1989;41:355–46.

Part II

Gamete Biology/
Ovarian Physiology

6
Sperm-Zona Pellucida Interaction in Macaques

JAMES W. OVERSTREET AND CATHERINE A. VANDEVOORT

The success of assisted reproduction technologies in treating human infertility suggests that many causes of infertility can be related to the interaction of spermatozoa with the female reproductive tract. Nonhuman primates are the best model, and may be the only relevant animal model, for human sperm transport and physiology in vivo (1). An understanding of the similarities and differences in gamete biology between primate species is an important step in developing the animal model. In this chapter we highlight our recent work on the physiology of spermatozoa from the laboratory macaque and, in particular, on sperm interaction with the zona pellucida of the oocyte.

Sperm Capacitation and the Acrosome Reaction

Sperm capacitation in vivo is thought to involve the gradual removal or alteration of proteins that coat the sperm surface (2). There is evidence in humans that this process is initiated, if not completed, during sperm penetration of the cervical mucus (3). One possible consequence of the surface alterations in capacitated spermatozoa is the exposure of receptor sites that allow the sperm to interact with receptors of the zona pellucida and undergo the acrosome reaction as a consequence of interaction with these receptors. Experiments in vitro have identified numerous changes in the sperm membranes that take place during capacitation (4–6). These changes include increased membrane fluidity and phosphorylation of membrane proteins and lipids. Alterations in the sperm surface may also be responsible for the initiation of hyperactivated motility, which is thought to be involved in penetration of the oocyte investments (7).

Sperm capacitation in vitro can be accomplished in most species by sperm washing and incubation in a balanced salt solution containing energy substrates and albumin (8). However, there are some requirements for in vitro sperm capacitation that are species specific. Macaque spermatozoa are unusual in that they require *activation* with caffeine and *dibutyryl cyclic AMP* (dbcAMP) to facilitate hyperactivated motility and *in vitro fertilization* (IVF) (9, 10). Both of these compounds act to increase intracellular cAMP, which is thought to be a mediator of the capacitation process (5). The acrosome reaction is a sequel to capacitation and involves fusion and vesiculation of the sperm plasma membrane and the outer acrosomal membrane. This morphological event appears to result from an influx of calcium that may act directly on membrane phospholipids, as well as through activation of membrane phospholipase (11).

In many species the trigger for the acrosome reaction appears to be a specific glycoprotein receptor for sperm on the zona pellucida of the oocyte. The mouse is the best characterized of these species, and the mouse zona pellucida has been shown to be composed of 3 sulfated glycoproteins, termed *ZP1, ZP2,* and *ZP3* (12). ZP3 appears to function as a sperm receptor (13), and after the mouse sperm binds to ZP3, the acrosome reaction is elicited, perhaps by aggregation of the sperm receptors for zona (14). Although the biology of sperm-zona pellucida interaction is less well characterized in primates, homologous intact zona pellucida has been shown to induce acrosome reactions in sperm of humans (15) and macaques (16).

Macaque Sperm Interaction with Homologous Zona Pellucida

The methodology for these experiments has been published in detail (16), and it will only be summarized here. Semen was collected from cynomolgus and rhesus macaques by electroejaculation and was washed by repeated centrifugation and dilution with the *Biggers Whitten and Whitingham* (BWW) culture medium supplemented with 3 mg/mL of *bovine semen albumin* (BSA). Following incubation for 1.5 h at 37°C, the sperm suspensions were activated by addition of 1 mM caffeine and 1 mM dbcAMP. Following an additional 30 min incubation, they were coincubated with ovarian oocytes obtained from the same macaque species and previously stored at −20°C.

During the *pulse* phase of the experiment, 6–8 oocytes were incubated in a 100-µL drop of the sperm suspension (10×10^6 sperm/mL) for 1 min. The oocytes were washed by pipetting to remove loosely bound sperm, and half were immediately fixed in absolute ethanol. The remaining oocytes were incubated for an additional hour in fresh culture medium at 37°C. The acrosomal status of sperm bound to the zona pellucida was assessed using an antisperm antiserum and indirect immunofluorescence; the sperm in suspension were similarly assessed using a fluoresceinated lectin that binds to the acrosomal contents (16).

TABLE 6.1. Sperm-zona pellucida interaction in cynomolgus macaques.

	Activated sperm		Control sperm	
	No. of sperm per zona	Percent acrosome reacted	No. of sperm per zona	Percent acrosome reacted
Pulse	77 ± 14	24 ± 15	5 ± 4	0
Chase	53 ± 15	30 ± 6	0	0

Data are from 6 experiments and are presented as means ± SEM.
Less than 1% of sperm in suspension were acrosome reacted.
Source: VandeVoort, Tollner, and Overstreet (16).

TABLE 6.2. Sperm-zona pellucida interaction in rhesus macaques.

	Activated sperm		Control sperm	
	No. of sperm per zona	Percent acrosome reacted	No. of sperm per zona	Percent acrosome reacted
Pulse	71 ± 28	30 ± 8	10 ± 4	11 ± 7
Chase	74 ± 34	43 ± 14	15 ± 7	23 ± 12[a]

Data are from 6 experiments and are presented as means ± SEM.
Less than 2% of sperm in suspension were acrosome reacted.
[a] Significant difference between pulse and chase (Tukey's Test, $P < 0.05$).
Source: VandeVoort, Tollner, and Overstreet (16).

The data in Tables 6.1 and 6.2 summarize the results of the experiments with cynomolgus gametes and rhesus gametes, respectively. In both species activated sperm were bound to the zona within 1 min, and many were acrosome reacted. Very few nonactivated (control) sperm from cynomolgus macaques bound to the zona, and none were acrosome reacted (Table 6.1). In contrast, rhesus sperm were able to bind and acrosome-react on the zona, although the numbers of bound and acrosome-reacted sperm were higher if the sperm had been previously treated with activators (Table 6.2).

Effect of Calcium on Macaque Sperm-Zona Pellucida Interaction

Although the number of acrosome-reacted sperm in suspension was low (footnotes, Tables 6.1 and 6.2) and the period of coincubation was brief, we could not rule out the possibility that sperm on the zona had undergone the acrosome reaction prior to binding rather than being induced to acrosome-react by the zona. Because calcium is required for the acrosome reaction, we evaluated macaque sperm-zona pellucida interaction in media with minimal calcium. We hypothesized that under these conditions induction of the acrosome reaction as a consequence of sperm-zona binding would be inhibited. Cynomolgus sperm were used in these

TABLE 6.3. Cynomolgus sperm-zona pellucida interaction in media with minimal calcium.

	Activated sperm with Ca^{++}		Activated sperm with no Ca^{++}	
	No. of sperm per zona	Percent acrosome reacted	No. of sperm per zona	Percent acrosome reacted
Pulse	60 ± 6	14 ± 2	74 ± 24	4 ± 1
Chase	63 ± 5	20 ± 2	70 ± 14	4 ± 1

Data are from 5 experiments and are presented as means ± SEM.
Less than 3% of sperm in suspension were acrosome reacted.

experiments and were prepared by centrifugation through a 60% Percoll suspension and incubation for 2.5 h at 37°C in BWW medium with 3-mg/mL BSA. The sperm were activated as in previous experiments, but after 30 min of incubation with the activators, the sperm were washed and diluted with BWW medium in which Sr^{++} was substituted for Ca^{++} (17). The results of the pulse-chase experiments were similar to previous experiments when Ca^{++} was present in the media, but in media with minimal Ca^{++}, sperm binding resulted in few acrosome reactions (Table 6.3). These data support the interpretation that macaque sperm acrosome reactions are induced as a consequence of sperm-zona binding.

Macaque Sperm-Zona Pellucida Interaction Following Sperm Recovery from the Uterus

Little is known in any species about the comparative physiology of sperm capacitation in vivo and in vitro. Sperm capacitation in the female reproductive tract has been demonstrated in many model species (18) and in humans (3). Because of the apparent requirement for chemical activators during in vitro capacitation of macaque sperm, it was of interest to evaluate the interaction of spermatozoa with the zona pellucida following capacitation in vivo. For these experiments cynomolgus sperm were recovered from the uterus 6 h after mating by ultrasound-guided, transabdominal aspiration (19). As a control, sperm were capacitated in vitro with caffeine and dbcAMP as previously described. Because of the low number of sperm recovered in vivo (2.5 × 10^5/mL in 15 µL), the coincubations were carried out for 1 h. Surprisingly, almost none of the uterine sperm bound to the zonae (Table 6.4), although the sperm that were capacitated in vitro were able to bind and acrosome-react on the zona even when present at low concentrations (data not shown). These results appear to indicate a difference in the physiology of macaque sperm capacitated in vivo and in vitro. Further experiments will be needed to better define the nature of these differences and these biological significances.

TABLE 6.4. Cynomolgus sperm-zona pellucida interaction following sperm recovery from the uterus 6 h after mating.

Source of sperm	No. of experiments	No. of oocytes	Mean no. of sperm per zona
Uterus	2	8	0.9
Activated	3	6	11.6

Sperm and zonae were coincubated in 15-µL drops for 1 h. Activated sperm were capacitated in vitro with caffeine and dbcAMP and diluted to the same concentration as the uterine sperm (2.5×10^5/mL).

Conclusions

These experiments have pointed to a number of differences in the physiology of sperm from humans and macaques. Our data support the previous findings that macaque sperm require activation with caffeine and dbcAMP during capacitation to promote optimal sperm-zona pellucida interaction (9, 10). Human sperm differ from macaque sperm in that they capacitate spontaneously in vitro in relatively simple media (8). Both macaque sperm and human sperm are induced to acrosome-react after binding to the zona pellucida, and in humans, as in macaques, this zona-induced acrosome reaction is inhibited in media with minimal calcium (17). Macaque sperm undergo the zona-induced acrosome reaction more rapidly than human sperm because few human sperm are acrosome reacted after 1 min of sperm-zona pellucida coincubation (15). In the pulse-chase experiments described in this chapter, there was relatively little change in the number of acrosome-reacted macaque sperm during the 1-h chase period, whereas the majority of zona-induced acrosome reactions of human sperm occur during this interval (15). These differences may be attributed to the requirement for chemical induction of capacitation in macaque sperm that may result in greater synchrony of capacitation in the sperm population and, thus, a more immediate and uniform response to a stimulus for induction of the acrosome reaction.

The inability of macaque uterine sperm to bind to the zona also appears to reflect a difference in the physiology of human and macaque sperm. Although comparable data are not available for human uterine sperm, sperm recovered from human cervical mucus between 1 and 80 h after coitus were able to bind and penetrate the human zona pellucida (20).

Collectively, the observations reported in this chapter support the supposition that there are significant species differences in sperm physiology between macaques and humans. Nevertheless, the data are insufficient for firm conclusions to be drawn on the utility of the laboratory macaque as an animal model for studies of sperm transport and physiology in the female tract. Additional studies with these and other species will be needed in order to identify the most appropriate nonhuman primate model for studies of the pathophysiology of human infertility.

Acknowledgments. These data were obtained with support from CONRAD (CSA-88-018) and the National Institutes of Health (RR-00169 and HD-29116). The technical assistance of Mr. Theodore Tollner is gratefully acknowledged.

References

1. Overstreet JW, VandeVoort CA. Sperm transport in the female genital tract. In: Asch RH, Balmaceda JP, Johnston I, eds. Gamete physiology. Norwell, MA: Serono Symposia, USA, 1990:43–52.
2. Yanagimachi R. Capacitation and the acrosome reaction. In: Asch RH, Balmaceda JP, Johnston I, eds. Gamete physiology. Norwell, MA: Serono Symposia, USA, 1990:31–42.
3. Lambert H, Overstreet JW, Morales P, Hanson FW, Yanagimachi R. Sperm capacitation in the human female reproductive tract. Fertil and Steril 1985;43:325–7.
4. Yanagimachi R. Mechanisms of fertilization in mammals. In: Mastroianni L, Biggers JD, eds. Fertilization and embryonic development in vitro. New York: Plenum Press, 1981:81–182.
5. Yanagimachi R. Mammalian fertilization. In: Knobil E, Neill JD, eds. The physiology of reproduction; vol 1. New York: Raven Press, 1988:135–85.
6. Saling PM. Mammalian sperm interaction with extracellular matrices of the egg. In: Milligan SR, ed. Oxford reviews of reproductive biology; vol 11. Oxford, UK: Oxford University Press, 1989:339–88.
7. Katz DF, Drobnis EZ, Overstreet JW. Factors regulating mammalian sperm migration through the female reproductive tract and oocyte vestments. Gamete Res 1989;22:443–69.
8. Rogers BJ. Mammalian sperm capacitation and fertilization in vitro: a critique of methodology. Gamete Res 1978;1:165–223.
9. Chan PJ, Reinhold JH, Dukelow WR. Nonhuman primate in vitro fertilization: seasonality, cumulus cells, cyclic nucleotides, ribonucleic acid, and viability assays. Fertil Steril 1982;38:609–15.
10. Boatman DE, Bavister BD. Stimulation of rhesus monkey sperm capacitation by cyclic nucleotide mediators. J Reprod Fertil 1984;84:357–66.
11. Fraser LR, Ahuja KK. Metabolic and surface events in fertilization. Gamete Res 1988;20:491–519.
12. Bleil JD, Wassarman PM. Mammalian sperm-egg interaction: identification of a glycoprotein in mouse egg zonae pellucidae possessing receptor activity for sperm. Cell 1980;20:873–82.
13. Bleil JD, Wassarman PM. Autoradiographic visualization of the mouse egg's sperm receptor bound to sperm. J Cell Biol 1986;102:1363–71.

14. Leyton L, Saling P. Evidence that aggregation of mouse sperm receptors by ZP3 triggers the acrosome reaction. J Cell Biol 1989;108:2163–8.
15. Cross NL, Morales P, Overstreet JW, Hanson FW. Induction of acrosome reactions by the human zona pellucida. Biol Reprod 1988;38:235–44.
16. VandeVoort CA, Tollner TL, Overstreet JW. Sperm-zona pellucida interaction in cynomolgus and rhesus macaques. J Androl 1992.
17. Morales P, Cross NL, Overstreet JW, Hanson FW. Acrosome intact and acrosome-reacted human sperm can initiate binding to the zona pellucida. Dev Biol 1989;133:385–92.
18. Bedford JM. Sperm capacitation and fertilization in mammals. Biol Reprod Suppl 1970;2:128–58.
19. VandeVoort CA, Tollner TL, Tarantal AF, Overstreet JW. Ultrasound-guided transfundal uterine sperm recovery from *Macaca fascicularis*. Gamete Res 1989;24:327–31.
20. Gould JE, Overstreet JW, Hanson FW. Assessment of human sperm function after recovery from the female reproductive tract. Biol Reprod 1984;31:888–94.

7
Nonhuman Primate Oocyte Biology: Environmental Influences on Development

DOROTHY E. BOATMAN, RALPH D. SCHRAMM,
AND BARRY D. BAVISTER

Proper oocyte maturation involves the remodeling of the nucleus from a diploid *germinal vesicle* (GV) configuration to the mature haploid *metaphase II* (MII) stage. At the same time the egg must already possess or create and store sufficient cytoplasmic constituents to carry the zygote through early embryonic development following activation by sperm. The functional distinction between nuclear (morphological) and cytoplasmic (developmental) oocyte maturation has been clearly revealed by application of *in vitro fertilization* (IVF) to *in vitro-matured* (IVM) oocytes. In spite of apparently normal nuclear maturation—that is, attainment of MII and extrusion of the *first polar body* (PB1)—IVM oocytes may have developmental defects (1–4). Some of these deficiencies include defects in blocks to polyspermy, inability to decondense sperm nuclei, underdevelopment of metaphase spindles, and inability of zygotes to progress beyond early cleavage stages.

Much of our current knowledge concerning proper oocyte maturation leading to developmental competence is derived from studies utilizing nonprimate animal models, particularly the mouse. However, the increasing interest in developing procedures applicable to the economically important domestic animal species, such as cow, pig, and sheep, has led to observations revealing significant species differences in the timing and regulation of cellular events leading to a developmentally competent egg. For example, in cattle oocytes, unlike those of the mouse, both protein and RNA synthesis are required for GV breakdown (5, 6), with additional protein synthesis occurring after GV breakdown (7).

This chapter is publication number 32-034 of the Wisconsin Regional Primate Research Center.

Now, nonhuman primate IVF and egg and embryo culture technologies are sufficiently evolved that we may begin addressing many of these experimental questions directly in primates. It has been hypothesized that primate follicles and oocytes differ from those of other mammals (3, 8). Primate eggs, at the GV stage when released from antral follicles prior to the LH surge, resumed spontaneous meiosis at a very low rate compared to expectations based on studies of other mammalian eggs (3, 8). In women, during superstimulated cycles in which hCG was administered, up to 25% of the oocytes were still at the GV stage at the time of oocyte aspiration (9). Although these oocytes resumed meiosis at a high rate and many fertilized, they had only one-tenth the developmental potential of eggs that were mature at the time of aspiration (9). The GV eggs represent a significant component of the current failure rate in IVF. Additionally, there is new interest in recovering GV eggs from unstimulated follicles during routine clinical infertility workups in order to minimize expense and numbers of procedures required.

Progress in oocyte maturation leading to full developmental potential of immature nonhuman primate oocytes should lead to substantial clinical improvement in human IVF. In the longer term, studies on primate oocyte IVM/IVF will open up new opportunities for developing specific models for human and simian developmental disorders and diseases, perhaps through application of genetic engineering procedures to create transgenic animals.

Yield and Quality of Eggs

Usually, in Old World primates only a single egg is produced during natural ovulatory cycles. By superstimulating with exogenous gonadotropins, 10–20 quality eggs may frequently be produced in a given cycle (10), and in women stimulation is repeatable. In monkeys, however, there is limited capacity for repeating exogenous hormone stimulation in the same animal (11–13). In monkeys, using excised ovaries from unstimulated animals, up to 100 antral follicular eggs of varying quality can be obtained per female, but, of course, this can only be achieved once (14–16). If the pool of preantral follicles and their oocytes can be tapped as in the mouse (17), a single individual could produce in excess of 1000 eggs (18–20).

This achievement will require new initiatives in culture technology. Since only a single mature egg is produced in a natural ovulatory cycle, the potential value of any artificial intervention designed to increase yield is quite high. There is great potential for increasing the numbers of available female gametes by application of IVM to eggs of both antral and preantral follicles. However, much work remains before that potential can be realized since our ability to duplicate in vitro the necessary conditions for achieving oocyte growth and development is still rudimentary.

Timing of Nuclear Changes

In the animal species that have been most studied, the timing of meiotic progression in vitro for oocytes derived from well-developed antral follicles is affected by the presence of gonadotropins in the culture medium and the integrity of the cumulus cell layers surrounding the oocytes (2, 21-32). Under optimal in vitro conditions, meiotic timing approximates that occurring in vivo following the LH surge. In primates in vivo ovulation occurs within 36-48 h post-LH (33). In vitro, in maturation studies using oocytes obtained from nonsuperstimulated primate ovaries (human and monkey), there is evidence for some asynchrony in meiotic progression. The GV breakdown or prophase appears to occur between 11 and 24 h, diakinesis/MI from 19 to 31 h, and MII from 22 to 50 h (15, 16, 34-38). Ongoing research in our laboratory and others will eventually determine the effects of in vitro or in vivo gonadotropin supplementation on the time course of these phenomena (39).

Cellular Markers for Maturational Changes

During oocyte maturation cellular changes leading to redistribution of cortical granules from centripetal to cortical locations are well documented (40-42). Other changes occur in cytoplasmic organelles, the cytoskeleton, and at the membrane surface. During maturation of human oocytes, the number of multivesicular bodies and vacuoles decreased, while the endoplasmic reticulum volume, aggregation of smooth endoplasmic reticulum tubuli, and clustering of mitochondria increased (43).

Tubulin stores underwent progressive rearrangements during maturation (44). In macaque GV oocytes antibodies to tubulin revealed a diffusely staining pattern with no microtubules present. In metaphase oocytes no diffuse stores were present, but microtubules were located at the spindle poles and the interpolar region. In GV oocytes, prominent foci of actin were located at the egg cell surface, perhaps delineating the foot processes of cumulus cells traversing the zona pellucida and communicating with the eggs. In MI eggs surface actin was associated with clusters of fine microvilli, whereas in MII eggs actin was associated with evenly dispersed microvilli and was also present in the PB1 near the cleavage furrow (44).

Variables Correlated with Oocyte Quality

The granulosa cells are critical to the development of the oocyte (2, 21-23, 32). Various steroid and protein hormones, as well as growth factors, promote granulosa cell function and may improve cytoplasmic maturation of oocytes, including decondensation of the sperm nucleus, and acquisition of full developmental competence. In vivo in follicular fluid high concentrations of steroids (µg/mL) and pituitary hormones bathe the maturing oocyte (24, 25) and may be needed for maturation and developmental competence (21–32). In vitro their beneficial effects may not be manifested until later in development.

In the rabbit, development to morula and blastocyst stages was synergistically enhanced by FSH, LH, *estradiol* (E_2) or *progesterone* (P_4), and prolactin present during IVM, but nuclear maturation, fertilization, or 2-cell cleavage were not affected (31). The role of FSH in stimulating the growth and development of follicles in vivo and oocyte maturation and cumulus expansion in vitro is well known. Fully grown oocytes obtained from hypogonadal mice are restricted in their developmental competence following IVM (27); PMSG or FSH priming improves the developmental competence of oocytes derived from normal cycling or hypogonadal animals (27, 29). In vivo, LH induces maturation of oocytes. LH present in oocyte maturation culture improved development in cows and rats, has led to birth of live calves (26, 28), and has been specifically implicated in promotion of male pronucleus formation (45). In the rhesus monkey normal fertilization and development were impaired if gonadotropins were omitted during IVM (16).

Prolactin prevents differentiation, spreading, and luteinization of granulosa and cumulus cells in culture (46), changes that may lead to uncoupling of cumulus-oocyte gap junctions and cause abnormalities in oocytes (21). High prolactin content in proestrous cow serum improved bovine IVM (30). Estradiol has been linked to normal fertilization of rabbit ova (47) and to developmental competence of IVF human ova (reviewed in 24). Progesterone produced during FSH stimulation in vitro or added to IVM cultures of rat oocytes increased their fertilizability (48). Increased P_4/E_2 correlated with normal embryo development in primates (reviewed in 24).

Effects of Superstimulation Regimen on Oocyte Competence

In the rhesus monkey we have examined the effects on oocyte maturation leading to embryonic competence produced by two superstimulation protocols using PMSG/hCG (49). The higher-dose and longer-duration scheme (average PMSG dosage, 2110 ± 219 IU), protocol I, resulted in a 2-fold greater E_2 output over the stimulation cycle than occurred with protocol II (average PMSG dosage 1380 ± 89 IU). In protocol I, per cycle, substantially more mature oocytes (2.6x) were recovered and more fertilized zygotes (3.9x) were produced, along with significantly increased numbers of early zonal blastocysts (9.7x) after 5 or more days in culture. When the mean development scores for the oocytes of each of the animals in the study were plotted against their normalized plasma E_2 levels for the stimulation cycles, it was evident that in higher E_2-producing cycles the majority of the embryos cleaved beyond the 2-cell stage, reaching morula and blastocyst stages, whereas oocytes derived from the lower E_2 cycles were penetrated by spermatozoa, but did not cleave beyond the 2-cell stage.

Developmental Competence of Oocytes: In Vivo versus In Vitro Maturation

We initially approached the question, Is a mature oocyte the same whether produced in vivo or in vitro? by examining, after superstimulation cycles, the developmental competence of rhesus eggs that had completed either all or part of their maturation time in vivo (24, 49, 50). For eggs that had matured and extruded PB1, we found no differences in their ability to become penetrated and fertilized by sperm whether maturation had been completed in vivo or in vitro (67% vs. 73%, respectively). However, eggs that had completed a substantial portion of their maturation in vitro were substantially less likely to cleave optimally—that is, reach 6- to 8-cells by 50 h postinsemination—than eggs that were initially mature at the time of oocyte aspiration (45.6% in vitro vs. 86% in vivo). Thus, in vitro maturation or culture fails to efficiently or totally support normal maturation leading to continued developmental competence of eggs.

Follicular fluid steroids were examined to determine the parameters associated with initially mature, and by inference optimally mature, eggs from superstimulation cycles (24, 51). Initially mature eggs originated in follicles whose follicular fluid P_4 levels were 38%–42% higher on average than those for eggs matured in vitro (i.e., that had not completed maturation at aspiration) or for dysmature and atretic eggs. No differences were observed for follicular fluid E_2 levels nor for the marker androgens chosen for assay. The P_4/E_2 ratios were also markedly higher for initially mature eggs compared to immature and dysmature types. From the data one cannot determine whether the steroid levels and ratios were merely markers for the successful completion of good maturation or actively contributed to the future health of oocytes/embryos subsequently cultured in vitro. However, recent findings in the rhesus monkey implicate in vivo steroids as supporting normal fertilization and development, but not follicle development or nuclear maturation of oocytes (52).

In Vitro Maturation of Oocytes from Excised Follicles of Unstimulated Rhesus Monkeys

Gonadotropins and culture medium were tested for their effects on nuclear (i.e., meiotic progression) and cytoplasmic (i.e., developmental capacity) maturation of oocytes obtained from excised follicles (0.2–

2 mm) from unstimulated rhesus monkeys (16). No significant differences were detected between culture in the relatively simple medium TALP containing *bovine calf serum* (BCS) and 4 amino acids and the more complex modified CMRL medium containing 21 amino acids and nucleosides plus BCS; therefore, those data were pooled for further analysis.

Gonadotropins did not significantly affect meiotic maturation as determined by *GV breakdown* (GVBD) (53.3% and 59.5%, minus and plus gonadotropins, respectively) and extrusion of PB1 (28.9% vs. 40.5%). However, both fertilization (i.e., sperm penetration and activation of eggs, 6.7% vs. 20.7%) and cleavage (2.2% vs. 13.5%) subsequent to sperm penetration were significantly enhanced in the treatments containing gonadotropins compared with the control. Nonetheless, even with gonadotropins the majority of the cultured oocytes (60%) did not mature sufficiently (i.e., reach PB1) such that fertilization was possible. Of the fertilized eggs only 44% reached 5- to 8-cells in culture, a substantial reduction in development compared to 86% of eggs matured in vivo, as described in preceding paragraphs (24, 49, 50).

Inhibition of Meiotic Progression by cAMP and Hypoxanthine

Mammalian eggs may be divided into two broad categories based on their in vitro responsiveness to inhibition of spontaneous meiotic progression by exogenous cAMP mediators and/or hypoxanthine. In the mouse, hypoxanthine, a phosphodiesterase inhibitor, is inhibitory to meiotic progression (53, 54), and follicular fluid levels of hypoxanthine are sufficiently high that it may be the elusive *oocyte maturation inhibitor* (OMI) much pursued by biochemists (53, 54). However, in the cow spontaneous oocyte maturation is highly resistant to inhibition by exogeous cAMP or phosphodiesterase inhibitors (55).

We tested the inhibition of rhesus nuclear maturation by cAMP and hypoxanthine (15). A crossover design was used to test whether or not inhibition—if it occurred—could be reversed by removal of the drugs following 40 h of culture. We found that rhesus oocyte maturation could be inhibited reversibly by hypoxanthine plus dbcAMP and, thus, regarding this control point of meiosis, resembled the murine more than the bovine system.

Follicular Diameter and Oocyte Properties

In the mouse at the time of antrum formation, the oocyte has essentially completed growth, and the nucleus, given the appropriate stimulus, has acquired the ability to complete meiotic progression (56). In domestic animals, such as the pig and cow, oocyte growth has not been completed in the smallest antral follicles (20, 57). Furthermore, such oocytes are deficient in their ability to complete meiotic progression (57) and/or after maturation lack the ability to support complete preimplantation development when penetrated by sperm (58, 59).

Equivalent data have been lacking for primate oocytes. In recent experiments in our laboratory, antral follicles of unstimulated rhesus monkey ovaries were divided into four size classifications: class I: 200–450 μm; class II: 451–700 μm; class III: 701–1000 μm; and class IV: >1000 μm (60). Samples of oocytes from each class were obtained at time = 0 h, their vitelline diameters measured, and nuclear configurations examined using a fluorescent DNA-binding dye. The remainder of the oocytes were cultured for 48 h in medium containing gonadotropins and serum, then scored for meiotic progression. Chromatin patterns of GV oocytes were classified according to the following criteria (61): GV1: nucleolus not rimmed by chromatin; GV2: partial rim of chromatin around nucleolus; and GV3: complete chromatim rim around nucleolus. In the mouse these three categories represent increasingly "mature" categories of nuclei, differing in competence to undergo meiotic progression in vitro (61). Our GV3 category corresponds to that normally found in well-developed antral follicles of the mouse (61).

The relationship between rhesus egg and follicle diameters and resumption of meiosis in culture is illustrated in Figure 7.1. The eggs in

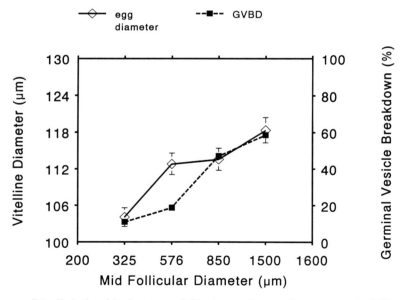

FIGURE 7.1. Relationship between follicular and egg diameters and ability of rhesus oocytes to resume meiosis in vitro. The midpoint diameter of each of the 4 follicle class sizes was plotted. GVBD represents resumption of meiosis.

class I follicles were significantly smaller than eggs in the other follicle classes. Those in class II were fully grown, but were less competent than eggs in classes III and IV to progress to metaphase stages. The relationship between follicle diameter and competence to resume meiosis was similar to that for marmoset monkey (New World primate) follicles divided into small, medium, and large categories (62). At time = 0, follicle classes I–III had a mixture of types of GV nuclei, whereas class IV follicles had no GV1, 20% GV2, and 80% GV3 (not shown). The change of oocyte nuclear distribution after 48 h of culture for the four follicle classes is illustrated by Figure 7.2. All follicle classes produced some MI and MII eggs. This appeared to occur predominantly by the progression of GV3 eggs to metaphase stages, but some involvement of GV2 eggs cannot be excluded.

Other Culture Variables

Results obtained from egg and embryo culture using nonprimate models, in which eggs are more plentiful, point out important experimental questions still to be addressed in primate research and illustrate the pitfalls of too readily generalizing between species. In our laboratory a large-scale experiment using over 700 bovine oocytes was conducted to test the effects of 7 common culture media on IVM and on embryonic

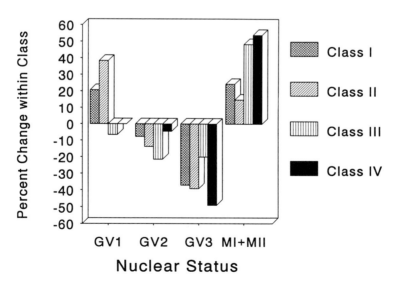

FIGURE 7.2. Change in distribution of nuclear configurations in rhesus oocytes after 48 h of culture. Within each follicle class, the percent (P) change was calculated: $P_{48h} - P_{0h} = P_{change}$.

development subsequent to IVM (63). Oocytes were exposed to the 7 different media only during the 24-h IVM culture. After IVM all eggs were transferred to standard IVF medium, TALP; then, zygotes were all transferred into the same culture medium for another 8–9 days of culture to permit embryo development to blastocyst stages.

Five of the media were equivalent, but 2 of them, Waymouth's and Ham's F12, reduced the ability of matured oocytes to be normally fertilized, as evidenced by decreased cleavage to 2-cells. When the data were corrected for the difference in fertilization, additional impairment of development was noted, as these 2 media resulted in a significantly lower percentage of 2-cells forming blastocysts. One might conclude that these media are bad for IVM in general, but Waymouth's is the preferred medium for mouse IVM and has been shown to promote male pronucleus formation in pig oocytes (64, 65).

Other work from our laboratory using hamster embryos implicates both carbon dioxide and oxygen gas tensions in promoting or inhibiting embryonic development (66, 67). For embryos obtained after in vivo fertilization, the gas level required for optimal development was significantly influenced by the stage of embryo when placed into culture. In general, developmentally younger stages required lower O_2 tensions (i.e., 5%) than did slightly older stages (10%) (67). Elevated CO_2 (10%) increased 2-cell development to blastocysts, possibly acting to modulate intracellular pH (66). We are just initiating similar experiments on gas tension for rhesus oocytes in light of our recent finding that oxygen levels in rhesus fallopian tubes are less than 10% (Fischer, Bavister, unpublished observations).

Summary and Conclusions

Primate oocyte biology is now at the crossroads of a new age of opportunity created largely by knowledge generated by the expanding data base of experimentation on an increasing number of mammalian species. With ample evidence that nonhuman primate oocytes are a high-fidelity model for human oocytes, it is anticipated that the ability to support IVM of nonhuman primate oocytes with full developmental competence will lead to substantial clinical improvement in human IVF. In the longer term, the application of genetic engineering procedures to create transgenic animals from IVM, IVF nonhuman primate oocytes will, in general, enhance the usefulness of nonhuman primates as models for human diseases and their amelioration.

Acknowledgments. We are grateful to Michael Tennier for skilled technical service, to past members of our research team, to the Gamete and Embryo Biology Training Grant (NIH-5-T32 HD-07342) for fellowship support (R.D.S.), and to the NIH for funding (HD-22023 to B.D.B. and RR-00167 to the Wisconsin Regional Primate Research Center).

References

1. Lanzendorf SE, Zelinski-Wooten MB, Stouffer RL, Wolf DP. Maturity at collection and the development potential of rhesus monkey oocytes. Biol Reprod 1990;42:703-11.
2. Staigmiller RB, Moor RM. Effect of follicle cells on the maturation and developmental competence of ovine oocytes matured outside the follicle. Gamete Res 1984;9:221-9.
3. Plachot M, Mandelbaum J. Oocyte maturation, fertilization and embryonic growth in vitro. Brit Med Bull 1990;46:675-94.
4. Zenzes MT, Belkien L, Bordt J, Kan I, Schneider HPG, Nieschlag E. Cytologic investigation of human in vitro fertilization failures. Fertil Steril 1985;43:883-91.
5. Hunter AG, Moor RM. Stage dependent effects of inhibitory ribonucleic acids and protein synthesis on meiotic maturation of bovine oocytes in vitro. J Dairy Sci 1987;70:1646-51.
6. Sirard MA, Leibfried-Rutledge ML, Barnes FL, Sims ML, First NL. Timing of nuclear progression and protein synthesis necessary for meiotic maturation of bovine oocytes. Biol Reprod 1989;40:1257-64.
7. Kastrop PMM, Hulshof SCJ, Bevers MM, Destree OHJ, Kruip TAM. The effects of Á-amanitin and cycloheximide on nuclear progression, protein synthesis, and phosphorylation during bovine oocyte maturation in vitro. Mol Reprod Dev 1991;28:249-54.
8. Lefevre B, Gougeon A, Testart J. Primate: a model for oocyte maturation study. Pathol Biol (Paris) 1990;38:166-9.
9. Flood JT, Chillik CF, van Uem JFHM, Iritani A, Hodgen GD. Ooplasmic transfusion: prophase germinal vesicle oocytes made developmentally competent by microinjection of metaphase II egg cytoplasm. Fertil Steril 1990;53:1049-54.
10. Zelinski-Wooten MB, Wolf DP, Hess DL, Stouffer RL. Individualized gonadotropin regimens for follicular stimulation in macaque IVF cycles [Abstract 22]. Serono Symposium on In Vitro Fertilization and Embryo Transfer in Primates, Beaverton, OR, 1992:37.
11. Bavister BD, Dees HC, Schultz RD. Refractoriness of rhesus monkeys to repeated ovarian stimulation by exogenous gonadotropins is caused by non-precipitating antibodies. Am J Reprod Immunol Microbiol 1986;11:11-6.
12. Wolf DP, Thomson JA, Zelinski-Wooten MB, Stouffer RL. In vitro fertilization-embryo transfer in nonhuman primates: the technique and its applications. Mol Reprod Dev 1989;27:261-80.
13. Iliff-Sizemore SA, Molskness TA, Stouffer RL. Effect of anti-human gonadotropin antibodies on fertility of female rhesus macaques [Abstract 11]. Serono Symposium on In Vitro Fertilization and Embryo Transfer in Primates, Beaverton, OR, 1992:32.

14. Smith DM, Conaway CH, Kerber ST. Influences of season and age on maturation in vitro of rhesus monkey oocytes. J Reprod Fertil 1978;54: 91–5.
15. Warikoo PK, Bavister BD. Hypoxanthine and cAMP maintain meiotic arrest of rhesus monkey oocytes in vitro. Fertil Steril 1989;51:886–9.
16. Morgan PM, Warikoo PK, Bavister BD. In vitro maturation of ovarian oocytes from unstimulated rhesus monkeys: assessment of cytoplasmic maturity by embryonic development after in vitro fertilization. Biol Reprod 1991;45:89–93.
17. Eppig JJ, Schroeder AC. Capacity of mouse oocytes from preantral follicles to undergo embryogenesis and development to live young after growth, maturation, and fertilization in vitro. Biol Reprod 1989;41:268–76.
18. Koering MJ. Ovarian architecture during follicle maturation. In: Dukelow SR, Erwin J, eds. Comparative primate biology; vol 3, reproduction and development. New York: Alan R. Liss, 1986:215–62.
19. Green SH, Zuckerman S. The number of oocytes in the mature rhesus monkey (*Macaca mulatta*). J Endocrinol 1951;7:194–202.
20. Motlik J, Crozet N, Fulka J. Meiotic competence in vitro of pig oocytes isolated from early antral follicles. J Reprod Fertil 1984;72:323–8.
21. Mattioli M, Galeati G, Seren E. Effect of follicle somatic cells during pig oocyte maturation on egg penetrability and male pronucleus formation. Gamete Res 1988;20:177–83.
22. Mattioli M, Galeati G, Bacci ML, Seren E. Male pronucleus formation depends on soluble factors produced by follicle somatic cells during pig oocyte maturation. Gamete Res 1988;21:223–32.
23. Moor RM, Mattioli M, Ding J, Nagai T. Maturation of pig oocytes in vivo and in vitro. J Reprod Fertil 1990;40(suppl):197–210.
24. Morgan PM, Boatman DE, Bavister BD. Relationships between follicular fluid steroid hormone concentrations, oocyte maturity, in vitro fertilization and embryonic development in the rhesus monkey. Mol Reprod Dev 1990;27:145–51.
25. McNatty KP, Hunter WM, McNeilly AS, Sawers RS. Changes in the concentration of pituitary and steroid hormones in the follicular fluid of human graafian follicles throughout the menstrual cycle. J Endocrinol 1975;64:555–71.
26. Brackett BG, Younis AI, Fayrer-Hosken RA. Enhanced viability after in vitro fertilization of bovine oocytes matured in vitro with high concentrations of LH. Fertil Steril 1989;52:319–24.
27. Schroeder AC, Eppig JJ. Developmental capacity of mouse oocytes that undergo maturation in vitro: effect of the hormonal state of the oocyte donor. Gamete Res 1989;24:81–92.
28. Shalgi R, Dekel N, Kraicer PF. The effect of LH on the fertilizability and developmental capacity of rat oocytes matured in vitro. J Reprod Fertil 1979;55:429–35.
29. Vanderhyden BC, Armstrong DT. Effects of gonadotropins and granulosa cell secretions on the maturation and fertilization of rat oocytes in vitro. Mol Reprod Dev 1990;26:337–46.
30. Younis AI, Brackett BG, Fayrer-Hosken RA. Influence of serum and hormones on bovine oocyte maturation and fertilization in vitro. Gamete Res 1989;23:189–201.

31. Yoshimura Y, Hosoi Y, Iritani A, Nakamura Y, Atlas SJ, Wallach EE. Developmental potential of rabbit oocytes matured in vitro: the possible contribution of prolactin. Biol Reprod 1989;41:26–33.
32. Vanderhyden BC, Armstrong DT. Role of cumulus cells and serum on the in vitro maturation, fertilization and subsequent development of rat oocytes. Biol Reprod 1989;40:720–8.
33. Weick RF, Dierschke DJ, Karsch FJ, Butler WR, Hotchkiss J, Knobil E. Periovulatory time courses of circulating gonadotropic and ovarian hormones in the rhesus monkey. Endocrinology 1973;93:1140–7.
34. Edwards RG. Maturation in vitro of human ovarian oocytes. Lancet 1965;2:926–9.
35. Klinger HP, Kava HW, Hachamovitch M. Chromosome studies in human female meiosis. In: Apgar V, ed. Down's syndrome (mongolism). Ann N Y Acad Sci 1970:431–9.
36. Jagiello GM, Ducayen MB, Miller WA, Lin JS, Fang JS. A cytogenetic analysis of oocytes from *Macaca mulatta* and *nemestrina* matured in vitro. Humangenetik 1973;18:117–22.
37. Thibault C, Gerard M, Menezo Y. Nuclear and cytoplasmic aspects of mammalian oocyte maturation in relation to follicle size and fertilization. Sperm Action Prog Reprod Biol 1976;1:233–40.
38. Suzuki S, Kitai H, Tojo R, et al. Ultrastructure and some biologic properties of human oocytes and granulosa cells cultured in vitro. Fertil Steril 1981;35:142–8.
39. Alak BM, Wolf DP. Rhesus monkey oocyte maturation in vitro: kinetics and effects of gonadotropin supplementation [Abstract 1]. Serono Symposium on In Vitro Fertilization and Embryo Transfer in Primates, Beaverton, OR, 1992:27.
40. Zamboni L. Ultrastructure of mammalian oocytes and ova. Biol Reprod 1970;2(suppl 2):44–63.
41. Szollosi D, Gerard M. Cytoplasmic changes in the mammalian oocytes during the preovulatory period. In: Beier HM, Lindner HR, eds. Fertilization of the human egg in vitro. Berlin: Springer-Verlag, 1983:35–55.
42. Thibault C, Szollosi D, Gerard M. Mammalian oocyte maturation. Reprod Nutr Dev 1987;27:865–96.
43. Sundstrom P, Nilsson O. Sequential changes in cytoplasmatic features during maturation of the human oocyte. In: Motta PM, ed. Developments in ultrastructure of reproduction: a celebrative symposium, the "opera omnia" of Marcello Malpighi. Proc VIIIth Internatl Symp Morphological Sciences, July 10–15, 1988. New York: Liss, 1989:327–33.
44. Johnson LD, Mattson BA, Albertini DF, et al. Quality of oocytes from superovulated rhesus monkeys. Hum Reprod 1991;6:623–31.
45. Mattioli M, Bacci ML, Galeati G, Seren E. Effects of LH and FSH on the maturation of pig oocytes in vitro. Theriogenology 1991;36:95–104.
46. Channing CP, Anderson LD, Hoover DJ, et al. The role of nonsteroidal regulators in control of oocyte and follicular maturation. Recent Prog Horm Res 1982;38:331–408.
47. Yoshimura Y, Hosoi Y, Bongiovanni AM, Santulli R, Atlas SJ, Wallach EE. Are ovarian steroids required for ovum maturation and fertilization? Effects of cyanoketone on the in vitro perfused rabbit ovary. Endocrinology 1987;120:2555–61.

48. Zhang X, Armstrong DT. Effects of FSH and ovarian steroids during in vitro meiotic maturation on fertilization of rat oocytes. Gamete Res 1989;23:267–77.
49. Boatman DE, Morgan PM, Bavister BD. Variables affecting the yield and developmental potential of embryos following superstimulation and in vitro fertilization in rhesus monkeys. Gamete Res 1986;13:327–38.
50. Boatman DE. In vitro growth of non-human primate pre- and peri-implantation embryos. In: Bavister BD, ed. The mammalian preimplantation embryo. New York: Plenum Press, 1987:273–308.
51. Bavister BD. Oocyte maturation and in vitro fertilization in the rhesus monkey. In: Stouffer RL, ed. The primate ovary. New York: Plenum Press, 1987:119–37.
52. Zelinski-Wooten MB, Hess DL, Wolf DP, Stouffer RL. Steroid depletion during ovarian stimulation impairs oocyte fertilization, but not folliculogenesis in rhesus monkeys. 48th annu meet American Fertility Soc, 1992.
53. Downs SM, Coleman DL, Ward-Bailey PF, Eppig JJ. Hypoxanthine is the principal inhibitor of murine oocyte maturation in a low molecular weight fraction of porcine follicular fluid. Proc Natl Acad Sci USA 1985;82:454–8.
54. Eppig JJ, Ward-Bailey PF, Coleman DL. Hypoxanthine and adenosine in murine ovarian follicular fluid: concentrations and activity in maintaining oocyte meiotic arrest. Biol Reprod 1985;33:1041–9.
55. Sirard MA, Coenen K, Bilodeau S. Effect of fresh or cultured follicular fractions on meiotic resumption in bovine oocytes. Theriogenology 1992;37:39–57.
56. Erickson GF, Sorensen RA. In vitro maturation of mouse oocytes from late, middle and preantral graafian follicles. J Exp Zool 1974;190:123–7.
57. Motlik J. Cytoplasmic aspects of oocyte growth and maturation in mammals. J Reprod Fertil 1989;38(suppl):17–25.
58. Fulka J, Pavlok A, Fulka J. In vitro fertilization of zona-free bovine oocytes matured in culture. J Reprod Fertil 1982;64:495–9.
59. Pavlok A, Lucas-Hahn A, Niemann H. Fertilization and developmental competence of bovine oocytes derived from different categories of antral follicles. Mol Reprod Dev 1992;31:63–7.
60. Schramm RD, Tennier MT, Boatman DE, Bavister BD. Chromatin configurations and meiotic competence of oocytes are related to follicular maturation in non gonadotropin-stimulated rhesus monkeys. Biol Reprod (in press).
61. Mattson BA, Albertini DE. Oogenesis: chromatin and microtubule dynamics during meiotic prophase [mouse]. Mol Reprod Dev 1989;25:374–83.
62. Adachi M, Yokoyama M, Tanioka Y. Culture of marmoset ovarian oocytes in vitro. Jpn J Anim Reprod 1982;28:51–5.
63. Rose TA, Bavister BD. Effect of oocyte maturation medium on in vitro development of in vitro fertilized bovine embryos. Mol Reprod Dev 1992;31:72–7.
64. van de Sandt JJM, Schroeder AC, Eppig JJ. Culture media for mouse oocyte maturation affect subsequent embryonic development. Mol Reprod Dev 1990;25:164–71.

65. Yoshida M, Ishigaki K, Pursel VG. Effect of maturation media on male pronucleus formation in pig oocytes matured in vitro. Mol Reprod Dev 1992;31:68–71.
66. Carney EW, Bavister BD. Regulation of hamster embryo development in vitro by carbon dioxide. Biol Reprod 1987;36:1155–63.
67. McKiernan SH, Bavister BD. Environmental variables influencing in vitro development of hamster two-cell embryos to the blastocyst stage. Biol Reprod 1990;43:404–13.

8
Stimulation of Follicle and Oocyte Development in Macaques for IVF Procedures

Richard L. Stouffer, Mary B. Zelinski-Wooten, Yasmin Aladin Chandrasekher, and Don P. Wolf

In many primate species the interactions between and within components of the hypothalamic-pituitary-ovarian axis ensure the maturation of a single follicle and the timely release of one oocyte capable of fertilization around the middle of the menstrual cycle. Knowledge of the processes involved in growth, selection, maturation, and ovulation of the dominant follicle has increased substantially in recent years, particularly from experimental studies in rhesus monkeys (1, 2). The importance of the pituitary gonadotropins, *follicle stimulating hormone* (FSH) and *luteinizing hormone* (LH), in the folliculo- and gametogenic functions of the ovary have been recognized for over 50 years (3, 4), but recent manipulations to stimulate follicular development have added new information on the processes and events controlled by FSH and LH (5, 6). It is clear that methods that elevate circulating levels of endogenous or exogenous gonadotropins (FSH and LH) override the mechanisms that select a single dominant follicle and stimulate the development of multiple follicles and their enclosed oocytes.

The large supply of fertilizable oocytes provided by follicular stimulation protocols should facilitate critical research on primate reproduction, particularly gametogenesis, fertilization, early embryogenesis, and implantation. The application of assisted reproductive techniques, such as IVF, to nonhuman primates will promote applied research to control human infertility or fertility and to preserve endangered species of primates. Oocytes have been recovered from maturing follicles during unstimulated, spontaneous cycles in monovular (7) and polyovular (8) primates for IVF-*embryo transfer* (ET) protocols. In certain scenarios (7) this may be the method of choice. However, the large numbers of oocytes and resultant embryos offered by follicular stimulation techniques make detailed experimentation possible. Although there are disadvantages as discussed later, follicular stimulation will be essential until techniques are perfected for the *in vitro maturation* (IVM) of oocytes collected from resting or immature (unstimulated) follicles to yield fertilizable eggs and normal embryos (9).

Follicle Stimulation Protocols

The authors recently reviewed the various methods and protocols used by investigators to stimulate multiple follicular growth in species ranging from great apes to Old and New World monkeys (10). The list continues to grow; for example, Verhage and coworkers (11) recently induced multiple follicular development and superovulation in the olive baboon (*Papio anubis*). Despite success in clinical practice (12) with clomiphene —an antiestrogen that inhibits the negative feedback of estrogen on pituitary gonadotropin secretion, thereby raising endogenous FSH and LH levels—its use for follicular stimulation in nonhuman primates has been rare (13). Researchers have preferred to administer exogenous gonadotropins of nonprimate (e.g., *pregnant mare serum gonadotropin* [PMSG]) (14) or primate (e.g., human *menopausal gonadotropins* [hMG]) (15) origin for follicular stimulation. With the realization that LH receptors in primates have a much higher affinity for primate, as opposed to nonprimate, LH (16) and that highly purified human hormones prepared for clinical protocols are readily available, most (if not all) groups are using regimens of human gonadotropins to stimulate the growth and maturation of multiple follicles in primate species (10).

Although a standardized regimen of human gonadotropins has not evolved, treatment generally begins in the early follicular phase and continues for 6–11 days (10). Gonadotropin doses range from 25–75 IU/day in macaques. Although the daily dose is often held constant throughout treatment, *step-up* protocols where the dose is increased during treatment have been successful, whereas *step-down* protocols have not (17, 18). Preparations of hMG containing FSH and LH activity are commonly used since both gonadotropins may be necessary for optimal growth and maturation of follicles. Prevailing concepts prescribe a predominant role for FSH in follicle growth, but assign an important role to FSH plus LH in the endocrine maturation of the preovulatory follicle (1, 19).

However, the need for exogenous LH to supplement endogenous levels during multiple follicular development has been questioned (6, 19); indeed, elevated LH levels may have detrimental effects, such as eliciting premature maturation of the follicle or oocyte (10, 20). Exogenous hFSH alone stimulates the growth of many follicles in women (21) and in nonhuman primates, such as cynomolgus (22) and rhesus (23) monkeys. Although IVF is common with human oocytes from FSH treatment protocols (21), reported success in nonhuman primate models with FSH alone has been limited to squirrel monkeys (24).

During efforts to develop a rhesus monkey model for IVF-ET related research at ORPRC, we compared the ability of three regimens of human gonadotropins to stimulate the development of moderate numbers of ovulatory follicles (18). Although a regimen of hMG (FSH + LH) met our criteria, a sequential regimen of hFSH followed by hFSH + hLH resulted in a greater frequency of classical endocrine responses (e.g., continually rising estradiol levels during treatment) and a greater cohort of maturing oocytes capable of IVF (10). In this protocol (plan A), rhesus monkeys exhibiting regular menstrual cycles receive (beginning at menses) twice-daily IM injections of 30-IU hFSH (Metrodin, Serono Laboratories, Inc.) for 6 days, followed by 30-IU hMG (hFSH + hLH, Pergonal, Serono Laboratories, Inc.) for 3 days. On day 10, animals receive a single IM injection of 1000-IU hCG (Profasi, Serono Laboratories, Inc.) to induce ovulatory maturation. Although bona fide

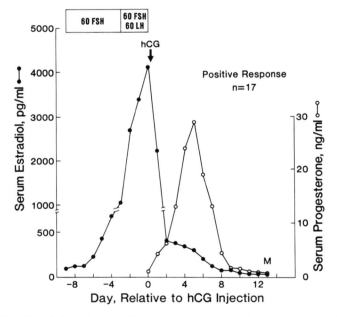

FIGURE 8.1. Circulating E- and P-levels in rhesus monkeys treated with human gonadotropins (plan A + hCG) to stimulate multiple follicular development. The daily dose (IU) of pituitary gonadotropins administered for 9 days beginning at menses is depicted in the horizontal bar. The arrow signifies the day of hCG injection to induce ovulatory maturation; data are normalized to the day of hCG injection (day 0). Values are the mean of 17 monkeys. See Table 8.1 for estimates of variance in the data group. Adapted with permission from Wolf, VandeVoort, Meyer-Haas, et al. (25).

ovulatory follicles develop (18), aspiration is typically performed ≥27 h after hCG injection to retrieve oocytes before ovulation (see below).

Figure 8.1 illustrates the patterns and levels of serum *estradiol* (E) and *progesterone* (P) in monkeys that responded positively to this follicle stimulation protocol. Estradiol levels rose slowly during the first days of FSH treatment, but increased markedly with addition of LH to the treatment regimen. The E-levels peaked on the day of hCG injection, then declined rapidly during the next 2 days and gradually thereafter in the luteal phase. The P-levels were at baseline (<0.5 ng/mL) during follicular stimulation, but rose rapidly after hCG administration. Levels peaked 5 days later and then declined to baseline within 10 days after hCG injection. While the steroid hormone patterns are similar to those observed in spontaneous menstrual cycles, peak levels were 10-fold higher than normal, and the length of the luteal phase was somewhat shortened (25). The high steroid levels reflected the development of numerous large (4–7 mm diameter) follicles on each ovary. An average of 18 oocytes were collected (Table 8.1) per animal after hCG injection, with 63% judged mature (based on achievement of metaphase II) at collection or within 18 h of in vitro culture. Linear regression analyses revealed that the number of ova collected correlated positively ($P < 0.05$) with peak levels of serum E. When peak E-levels were expressed per mature oocyte, a value of 514 pg/mL was obtained, which is similar to that in spontaneous monovular cycles.

Despite our success, it was clear that the response of normal cycling monkeys to follicular stimulation protocols was variable. Factors reportedly influencing the stimulation response include animal age, season, and ovarian status or cycle regularity (10, 26). When characterizing the patterns and circulating levels of E during stimulation protocols, we noted four types of responders: (i) *classical responders* with continuously rising E-levels throughout treatment; (ii) *biphasic*

TABLE 8.1. Endocrine and oocyte characteristics from gonadotropin-stimulated (plan A + hCG) cycles in rhesus monkeys.

Estradiol peak (pg/mL)	4480 ± 1012[a]
Estradiol peak (day)	10
Oocytes retrieved (number)	17.6 ± 3.2
Oocytes for IVF (% mature)[b]	63 ± 5
Progesterone peak (ng/mL)	32.6 ± 6.4
Progesterone peak (days post-hCG)	5.2 ± 0.3
Luteal phase (days)	11.8 ± 0.7

[a] Mean ± SEM; n = 17.
[b] Metaphase II within 18 h of follicle aspiration.
Source: Adapted with permission from Wolf, VandeVoort, Meyer-Haas, et al. (25).

responders with E-levels declining by >20% after more than 5 days of treatment, but rebounding thereafter; (iii) *abbreviated responders* with E-levels declining continuously after more than 5 days of treatment; and (iv) *nonresponders* with E-levels never rising above those observed in spontaneous cycles. A sequential regimen of FSH followed by FSH + LH (plan A) resulted in a greater frequency of classical responders than continual FSH + LH (67% vs. 36% of treated animals) (10). However, a significant number of animals fell into categories ii–iv (33% in plan A) and either did not reach follicle aspiration or if they did, the percentage of collected oocytes that fertilized and cleaved in vitro (13% of 31 oocytes) was less than those from classical responders (41% of 179 oocytes).

Since E-levels tended to plateau in some classical responders or to decline in biphasic/abbreviated responders during the last 3 days of treatment, we questioned whether the final days of gonadotropin exposure contribute to follicular maturation. Moreover, we hypothesized that it may be advantageous to individualize the treatment regimen, as in clinical IVF protocols (21), to optimize follicular stimulation. To test these hypotheses, monkeys received 6 days of hFSH (30 IU b.i.d.) followed by 1, 2, or 3 days of hFSH + hLH (30 IU each, b.i.d.). Treatment ended when serum E was >2 ng/mL and large follicles (≥4 mm diameter) were detected by transabdominal ultrasonography. The next day, hCG (1000 IU) was administered.

Table 8.2 summarizes the endocrine and oocyte characteristics from individualized protocols for follicular stimulation in rhesus monkeys (unpublished observations). Peak E-levels on the day of hCG injection

TABLE 8.2. Endocrine and oocyte characteristics from individualized, gonadotropin-treatment regimens in rhesus monkeys.

Parameters	Interval of FSH ± LH treatment		
	7 days	8 days	9 days
Regimens (n)	5	23	8
Estradiol peak (pg/mL)*	3090 ± 460[a]	3548 ± 350[a]	4979 ± 1230[b]
Progesterone peak (ng/mL)**	56 ± 15	32 ± 5	53 ± 11
Follicle size (% ≥ 4 mm)	56%	65%	60%
Oocytes retrieved (n)	20 ± 6	20 ± 2	20 ± 6
Oocyte maturity (% MI + MII)	82%	81%	83%
Fertilization (% of inseminated)***	40%	44%	31%

The "Parameters" are $\bar{x}$ ± SEM; unpublished data.
* Classical responders; peak levels on the day of hCG injection.
** Peak levels 5 days after hCG injection.
*** Data base includes IVF experiments both before and after sperm processing was modified to better remove decapacitation factors in macaque semen (see Chapter 5, this volume).
Significant differences ($P < 0.05$) between treatment intervals are noted by a and b superscripts between columns.

were comparable among 7- and 8-day treatment groups, but somewhat higher in the 9-day group. Nevertheless, there were no differences among groups in the size distribution of 2- to 7-mm follicles by day 6–7 of treatment or in the number of oocytes collected per animal. Although the percentage of oocytes collected at metaphase I and II was equally high in all groups (81%–83%), the percentage of inseminated eggs that fertilized in vitro tended to be higher in the 7- and 8-day treatment groups (40%–44% fertilization rate). Finally, peak levels and the pattern of circulating P in the luteal phase did not differ significantly between groups.

We conclude that reducing the gonadotropin treatment interval from 9 to 7 days is not detrimental to follicular growth or maturation, reinitiation of oocyte meiosis, or the subsequent development and function of corpora lutea. Reducing the interval of hMG (FSH + LH) exposure in the sequential protocol to as little as 1 day provides further evidence that exogenous LH may not be essential for follicular stimulation to yield normal oocytes in macaque species. The need for exogenous or endogenous LH in follicular stimulation protocols and subsequent IVF or embryogenesis in monkeys awaits clarification (6, 19). Nevertheless, the use of sequential regimens of human gonadotropins can proceed in macaques on an individualized basis to shorten the treatment interval without compromising follicle or gamete quality. We have yet to attempt longer treatment intervals in poor responders/nonresponders to accentuate follicular development; efforts in this direction are currently discouraged by the production of neutralizing antibodies to human gonadotropins (10, 25).

While a fixed-treatment protocol (e.g., plan A) may be required in experiments testing the effects of various factors on follicular or oocyte development, an individualized protocol that regularly monitors follicle growth (via ultrasound) and maturation (via serum E radioimmunoassay) permits adjustments to achieve an optimal response within each monkey. An individualized response may more efficiently provide multiple follicles and oocytes for IVF research and its applications to propagating endangered species.

Nevertheless, an individualized approach does not eliminate the occurrence of nonclassical responders. For example, 9 of 51 monkeys (18%) in our individualized protocols exhibited biphasic responses. Interestingly, fertilization of oocytes collected from biphasic animals during 8-day protocols (39% of inseminated eggs, n = 5 monkeys) was comparable to that in classical responders (Table 8.2) and greater ($P < 0.05$) than that from biphasic animals during 9-day protocols (18%, n = 3). Perhaps the earlier collection provides oocytes that are less affected by the processes underlying the transient drop in serum E levels. More efficient methods are needed to (i) screen animals prior to follicular stimulation to identify potential nonresponders (27) and (ii) promote follicular stimulation and reduce response variation.

Many of the abbreviated and biphasic E-responses in monkeys are associated with a spontaneous LH surge (>100 ng/mL) or "mini-surge" (<100 ng/mL) on the day before declining E-levels (10, 25). The addition of adjuvant drugs, such as GnRH agonists commonly used in clinical IVF programs (28), may prevent endogenous LH surges and/or provide a more homogeneous ovarian pool of growing follicles for the stimulation protocol. Since use of GnRH analogs to down-regulate endogenous gonadotropin secretion can increase the dosage and duration of gonadotropin treatment required for follicular stimulation in clinical programs (28), their use in macaques (29, 30) must again be weighed against further antigen exposure and production of antibodies to human gonadotropins. Alternatively, acute treatment with GnRH antagonist (31) even as late as 5–7 days after initiating gonadotropin regimens may prevent endogenous surges while permitting efficient follicular development in macaques.

Initiation of Ovulatory Events

The LH surge of the normal menstrual cycle initiates a series of events within the preovulatory follicle that lead to resumption of meiosis and ultimately to ovulation with release of a metaphase II oocyte capable of fertilization in the oviduct (32). A timely LH surge of normal magnitude and duration does not typically occur during follicular stimulation protocols, and when one unexpectedly occurs, oocyte collection may be disrupted and canceled (10). Because of its availability and LH-like activity, hCG has been the hormone of choice for inducing ovulatory events in follicle stimulation protocols for women and nonhuman primates (10). Although some clinical programs have permitted follicles to "coast" for 48 h before giving hCG, more typically, hCG is administered within 24 h of the last hMG/hFSH injection (10). Oocytes are then collected at a prescribed interval after hCG administration, usually between 24 (8) and 38 (15) h in monkeys. This time interval ideally would permit completion of oocyte maturation before ovulation begins to reduce the number of eggs available by follicle aspiration. We hypothesized that as the preovulatory maturation interval (hours post-hCG administration) increases, the percentage of oocytes maturing to metaphase II before harvest would increase. To evaluate this possibility and the IVF potential of collected oocytes, rhesus monkeys (n = 38) were treated with a sequential regimen of human gonadotropins (FSH followed by FSH + LH) (25), and the interval from hCG injection to follicle aspiration varied between 27 and 36 h (unpublished data).

The differences in the percentage of collected oocytes that reinitiated meiosis (% at *metaphase I* [MI] + *metaphase II* [MII] + *ovulation sites*

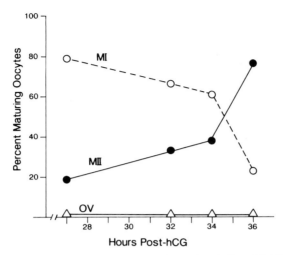

FIGURE 8.2. Percent of maturing oocytes collected at MI and MII or lost due to ovulation (OV) when follicles were aspirated from gonadotropin-treated monkeys either 27 (n = 12), 32 (n = 3), 34 (n = 20), or 36 (n = 3) h after hCG injection. Unpublished data.

[OV]) over the interval of 27–36 h post-hCG (70%–86%) were not significant. However, the percentage of maturing oocytes reaching MII doubled between 27 and 34 h, and then doubled again by 36 h post-hCG (Fig. 8.2). There was an inverse relationship between the percentage of MI vs. MII oocytes, suggesting that few oocytes entered the maturing pool (i.e., reinitiated meiosis) after 27 h, and few oocytes escaped the pool via ovulation. The latter was corroborated by the presence of only a few ovulatory stigma on ovaries through 36 h post-hCG. In addition, the time required for >90% of MI oocytes to mature to MII during in vitro culture (33) declined from 7 to 3 h as the hCG exposure time increased from 27 to 36 h.

The data are consistent with the concept that extending the interval between hCG administration and follicle aspiration up to 36 h (i) increases the yield of mature (MII) oocytes at collection and (ii) increases the rate at which meiotically active oocytes reach maturity in vitro. Nevertheless, IVF of oocytes collected 27 h post-hCG was as great or better (73% of inseminated [n = 95] eggs) than that for oocytes collected at 34 h (52% of 216 eggs). Since oocyte loss due to ovulation was not significant, follicle aspiration between 27 and 36 h post-hCG yields oocytes that fertilize, but longer culture for IVM of MI oocytes is required at the shorter collection times. Whether this longer, but not excessive, culture interval is beneficial for successful IVF of MI oocytes requires further study.

Alternative Ovulatory Stimuli

Although hCG has LH-like activity, there is concern that (i) hCG action does not entirely mimic the effects of LH on target cells, at least in nonprimate species (34); and (ii) the long half-life of hCG leads to pharmacologic gonadotropin levels through the midluteal phase that have detrimental (e.g., desensitization) effects on the development and function of the corpus luteum (35). A recent study on macaque granulosa lutein cells (36) does not support disparate actions of LH and CG in primate species where they are both endogenous hormones.

However, the long half-life of hCG is a major factor in the antigenicity of the molecule and antibody production in macaques during gonadotropin treatment protocols (10). It should be possible to replace the ovulatory hCG stimulus in IVF protocols with surge levels of hLH or with an induced surge of endogenous LH. The spontaneous LH surge (37) in rhesus monkeys and women is much longer (48–50 h) than that in many small laboratory animals (e.g., rats and rabbits: 4–8 h) and domestic animals (e.g., sheep and cows: 10–16 h). Whether the lengthy LH surge is necessary for ovulatory changes in the follicle and its enclosed oocyte in primates is not known. Therefore, studies were performed to titrate the periovulatory LH requirements in macaques during IVF-related cycles in which multiple-follicle development was promoted with a fixed gonadotropin-treatment regimen (37–39).

After 9 days of sequential FSH ± LH treatment (plan A), monkeys received either: (i) no ovulatory stimulus; (ii) 1–3 sc injections of 100-μg GnRH or 2 injections of 50-μg GnRH agonist (Lupron, TAP Pharmaceuticals) to stimulate endogenous LH release; (iii) 1 or 2 IM injections of hLH (2500 IU); or (iv) 1 IM injection of hCG (1000 IU). Multiple blood samples were collected during the next 72 h to measure LH-like bioactivity and the length of the LH surge (defined as the interval from injection to the subsequent fall in bioactive LH-like levels to <100 ng/mL). Twenty-seven hours after the first injection, follicles were aspirated and indexes of oocyte maturity and granulosa cell luteinization evaluated.

As examples, Figures 8.3 and 8.4 illustrate the hormonal patterns in monkeys receiving 1 injection of GnRH or hLH, respectively, to induce periovulatory events. Serum E rose to peak levels around the time of the ovulatory stimulus in both groups and was similar to those shown previously for plan A + hCG-treated monkeys (Fig. 8.1). Following GnRH injection, bioactive LH increased within 1–2 h to peak levels ($\bar{x}$ = 433 ng/mL) and then returned to baseline after 6 h (Fig. 8.3); nevertheless, the steroid pattern was reminiscent of that in animals receiving *no* ovulatory stimulus—serum E levels declined rapidly to baseline and P-levels remained low. In concordance with the lack of a

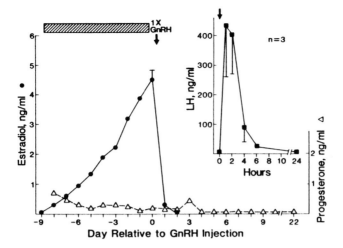

FIGURE 8.3. Serum E-, P-, and bioactive LH (inset) levels in monkeys (n = 3) receiving a single injection of 100-μg GnRH (Relisorm, Serono) after gonadotropin treatment (horizontal bar; plan A) for promotion of multiple follicular development. Mean steroid and LH levels are presented, with the SEM indicated for the E-peak and LH surge. Adapted with permission from Zelinski-Wooten, Lanzendorf, Wolf, Aladin Chandrasekher, and Stouffer (37), © by The Endocrine Society, 1991.

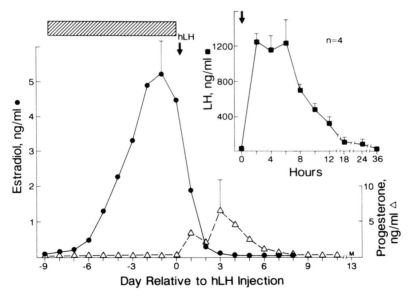

FIGURE 8.4. Steroid and bioactive LH (inset) patterns in monkeys (n = 4) receiving 1 injection of 2452-IU hLH for ovulatory maturation after gonadotropin treatment (horizontal bar; plan A). An M depicts the mean day of menses. Adapted with permission from Zelinski-Wooten, Hutchison, Aladin Chandrasekher, Wolf, and Stouffer (39), © by The Endocrine Society.

functional luteal phase, menses was either absent or delayed (22–43 days) after GnRH injection. Following LH injection (Fig. 8.4), bioactivity increased markedly and remained at peak levels ($\bar{x} = 1200\,\mathrm{ng/mL}$) for 6 h before declining to baseline after 24 h. Within 24 h of LH administration, serum P levels increased, but prematurely returned to baseline in 1–6 days (n = 4 monkeys); peak levels were markedly less than those observed in hCG-treated monkeys (e.g., Fig. 8.1). Menses was typically delayed until several days after P-levels returned to baseline.

In monkeys receiving no ovulatory stimulus or GnRH/GnRHa regimens that produced LH surges lasting up to 14+ h (Table 8.3), very few oocytes reinitiated meiotic maturation, and granulosa cells did not display immunocytochemical staining for *P-receptor* (PR) or an enhanced ability to secrete P in vitro. Consistent with minimal evidence for granulosa cell luteinization, these animals (except for one; see footnote in Table 8.3) did not display a functional luteal phase. In contrast, extending the interval of the LH surge to $\geq$24 h produced maturing oocytes; the percentage of MI + MII eggs at collection was similar at surge durations of 24 and 50 h (LH-treated monkeys), as well as 96 h (hCG-treated animals). However, PR expression was only detected in some animals, and P-production in vitro was less in monkeys experiencing 24-h LH surges compared to the greater responses observed during 96-h hCG surges. Notably, the length of the functional luteal phase and serum P

TABLE 8.3. Comparison of protocols that produce LH surges of various intervals during IVF-related cycles in rhesus monkeys: subsequent periovulatory events.

Ovulatory stimulus (n = 3–6/group)	Interval of gonadotropin surge (h)	Oocyte maturation (%)	Granulosa cells		Luteal phase (days)
			PR expression (−/+)	Progesterone production (ng/ml)	
None	0	13	−	19	0
GnRH (100 µg × 1)	6	0	−	35	0
GnRH (100 µg × 3)[a]	10	20	−	61	0
GnRHa (50 µg × 2)	+14	12	−	105	0
hLH (2452 IU × 1)	18–24	76	2(−),2(+)	290	1–6
hLH (2500 IU × 2)	36–48	69	+	DI[b]	8–18
hCG (1000 IU)	96	86	+	680	12

"Oocyte maturation" represents percent oocytes at metaphase I or II by the time of follicle aspiration. In "Granulosa cells" cells were stained immunocytochemically for PR and incubated in vitro to assess P-production (ng/mL/24 h). "Luteal phase" represents the interval of progesterone levels above baseline or from day 10 until menses.
[a] Excludes one animal with periovulatory events similar to those in monkeys with 18–24 h surges.
[b] Data incomplete.
Source: Adapted with permission from Zelinski-Wooten, Lanzendorf, Wolf, Chandrasekher, and Stouffer (37); Aladin Chandrasekher, Brenner, Molskness, Yu, and Stouffer (38); Zelinski-Wooten, Hutchison, Aladin Chandrasekher, Wolf, and Stouffer (39), © by The Endocrine Society; and unpublished data.

levels (39) were less in monkeys with 24-h LH vs. 96-h CG surges. Although the length of the luteal phase and serum P levels in some animals approached those observed following 96-h CG surges, these parameters were more variable after 48-h LH surges.

Collectively, the data indicate that short (≤ 14 h) LH surges reminiscent of those occurring in rodents and domestic animals are insufficient to reinitiate meiotic maturation of oocytes or to support the development and function of the corpus luteum in primates. However, attenuated (24 h) surges approximately one-half the duration of spontaneous surges in normal menstrual cycles induce early ovulatory changes in primate follicles—particularly, oocyte meiosis—but fail to sustain the development and/or function of the corpus luteum. *Pulsatile* administration of hLH (200 IU t.i.d.) following an attenuated LH surge produces luteal function and lifespan comparable to that in normal cycles (39).

These data support the concept that the threshold of gonadotropin exposure for periovulatory events varies (i) between species and also (ii) between different events within the ovulatory follicle. Thus, the reinitiation of oocyte maturation requires an LH surge of lesser amplitude or duration than luteinization events in rabbit (40) and macaque (39) follicles. Even larger gonadotropin surges may be required for actual ovulation (40). The amplitude or duration of the LH surge required for ovulation of primate follicles is unknown, but there is a report that a single injection of recombinant hLH following a follicular stimulation protocol produced a 24-h surge and ovulation of at least 2 follicles in one monkey (41).

It remains to be determined whether the threshold for reinitiation of meiosis is optimal for oocyte maturation and early embryogenesis. Although fertilization rates were comparable between LH- and hCG-treated groups, the percentage of fertilized eggs that underwent early cleavage after an 18- to 24-h LH surge was half that of eggs experiencing a 96-h hCG surge (39). Whether this is an artifact of the small sample size or an indicator of suboptimal embryo quality following an attenuated surge is unknown. The ability of embryos originating from MII oocytes of LH-treated macaques to implant and develop to live offspring awaits examination. However, oocyte exposure to an endogenous LH surge elicited by GnRH/GnRH agonist has occasionally led to clinical pregnancies in women (42) and in the rhesus monkey (33).

Limitations and Improvements of Follicular Stimulation Protocols

The most important limitation of current protocols is the lack of nonhuman primate hormones that forces the use of foreign (i.e., antigenic) gonadotropins. Nonprimate (43) and human (10) gonadotropins elicit production of neutralizing antibodies in macaques, re-

TABLE 8.4. Antihuman gonadotropin antibodies in macaque serum as a function of the number of follicular stimulation protocols.

Protocol	n	Percent ^{125}I-hCG bound		Estradiol response
		Before	After	
First	21	1.1 ± 0.1	1.4 ± 0.1	Classical
	3	1.0 ± 0.2	20.3 ± 2.0	Classical
Second	1	1.6	1.6	Classical
	8	1.3 ± 0.1	21.3 ± 4.6	Classical/abbrev./none
	3	18.5 ± 4.3	31.4 ± 9.4	Abbrev./none
Third	2	21.4	39.9	Abbrev./none

"Percent ^{125}I-hCG bound" represents protein A precipitation of antibody-bound ^{125}I-hCG in serum before and after completion of human gonadotropin (e.g., plan A + hCG) protocols.
"Estradiol response" represents patterns of serum E in response to gonadotropin treatment, as described in the text and reference 10.
Source: Adapted with permission from Wolf, VandeVoort, Meyer-Haas, et al. (25) and unpublished data.

sulting in ovarian refractoriness to further gonadotropin therapy. In our laboratory, the first regimen of human gonadotropins (plan A + hCG) typically does not lead to the presence of antihuman gonadotropin antibodies (as detected by protein A-precipitable ^{125}I-hCG from serum) (Table 8.4). However, after 2 treatment regimens more than 90% of the monkeys have circulating antibodies that are associated with a high incidence of failure to promote multiple follicular development. Importantly, these antibodies do not disrupt normal menstrual cyclicity, fertility, and successful pregnancy in rhesus monkeys (44), but they prevent further use of the animals in IVF or other protocols employing human gonadotropins. To minimize antibody production we routinely perform only one follicular stimulation protocol on each monkey; this is possible since we have a large colony of macaques for research purposes at ORPRC.

The availability of ample supplies of nonhuman primate gonadotropins would overcome this problem and greatly facilitate experimentation. For example, monkeys that are classical responders and yield large numbers of high-quality oocytes in the first treatment protocol could be reused, thereby increasing the success rate and permitting sequential experiments on the same animal. Also, nonhuman primate gonadotropins would permit IVF-related procedures to preserve the genetic diversity and maintain endangered species of macaques or other primates. We recently characterized the menstrual cycle in *Macaca nigra*, a species that is declining rapidly in the wild and is the only nonhuman primate model for type II diabetes mellitus (45). The menstrual cycle in this species is 10 days longer than in rhesus macaques due to a period of ovarian quiescence (circulating E $\leq$ 50 pg/mL) during the early follicular phase.

Novel protocols will be required to optimize follicular development in *M. nigra* cycles, and we await macaque hormones that would allow repetitive experimentation on representative females from a limited population.

Until macaque gonadotropins are available, efforts such as those described in previous sections will continue to reduce or eliminate the need for foreign gonadotropins as promoters of follicle development or ovulatory maturation. Clinical studies in women suggest that exogenous growth hormone (46) or its releasing factor GRF (47) reduces the gonadotropin treatment interval and the amount of gonadotropin required for multiple follicular development. Whether similar adjuvant therapy would reduce gonadotropin requirements in macaques is unknown. Novel methods that increase the circulating levels of endogenous gonadotropins, such as the pulsatile administration of GnRH (48) or systemic administration of activin (49), also warrant testing.

Summary

We have developed fixed and individualized treatment regimens of human gonadotropins that promote the development of multiple large follicles in rhesus monkeys and provide high-quality oocytes for IVF-ET protocols and basic research in primates. The sequential treatment regimens of FSH followed by FSH + LH are based on the proposed predominant role for FSH in follicular growth, but include a possible action for LH in later follicular maturation. The combined use of ultrasonography and E-assays permits regular monitoring of the follicle growth and maturation, respectively. The maturational status of oocytes collected at follicle aspiration is dependent on the duration of the ovulatory stimulus and the interval between the LH/CG surge and oocyte collection. In the past two years, more than 50 gonadotropin treatment regimens were performed annually that culminated in the recovery of over 1000 oocytes per year for IVF-related research in the rhesus monkey. Nevertheless, further improvements are needed to reduce the variability in animal response to treatment regimens and to optimize the numbers of mature follicles/oocytes in responders. Researchers eagerly await the availability of nonhuman primate gonadotropins made through recombinant DNA technology (41) to overcome current limitations due to use of antigenic preparations of foreign gonadotropins. Until then, efforts continue to reduce or replace the need for human FSH, LH, or CG by testing adjuvant or alternative treatment protocols.

Acknowledgments. The authors express their appreciation to all current and past members of the IVF-ET team at ORPRC for their contributions to these studies. A special thanks to Dr. Vaughn Critchlow, Director of

ORPRC, for his vision and support in developing a nonhuman primate model for IVF-related research; to Dr. David Hess and his associates in the Hormone Assay Laboratory; to Mr. William Baughman and his assistants in the Surgical Unit; and to Ms. Patsy Kimzey for her outstanding editorial and secretarial assistance. We also thank Dr. Jim Hutchison and Serono Laboratories, Inc., for their generous supply of hormone preparations. This work was supported by NIH Grants RR-00163, HD-18185, HD-20869 (R.L.S.), HD-22408 (R.L.S.), HD-28484 (D.P.W.), and Serono Laboratories, Inc. This chapter is publication number 1846 of the Oregon Regional Primate Research Center.

References

1. Hodgen GD. The dominant ovarian follicle. Fertil Steril 1982;38:281–300.
2. Zeleznik AJ. Control of follicular growth during the primate menstrual cycle. In: Mashiach S, Ben-Rafael Z, Laufer N, Schenker JG, eds. Advances in assisted reproductive technologies. New York: Plenum Press, 1990:83–91.
3. Hisaw FL, Greep RO, Fevold HL. Experimental ovulation of *Macacus rhesus* monkeys [Abstract 56]. Anat Rec 1935;61(suppl):24–5.
4. Hartman CG. Use of gonadotropic hormones in adult rhesus monkey. Bull Johns Hopkins Hosp 1938;63:351–71.
5. Baird DT. A model for follicular selection and ovulation: lessons from superovulation. J Steroid Biochem 1987;27:15–23.
6. Hodgen GD, Kenigsberg D. New developments in ovulation induction. In: Besch PK, ed. The biochemistry of the reproductive years. Washington, DC: American Assoc for Clinical Chemistry, 1985:201–26.
7. Cranfield MR, Schaffer N, Bavister BD, et al. Assessment of oocytes retrieved from stimulated and unstimulated ovaries of pig-tailed macaques (*Macaca nemestrina*) as a model to enhance the genetic diversity of captive lion-tailed macaques (*Macaca silenus*). Zoo Biol 1989;suppl 1:33–46.
8. Lopata A, Summers PM, Hearn JP. Births following the transfer of cultured embryos obtained by in vitro and in vivo fertilization in the marmoset monkey (*Callithrix jacchus*). Fertil Steril 1988;50:503–9.
9. Cha KY, Koo JJ, Ko JJ, Choi DH, Han SY, Yoon TK. Pregnancy after in vitro fertilization of human follicular oocytes collected from nonstimulated cycles, their culture in vitro and their transfer in a donor oocyte program. Fertil Steril 1991;55:109–13.
10. Wolf DP, Thomson JA, Zelinski-Wooten MB, Stouffer RL. In vitro fertilization-embryo transfer in nonhuman primates: the technique and its applications. Mol Reprod Dev 1990;27:261–80.
11. McCarthy TJ, Fortman JD, Boice ML, Fazleabas AT, Verhage HG. Induction of multiple follicular development and superovulation in the olive baboon, *Papio anubis*. J Med Primatol 1991;20:308–14.
12. Adashi EY. Clomiphene citrate: mechanism(s) and site(s) of action—a hypothesis revisited. Fertil Steril 1984;42:331–44.
13. Fourie FR, Snyman E, van der Merwe JV, Grace A. Primate in vitro fertilization research: preliminary results on the folliculogenic effects of three different ovulatory induction agents on the chacma baboon, *Papio ursinus*. Comp Biochem Physiol [A] 1987;87:889–93.

14. Bavister BD, Boatman DE, Collins K, Dierschke DJ, Eisele SG. Birth of rhesus monkey infant after in vitro fertilization and nonsurgical embryo transfer. Proc Natl Acad Sci USA 1984;81:2218-22.
15. Balmaceda JP, Pool TB, Arana JB, Heitman TS, Asch RH. Successful in vitro fertilization and embryo transfer in cynomolgus monkeys. Fertil Steril 1984;42:791-5.
16. Cameron JL, Stouffer RL. Comparison of the species specificity of gonadotropin binding to primate and nonprimate corpora lutea. Biol Reprod 1981;25:568-72.
17. Abbasi R, Kenigsberg D, Danforth D, Falk RJ, Hodgen GD. Cumulative ovulation rate in human menopausal/human chorionic gonadotropin-treated monkeys: "step-up" versus "step-down" dose regimens. Fertil Steril 1987;47:1019-24.
18. VandeVoort CA, Baughman WL, Stouffer RL. Comparison of different regimens of human gonadotropins for superovulation of rhesus monkeys: ovulatory response and subsequent luteal function. J In Vitro Fertil Embryo Transfer 1989;6:85-91.
19. Chappel SC, Howles C. Reevaluation of the roles of luteinizing hormone and follicle-stimulating hormone in the ovulatory process. Hum Reprod 1991;6:1206-12.
20. Moon YS, Yun YW, King WA. Detrimental effects of superovulation. Semin Reprod Endocrinol 1990;8:232-41.
21. Jones GS, Acosta AA, Garcia JE, Bernardus RE, Rosenwaks Z. The effect of follicle-stimulating hormone without additional luteinizing hormone on follicular stimulation and oocyte development in normal ovulatory women. Fertil Steril 1985;43:696-702.
22. Schenken RS, Williams RF, Hodgen GD. Ovulation induction using "pure" follicle-stimulating hormone in monkeys. Fertil Steril 1984;41:629-34.
23. Stouffer RL, Hodgen GD, Graves PE, Danforth DR, Eyster KM, Ottobre JS. Characterization of corpora lutea in monkeys after superovulation with human menopausal gonadotropin or follicle-stimulating hormone. J Clin Endocrinol Metab 1986;62:833-9.
24. Dukelow WR, Vengesa PN. Primate models for fertilization and early embryogenesis. In: Benirschke K, ed. Primates: the road to self-sustaining populations. New York: Springer-Verlag, 1986:445-61.
25. Wolf DP, VandeVoort CA, Meyer-Haas GR, et al. In vitro fertilization and embryo transfer in the rhesus monkey. Biol Reprod 1989;41:335-46.
26. Armstrong DT, Leung PCK. The physiological basis of superovulation. Semin Reprod Endocrinol 1990;8:219-31.
27. Scott RT, Toner JP, Muasher SJ, Oehninger S, Robinson S, Rosenwaks Z. Follicle-stimulating hormone levels on cycle day 3 are predictive of in vitro fertilization outcome. Fertil Steril 1989;51:651-4.
28. Loumaye E. The control of endogenous secretion of LH by gonadotrophin-releasing hormone agonists during ovarian hyperstimulation for in-vitro fertilization and embryo transfer. Hum Reprod 1990;5:357-76.
29. Thibodeaux JK, Roussel JD, Menezo Y, et al. GnRH down-regulation in a superovulation regime for nonhuman primates [Abstract]. Theriogenology 1990;33:337.
30. Lefèvre B, Gougeon A, Nomé F, Testart J. Effect of a gonadotropin-releasing hormone agonist and gonadotropins on ovarian follicles in

cynomolgus monkey: a model for human ovarian hyperstimulation. Fertil Steril 1991;56:119–25.
31. Byrd S, Itskovitz J, Chillik C, Hodgen GD. Flexible protocol for administration of human follicle-stimulating hormone with gonadotropin-releasing hormone antagonist. Fertil Steril 1992;57:209–14.
32. Yoshimura Y, Wallach EE. Studies of the mechanism(s) of mammalian ovulation. Fertil Steril 1987;47:22–34.
33. Lanzendorf SE, Zelinski-Wooten MB, Stouffer RL, Wolf DP. Maturity at collection and the developmental potential of rhesus monkey oocytes. Biol Reprod 1990;42:703–11.
34. Niswender GD, Roess DA, Barisas BG. Receptor-mediated differences in the actions of ovine luteinizing hormone vs. human chorionic gonadotropin. In: Stouffer RL, ed. The primate ovary. New York: Plenum Press, 1987:237–48.
35. VandeVoort CA, Hess DL, Stouffer RL. Luteal function following ovarian stimulation in rhesus monkeys for in vitro fertilization: atypical response to human chorionic gonadotropin treatment simulating early pregnancy. Fertil Steril 1988;49:1071–5.
36. Molskness TA, Zelinski-Wooten MB, Hild-Petito SA, Stouffer RL. Comparison of the steroidogenic response of luteinized granulosa cells from rhesus monkeys to luteinizing hormone and chorionic gonadotropin. Biol Reprod 1991;45:273–81.
37. Zelinski-Wooten MB, Lanzendorf SE, Wolf DP, Aladin Chandrasekher Y, Stouffer RL. Titrating luteinizing hormone surge requirements for ovulatory changes in primate follicles, I. Oocyte maturation and corpus luteum function. J Clin Endocrinol Metab 1991;73:577–83.
38. Aladin Chandrasekher Y, Brenner RM, Molskness TA, Yu Q, Stouffer RL. Titrating luteinizing hormone surge requirements for ovulatory changes in primate follicles, II. Progesterone receptor expression in luteinizing granulosa cells. J Clin Endocrinol Metab 1991;73:584–9.
39. Zelinski-Wooten MB, Hutchison JS, Aladin Chandrasekher Y, Wolf DP, Stouffer RL. Administration of human luteinizing hormone (hLH) to macaques following follicular development: further titration of LH surge requirements for ovulatory changes in primate follicles. J Clin Endocrinol Metab (in press).
40. Bomsel-Helmreich O, Huyen LVN, Durand-Gasselin I. Effects of varying doses of hCG on the evolution of preovulatory rabbit follicles and oocytes. Hum Reprod 1989;4:636–42.
41. Simon JA, Danforth DR, Hutchison JS, Hodgen GD. Characterization of recombinant DNA derived-human luteinizing hormone in vitro and in vivo. JAMA 1988;259:3290–5.
42. Gonen Y, Balakier H, Powell W, Casper RF. Use of gonadotropin-releasing hormone agonist to trigger follicular maturation for in vitro fertilization. J Clin Endocrinol Metab 1990;71:918–22.
43. Bavister BD, Dees C, Schultz RD. Refractoriness of rhesus monkeys to repeated ovarian stimulation by exogenous gonadotropins is caused by nonprecipitating antibodies. Am J Reprod Immunol Microbiol 1986;11:11–6.
44. Iliff-Sizemore SA, Molskness TA, Stouffer RL. Effect of anti-human gonadotropin antibodies on fertility of female rhesus macaques [Abstract].

Serono Symposium on In Vitro Fertilization and Embryo Transfer in Primates, Beaverton, OR, May 28–31, 1992.
45. Thomson JA, Hess DL, Dahl KD, Iliff-Sizemore SA, Stouffer RL, Wolf DP. The Sulawesi crested black macaque (*Macaca nigra*) menstrual cycle: changes in perineal tumescence and serum estradiol, progesterone, follicle-stimulating hormone, and luteinizing hormone levels. Biol Reprod 1992;46:879–84.
46. Owen EJ, Shoham Z, Mason BA, Ostergaard H, Jacobs HS. Cotreatment with growth hormone, after pituitary suppression, for ovarian stimulation in in vitro fertilization: a randomized, double-blind, placebo-control trial. Fertil Steril 1991;56:1104–10.
47. Volpe A, Coukos G, Barreca A, Giordano G, Artini PG, Genazzani AR. Clinical use of growth hormone-releasing factor for induction of superovulation. Hum Reprod 1991;6:1228–32.
48. Braat DDM, Berkhout G, ten Brug CS, Schoemaker J. Multiple follicular development in women with normal menstrual cycles treated with pulsatile gonadotrophin-releasing hormone. Hum Reprod 1991;6:1379–83.
49. Stouffer RL, Woodruff TK, Dahl KD, Hess DL, Mather JP, Molskness TA. Disparate actions of exogenous activin-A on follicular development and steroidogenesis during the menstrual cycle in rhesus monkeys [Abstract]. Serono Symposia Ninth Ovarian Workshop on Ovarian Cell Interactions: Genes to Physiology, Chapel Hill, NC, July 9–11, 1992.

Part III

Preimplantation/Implantation Biology

9

Overview of the Morphology of Implantation in Primates

ALLEN C. ENDERS

Although the great apes are considered the species of primate that show the greatest structural similarity to human beings during the implantation stage, early stages are available only from the chimpanzee (1). Furthermore, the most informative implantation site was described originally over 50 years ago (2). Since the possibility of obtaining more than one or two additional specimens is remote, the information on the chimpanzee is not very useful. Whether considering periimplantation signaling (3) or morphological development, the only meaningful comparisons are between species that have been studied more extensively.

Many aspects of early development of Old World primates show extensive similarities; for example, the development of a secondary yolk sac (1) and the derivation of extraembryonic mesoderm from parietal endoderm (4). Similarly the development of polarity within the inner cell mass leading to amnion formation by cavitation appears common to these species (5). It is not surprising, therefore, that despite the different form and structure of implantation sites and early placentas in different primates, the same tissues and processes appear to be involved in implantation in all cases. One can view the different forms taken by the chorioallantoic placenta (superficial and interstitial, bidiscoidal) as being the result of the relative contributions of different trophoblast populations during the various stages of implantation. This chapter uses information drawn from studies of the best-known Old World primates—the macaque, baboon, and human—and the only well-known New World primate—the marmoset—to consider the way in which these factors affect the differences.

Apposition and Adhesion

Implantation is not a single event but a series of events in the continued development and association of the blastocyst and the endometrium

TABLE 9.1. Relative time sequence of some developmental sites in the periimplantation period.

Day	Human	Baboon	Macaque	Marmoset
4	Morula to blastocyst			
5	Expanding blastocyst	Morula to blastocyst	Morula to blastocyst	
6	Expanded blastocyst	Blastocyst maturing	Blastocyst maturing	
7	Hatching blastocyst	Hatching blastocyst	Expanding blastocyst	
8			Hatching blastocyst	
9	Attachment	Attachment	Attachment, penetration to basal lamina	
10	Epithelial penetration	Penetration to basal lamina	Trophoblastic plate	
11	Expanding site, amnion distinct	Penetration of maternal vessels	Penetration of maternal vessels, secondary implantation	Attachment to uterus
12	Recruitment of parietal trophoblast	Trophoblast cleft and lacuna formation	Expansion of lacunae	
13	Penetration of vessels, cleft formation	Expansion of lacunae	Primary villi, primitive streak	Penetration to basal lamina
14	Primary villi	Primary villi		
15	Primitive streak	Secondary villi	Secondary villi	Trophoblastic plate
16	Secondary villi			
18	Notocordal process			Fusion of parietal trophoblast
19	Primitive pit			
21	Somites	Tertiary villi	Tertiary villi	
22	Heartbeat, tertiary villi			
30	⋮			
45	⋮			Maternal vessels surround Tertiary villi
60			Spread of trophoblast shell	Penetration of maternal vessels

(Table 9.1). Even if we define implantation as starting when the blastocyst assumes a relatively fixed position in relation to the uterus and initiates a more intimate relationship, there would nevertheless be some chronological slack between the time when apposition of an adhesive blastocyst to the uterus begins and the time that adhesion of trophoblast to uterine surface can be demonstrated. As can be shown by ultrasound, the lumen of the simplex uterus of primates is virtually closed at the time of implantation, consisting of a horizontal slot. Presumably, this condition is essential to bring about apposition of the endometrial epithelium to trophoblast of the blastocyst and is thought to be the result of the shape of the uterus, muscular activity, paucity of luminal fluid, and endometrial edema. The supposition of mechanical positioning of blastocysts within the uterus could be examined using blastocyst-sized beads of appropriate density introduced into the uterine lumen on day 5–6 after ovulation, with perfusion fixation on day 8–9.

Because of the orientation of the inner cell mass toward the uterine surface, it is evident that in all four species initial adhesion to the uterine luminal epithelium occurs in the peri-inner cell mass region. In all but the human, it has been shown by ultrastructural studies that syncytial trophoblast developing in the peri-inner cell mass region invades the uterine epithelium (6), and at the light microscope level, the earliest human implantation sites appear to have had the epithelium penetrated by syncytial masses. Syncytial trophoblast has been seen in blastocysts flushed from the uterus of baboons (7), but fine structural studies of initial adhesion of trophoblast to epithelium have not yet been accomplished in any species.

Formation of syncytial trophoblast can occur in vitro (8). Therefore, in addition to obtaining normal periimplantation blastocysts for studying adhesion mechanisms, blastocysts from *in vitro fertilization* (IVF) could be used in a procedure similar to that of Lindenberg et al. (9), but using a polarized epithelium to examine adhesion and junctional complex penetration. It should be possible to use the marmoset to study adhesion since this animal has multiple blastocysts without the necessity of inducing extra follicle development, and the trophoblast does well in vitro (10). With the little information currently available, it appears that the adhesion stage prior to epithelial penetration would be a short stage in all species; there is little reason to expect differences between primate species at this stage, but the absence of information always makes the situation appear simpler.

Invasion of Endometrial Epithelium

More information is available concerning invasion of the uterine epithelium by trophoblast. The initial invasion has been suggested to occur in the macaque and marmoset by intrusion of processes of syncytial

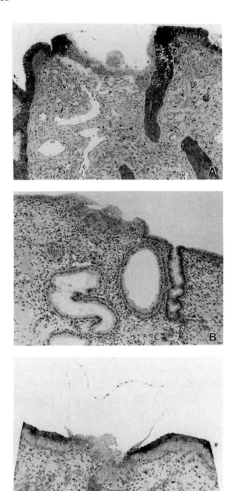

FIGURE 9.1. Trophoblastic plate stages 1–2 days after implantation in the human (*A*), baboon (*B*), and macaque (*C*). All 3 implantation sites have expanded in the plane of the luminal epithelium. The baboon implantation site appears to have penetrated farthest and has the most highly dilated maternal vessels below it (35×).

trophoblast between uterine epithelial cells (6, 11). This stage is followed by a consolidation of the association with the epithelium by penetration of additional processes and by the coalescence of the areas of penetration to form a trophoblastic plate, which usually spreads in the plane of the epithelium (Fig. 9.1). Epithelial invasion occurs within a day of adhesion in the baboon (12), 1–2 days in the macaque (13), and probably a similar 1–2 days in the marmoset (3). No information is available on the human at this stage, suggesting that epithelial penetration takes place in only a day or two.

Trophoblast-Epithelial Cell Fusion

There is now considerable evidence that during the trophoblastic plate stage in the macaque and baboon, some fusion of syncytial trophoblast with maternal epithelial cells takes place. There is no evidence that trophoblast fuses with the apical ends of epithelial cells in the initial implantation as it does in the rabbit (14). Three lines of evidence suggest that heterokaryons are formed between fetal syncytial trophoblast and maternal epithelium or symplasma. First, in both the baboon and macaque, at the margin of the implantation site, cytoplasm that is confluent with syncytial trophoblast has apical vesicles similar to those in maternal epithelial cells (15). Second, in macaques some masses of syncytium contain nuclei that have intranuclear channels similar to those seen in epithelial plaque cells. These channels are also seen in the human endometrial epithelial cells on day 19 of the cycle (16). Third, near the basal lamina between nuclei, whorls and pockets of cell membrane suggest fusion of lateral cell membranes between cells.

The association of clusters of small nuclei and large nuclei in human implantation sites has long been noted. Hertig and Rock (17) considered that the stromal nuclei were ingested by trophoblast. Later, Falck Larsen (18) suggested that the small nuclei represented areas of fusion of epithelial cells with trophoblast. The large nuclei of the syncytium (apparently polyploid) are often found in older syncytium or syncytium that is farther from the maternal surface. Such cells have been seen in the macaque and baboon as well as in humans, and in the baboon these are the principal nuclear type in a failing early implantation site. Consequently, it is unclear whether the differences in nuclear size relate to age changes in syncytium per se or to fusion of syncytial trophoblast with uterine cells. This question is clearly a target for in vitro investigation should the use of spare blastocysts from IVF be considered ethical.

Stromal Invasion and Invasion of Maternal Vessels

With expansion of the trophoblastic plate, the epithelial basal lamina is penetrated, as is the underlying stroma, followed by penetration of subepithelial maternal vessels. During these processes variation between the species increases. Trophoblast penetrates maternal blood vessels most rapidly in the baboon (12). Subepithelial vessels are dilated, and ectoplasmic processes from the syncytial trophoblast penetrate the vessels almost as soon as they have passed through the epithelial basal lamina. At the other extreme is the marmoset, in which maternal vessels are slowly surrounded by trophoblast and penetration of trophoblast into vessels does not occur until 45 days (19) to 60 days (20) of gestation. In the macaque penetration of vessels occurs almost as soon as epithelial basal lamina penetration, but a day or so later than in the baboon.

No human material at this stage has been studied by electron microscopy. However, examination of paraffin-embedded material in the Carnegie collection indicates that maternal vessels are initially surrounded and that vessel penetration probably takes place a day or so later than in the baboon or macaque. It appears, therefore, that the trophoblast penetrates deeper into the stroma before breaching maternal

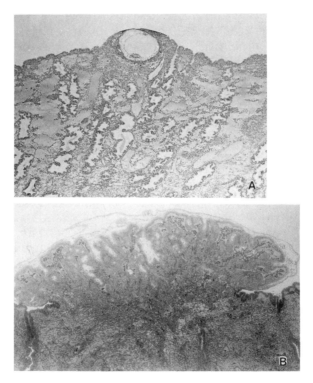

FIGURE 9.2. Human (A) and baboon (B) implantation sites 4–5 days after implantation. A: Although trophoblast of the circumference of the conceptus is within the endometrium, the actual invasion into the stroma is slight. This specimen shows extreme dilation of the venules. B: Growth of the lacunae has expanded the implantation site above the surface of the endometrium (25×).

vessels. In other words, the type of syncytium that actively invades the stroma persists for a longer time, which results in deeper invasion into the endometrium and progressively involves more of the trophoblast circumference (Fig. 9.2).

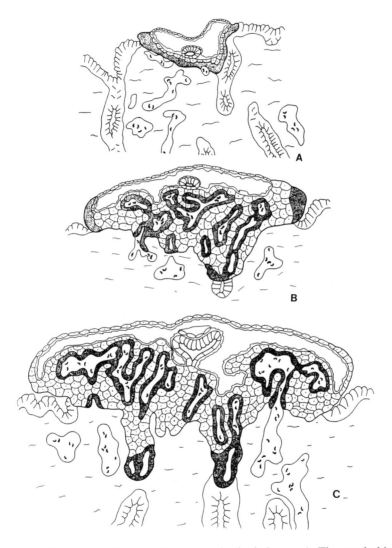

FIGURE 9.3. Drawings of implantation stages in the baboon. *A:* The trophoblastic plate stage is shown. Syncytial trophoblast (stippled) has tapped the enlarged maternal vessels. *B:* Once lacunae are formed, expansion of the implantation site is primarily internal, and most of the lacunae lie above the surface of the endometrium. *C:* The site has expanded, and abundant cytotrophoblast is becoming organized into primary villi.

Trophoblast Differentiation

At the same time that trophoblast is tapping maternal vessels, there are profound changes occurring in syncytial trophoblast in the macaque, baboon, and human. The majority of this trophoblast converts from a multinucleate invasive tissue to a unilaminar microvillous absorptive tissue (21). This conversion occurs toward the end of the initial invasion of endometrium and during the early stages of formation of blood-filled spaces within the trophoblast. The formation of an extensive series of clefts facilitates the subsequent expansion of these spaces into blood-filled intrasyncytial lacunae. It is important to note that in the human this conversion corresponds to the time of change in the rate of hCG production (22). Thus, it marks a major change in the trophoblast and if not the end of the implantation period, at least the conclusion of the initial series of implantation events. It does not mark the end of changes in trophoblast nor the end of invasion of the endometrium, but does denote a change in the way in which these are accomplished.

Expansion of the Implantation Site

Once maternal blood vessels are tapped, there is again variation among species. In the baboon, which is earliest to tap vessels, clefts in syncytial trophoblast develop rapidly into lacunae, and rapid growth of trophoblast lifts the placenta off the surface of the endometrium (23, 24). During this time period most of the growth is within the placenta; little is within the endometrium.

In the human all of the trophoblast, including the abembryonic trophoblast, is converted into syncytial and cytotrophoblast with clefts and lacunae (21). This expansion of the lacunar area is largely contained within the endometrium (Figs. 9.2 and 9.3).

In the macaque abembryonic trophoblast forms a secondary implantation site on the opposite surface of the endometrium. Consequently, both placental discs lift from the surface of the endometrium (Fig. 9.4). In the marmoset the abembryonic trophoblast of twins and triplets fuses so that there comes to be a shared exocelom (11).

Therefore, if we compare the implantation site of the baboon and human at the close of the previllous stage, the most superficial site is that of the baboon, but it is superficial because trophoblast invades maternal vessels rapidly and only trophoblast near the inner cell mass is involved. The human previllous stage is more deeply embedded in the endometrium because trophoblast grows within stroma for a longer period of time before tapping vessels and because all of the trophoblast becomes involved. Thus, both arrive at extensive lacunar spaces, but by different processes.

9. Overview of the Morphology of Implantation in Primates

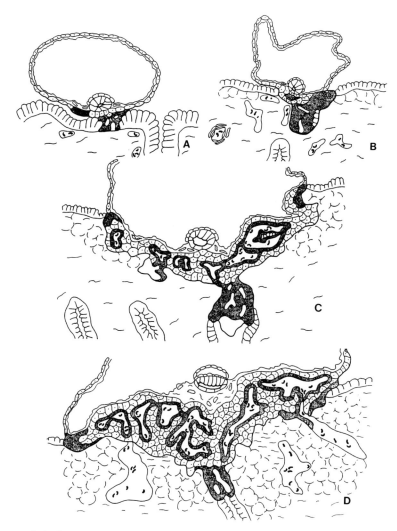

FIGURE 9.4. Drawings of implantation stages in the macaque. *A:* Syncytial trophoblast (stippled) is invading uterine epithelium. *B:* The trophoblastic plate is shown. Trophoblast has extended along the epithelium and into a gland. *C:* Trophoblast has developed clefts and blood-filled lacunae. Note that there is a mixture of syncytial and cellular trophoblast (dots). *D:* Toward the end of the lacunar stage, cytotrophoblast accumulates between lacunae. These columns of cytotrophoblast are precursors of primary villi.

The trophoblastic plate of the macaque and baboon is a mixture of syncytium and cytotrophoblast. Vessel penetration is initially by syncytial trophoblast, but as lacunae form, maternal blood comes in contact with both syncytium and cytotrophoblast. Although there is a nearly complete

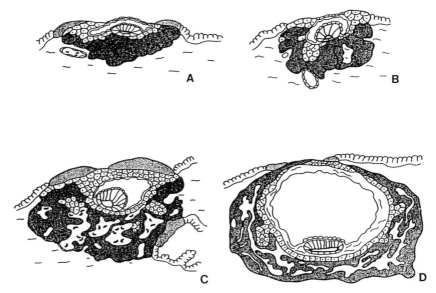

FIGURE 9.5. Drawings of implantation stages in the human. *A and B:* Shown are implantation sites about a day after implantation. Syncytial trophoblast has not yet invaded maternal vessels. In *A* trophoblast has spread in the plane of the epithelium. In *B* trophoblast has extended further into the stroma. *C:* Two of three days after implantation, syncytial trophoblast has penetrated maternal vessels and no longer surrounds regions of maternal tissue. *D:* Four to five days after implantation, syncytial trophoblast now surrounds the conceptus, and primary villi project from the organized inner layer of cytotrophoblast.

layer of cytotrophoblast lining syncytium toward the embryonic surface, there are also pockets of cytotrophoblast cells at the maternal surface (12). (In the human the implantation site "cleans up" at the time that maternal vessels are tapped in that the previously surrounded maternal cells disappear, the border of trophoblast on the endometrium becomes smoother, and the cytotrophoblast cells become segregated largely to the embryonic side of the trophoblastic plate (Figs. 9.2 and 9.5). In the macaque and baboon, cytotrophoblast cells from the trophoblastic plate enter maternal vessels shortly after initial vessel penetration and become situated in the superficial arterioles that they at least partially plug (12).

The anastomosing venules that are confluent with placental lacunae remain patent. In these species and the human, veins in the area of the implantation site are dilated (12). This dilation should provide decreased resistance and therefore encourage an increase in blood flow. Plugging of arterioles adjacent to the trophoblast may diminish blood pressure in the rhesus and baboon, protecting the extraembryonic tissues that are not yet adequately supported by connective tissue. The human implantation site confronts the same blood pressure situation by containing lacunae within the endometrium. However, it should be noted that although all three species show some bleeding into endometrial glands and occasionally into the lumen, the extent of such hemorrhage is greater in the human material than in the macaque and baboon. Difference in the conditions of obtaining tissues must make us cautious in interpretation of this observation, however.

At the end of the lacunar (previllous) stage, proliferation of cytotrophoblast cells in clusters adjacent to the blastocyst cavity (human) or between lacunae (macaque and baboon) results in formation of the primary villi. Almost immediately, mesenchyme begins to grow into these villi, converting them into secondary villi. Although vessels form rapidly within the villi, no blood circulates in the vessels until embryonic heartbeat is established. Thus, there is a period of placenta formation following implantation of about a week. The reorganization of trophoblast at the junction with the endometrium results in the establishment of structures that persist for weeks or months rather than days. These structures—the anchoring villi and their distal portion, the trophoblastic shell—issue in the period of placental expansion. It is only at this stage (that of formation of trophoblastic shell) that, in the human, trophoblast cells invade as individual cells or as clusters into the endometrium beyond the front of the trophoblastic shell. Invasion of maternal arterioles has not been reported in the human prior to this time.

Anchoring Villi

Anchoring villi are fascinating structures that are currently under investigation in a number of laboratories including our own. Since they will be the subject of a subsequent presentation and are part of the stage of placental expansion rather than implantation, only a few comments are included here. In the baboon and macaque, the initial anchoring villi undergo further changes, including elongation and accumulation of extracellular matrix between trophoblast cells. This material, which was originally called fibrinoid, is a complex mixture of substances, many of which are normally associated with basal laminas (25). Furthermore, in both macaque and baboon, the border of cytotrophoblast of the shell with the endometrium is relatively even, and fewer cells are seen in the maternal stroma in these species than in the human. The invasion of arteries, however, is extensive. The distal ends of anchoring villi contribute cells and matrix to the trophoblastic shell. The accumulation of matrix material is readily visible in plastic-embedded material.

As placental expansion proceeds, differences in cells both within anchoring villi and between anchoring villi in the same placenta become increasingly apparent. It seems reasonable that the cells of these structures respond to a large number of factors to control their growth, differentiation, and life span, whereas some of the shorter-lived early trophoblast types may have more prescribed development. It will be interesting to see what factors influence the replication, migration, and differentiation of cytotrophoblast cells as they move from the proximal region of anchoring villi to more distal aspects of the trophoblastic shell.

Acknowledgments. The author would like to thank Dr. Barry F. King for reading the manuscript, Dr. Thomas Blankenship for numerous discussions in relation to these studies, and Dr. Andrew Hendrickx, Dr. Cathy VandeVoort, and Pam Peterson for assistance in acquiring the implantation stages. I also thank Katy Lantz for the preparation of tissues for light and electron microscopic examination and Carrie Beth Mattos for the excellent drawings. These studies were supported by Grants HD-10342 (A.E.) and HD-24491 (B.F.K.) from the National Institute of Child Health and Human Development and RR-00169 to the California Primate Research Center.

References

1. Luckett WP. Comparative development and evolution of the placenta in primates. Contrib Primatol 1974;3:142–234.
2. Heuser CH. The chimpanzee ovum in the early stages of implantation (10½ days). J Morphol 1940;66:155–74.
3. Hearn JP, Webley GE, Gidley-Baird AA. Chorionic gonadotrophin and embryo-maternal recognition during the peri-implantation period in primates. J Reprod Fertil 1991;92:497–509.
4. Enders AC, King BF. Formation and differentiation of extraembryonic mesoderm in the rhesus monkey. Am J Anat 1988;181:327–40.
5. Enders AC, Schlafke S, Hendrickx AG. Differentiation of the embryonic disc, amnion and yolk sac in the rhesus monkey. Am J Anat 1986;177:147–69.
6. Enders AC, Hendrickx AG, Schlafke S. Implantation in the rhesus monkey. Am J Anat 1983;176:275–98.
7. Enders AC, Lantz KC, Schlafke S. Differentiation of trophoblast of the baboon blastocyst. Anat Rec 1989;225:329–40.
8. Enders AC, Boatman D, Morgan P, Bavister BD. Differentiation of blastocysts derived from in vitro fertilized rhesus monkey ova. Biol Reprod 1989;41:715–27.
9. Lindenberg S, Hyttel P, Lenz S, Holmes PV. Ultrastructure of the early human implantation in vitro. Hum Reprod 1987;1:533–8.
10. Summers PM, Campbell JM, Miller MW. Normal in-vivo development of marmoset monkey embryos after trophectoderm biopsy. Hum Reprod 1988;3:389–93.
11. Smith CA, Moor HDM, Hearn JP. The ultrastructure of early implantation in the marmoset monkey (*Callithrix jacchus*). Anat Embryol 1987;175:399–410.

9. Overview of the Morphology of Implantation in Primates 157

12. Enders AC, King BF. Early stages of trophoblastic invasion of the maternal vascular system during implantation in the macaque and baboon. Am J Anat 1991;192:329–65.
13. Enders AC, Schlafke S. Implantation in nonhuman primates and in the human. In: Dukelow WR, Erwin J, eds. Comparative primate biology, reproduction and development. New York: Alan R Liss, 1986:291–310.
14. Enders AC, Schlafke S. Penetration of the uterine epithelium during implantation in the rabbit. Am J Anat 1971;132:219–40.
15. Enders AC. The role of different trophoblast types in implantation in primates. In: Implantation in mammals. Raven Press, 1992.
16. Dockery P, Li TC, Rogers AW, Cooke ID, Lenton EA. The ultrastructure of the glandular epithelium in the timed endometrial biopsy. Hum Reprod 1988;3:826–34.
17. Hertig AT, Rock J. Two human ova of the pre-villous stage, having a developmental age of about eight and nine days respectively. Contrib Embryol Carnegie Inst 1949;33:169–86.
18. Falck Larsen J. Human implantation and clinical aspects. Prog Reprod Biol 1980;7:284–95.
19. Merker H-J, Bremer D, Csato W, Heger W, Gossrau R. Development of the marmoset placenta. In: Neubert D, Merker H-J, Hendrickx AG, eds. Non-human primates—developmental biology and toxicology. Berlin: Ueberreuter Wissenschaft, 1988:245–72.
20. Smith CA, Moore HDM. The morphology of early development and implantation in vivo and in vitro in the marmoset monkey. In: Neubert D, Merker H-J, Hendrickx AG, eds. Non-human primates—developmental biology and toxicology. Berlin: Ueberreuter Wissenschaft, 1988:171–90.
21. Enders AC. Trophoblast differentiation during the transition from trophoblastic plate to lacunar stage of implantation in the rhesus monkey and human. Am J Anat 1989;186:85–98.
22. Lenton EA, Woodward AJ. The endocrinology of conception and implantation in women. J Reprod Fertil 1988;suppl 36:1–15.
23. Houston ML. The placenta. In: Hendrickx AG, ed. The embryology of the baboon. University of Chicago Press, 1971:153–72.
24. Tarara R, Enders AC, Hendrickx AG, et al. Early implantation and embryonic development of the baboon: stages 5, 6 and 7. Anat Embryol 1987;176:267–75.
25. Damsky CH, Fitzgerald ML, Fisher SJ. Distribution patterns of extracellular matrix components and adhesion receptors are intricately modulated during first trimester cytotrophoblast differentiation along the invasive pathway in vivo. J Clin Invest 1992;89:210–22.

10
Physiology of Implantation in Primates

J.P. HEARN, P.B. SESHAGIRI, AND G.E. WEBLEY

Embryo implantation in primates relies on endocrine mechanisms that are distinct from those in nonprimate species. In all primate species studied to date, the embryo enters the uterus as a morula on day 3–4 after ovulation, hatches from the zona pellucida on day 6–8 and attaches to the maternal endometrial epithelium on day 8–9 in Old World species (1, 2); day 11–12 in the marmoset monkey (3, 4) and probably day 6 in the squirrel monkey (5). The morphology of implantation is now well described for a few primate species, including the rhesus monkey (1, 6), baboon (7), and marmoset (8, 9). Comparisons with human implantation still depend largely on the classic studies of Hertig and Rock (10).

In contrast, there is still relatively little known of the cellular and molecular processes that regulate the activation and sequential expression of embryonic signals, the attachment of the embryo to the endometrial epithelium, the rapid invasion of trophoblast, the establishment of embryo-maternal vascular links, and the initial maternal responses to implantation by the endometrium. For example, the genes for *chorionic gonadotropin* (CG) are thought to be expressed only in primates (11), yet the mechanisms controlling periimplantation CG gene expression remain undefined. The capacity of the primate embryo to secrete CG from about the time of attachment (12–14) allows the embryo in primates to initiate and regulate implantation and at the same time to revitalize *corpus luteum* (CL) function over the short time span of 2–3 days allowed between embryo attachment, the establishment of trophoblast-endometrial vascular contact, and the demise of the CL at the end of the ovarian cycle.

The expansion of research on human IVF has advanced our knowledge of preimplantation embryonic development in the human species, in some cases beyond that in nonhuman primates (15). However, the human embryos studied are often those rejected from IVF procedures as unsuitable for transfer, so that carefully controlled experimental work in embryonic differentiation is likely to require nonhuman material from primates that share similar developmental mechanisms with the human. As development of the embryo moves to hatching of the blastocyst, attachment, and invasion, human material is even less available and is often in poor condition from having been cultured from IVF. In addition, the ethical constraints in studying human embryo differentiation increase the needs for experimental, controlled, nonhuman primate studies as differentiation proceeds.

While implantation has been monitored for many years through the study of endocrine changes in the peripheral circulation, until recently, it has not been possible to obtain sufficient material for detailed studies of cell signals and cellular physiological interactions at the embryo-maternal interface. Nor has it been easy to clarify the dynamics of embryo-endometrial relationships with the early recognition of embryo signals by the CL; a dialogue that must be established rapidly if pregnancy is to proceed.

This chapter examines our current understanding of periimplantation embryonic development in the marmoset monkey, rhesus monkey, and human. In doing so, we consider the onset of activation of the embryonic genome; the timing of first secretion of CG and its possible role in implantation; the effects of antisera to CG on the progression of implantation and early pregnancy; and the possible significance of other putative embryonic messages that have been described, but not as yet been assigned a clear physiological role.

Regulation of Implantation

Species Differences

The three species considered in this chapter all show ovarian cycles of approximately 28 days. Yet many Old World primates show somewhat longer cycles, while numerous New World species show much shorter ones; for example, the 9 days reported in the squirrel monkey (5) and 16 days reported in the cotton-top tamarin (16). The common factor in all these species is the necessity for CG secretion in time to "rescue" the CL of the cycle and transform it into the CL of pregnancy. It is likely, though yet unproven, that CG might act on the CL before embryo-maternal vascular connections are fully developed, but in all species studied these connections are formed within 3–4 days, so that the circulating levels of CG increase exponentially in vivo at that time.

The morphology of implantation is reviewed by Enders in Chapter 9. The hominoidea, including the apes, probably all share a similar interstitial and highly invasive implantation as seen in the human, with a strong decidual reaction. In contrast, the cercopithecoidea and the ceboidea show a relatively superficial implantation, with the embryo remaining above the epithelial layer and not burrowing under and enclosed by the epithelium as seen in the human. These morphological

TABLE 10.1. Activation of the embryonic genome in mammals.

Species	Stage	Reference
Mouse	2-cell	20, 21
Rabbit	8-cell	22
Sheep	8-cell	23
Cow	8- to 16-cell	24
Human	4- to 8-cell	12
	6- to 8-cell	25

differences are probably academic with regard to the needs of all primates to make rapid vascular contact in order to ensure nutrients for the embryo and luteotropic support of the CL. The variations in reproductive capacity, endocrine patterns, and aspects of morphology seen in several primate species were reviewed in a previous report (17).

Activation of the Embryonic Genome

The proteins that control embryonic development are coded for by two genomic sources: the oocyte (maternal genome) and the transcription products of the embryonic genome proper. Cytoplasmic factors derived from the oocyte regulate events to the blastocyst stage (18), but neither the male nor the female genome is totipotent by itself (19). At later stages of preimplantation embryo development, the time of establishment of trophoblast cell lineage-specific gene expression is not yet known. The first step in identifying and characterizing embryo-specific signals and the ontogeny of their expression is the clarification of their first transcription. Table 10.1 gives a summary of the activation of the embryonic genome in 5 mammalian species studied to date.

With regard to CG, the presence of β-subunit transcription was detected in 3 of 7 triploid embryos (25). Although their normality may be doubtful, transcription at this early stage, with equal expression among blastomeres, does raise questions about the cell-specific expression of CG in preimplantation embryos. At the 6- to 8-cell stage, cell lineage specificity is not yet established, and trophectoderm lineage is not thought to have occurred. Although these results suggest early transcription of the β-subunit, attempts to measure CG in preimplantation embryonic tissue or secreted into culture media have not shown a presence of biologically active CG until it is measured in low levels immediately before embryonic attachment; the secretion levels of CG by the embryo quickly rise exponentially once attachment to the endometrium has occurred (14, 26).

Timing of Implantation

Table 10.2 shows the estimated day after ovulation of embryonic attachment and the first day of measurement of CG in the peripheral circulation for 5 primate species. The expression and secretion of CG is

TABLE 10.2. Estimated day, after ovulation, of embryo attachment to uterine endometrial epithelium in vivo and the first day of measurement of CG in peripheral plasma in 5 primate species.

Species	Embryo attachment (day)	CG first detected (day)	Reference
Human	7–9	9–12	2
Chimpanzee	7–9?	10–11	27
Rhesus	8–10	11–12	28
Baboon	8–10	11–12	29
Marmoset	11–12	14–17	4, 14

the first clear signal from embryo to mother, indicating the embryo's presence and, perhaps, its viability. The CG is measured approximately 3 days after embryo attachment, giving time for vascular connections to be formed.

The Old World species studied so far show embryo attachment on day 7–10. Earlier estimates of human embryo attachment on day 5–6 were probably inaccurate and depended on the necessarily indistinct timing of Hertig and Rock (10). More recent estimates (2) indicate that the human is more aligned to the Old World primate timing of 8–10 days reported for the rhesus monkey and the baboon. The marmoset monkey, the only species studied so far in detail from the ceboidea, has embryonic attachment at day 11–12 after ovulation. It is likely that the variation of days from ovulation to embryo attachment will be greater in the ceboidea. The New World primates show a far broader variation in their reproductive patterns than is seen in the Old World species, and the length of the luteal phase varies quite widely. For example, the marmoset (*Callithrix jacchus*) shows a luteal phase of 18–20 days in a cycle of 28–30 days. The cotton-top tamarin has a luteal phase of approximately 8–10 days in a cycle of 16–18 days. The squirrel monkey, one of the first of the New World primates to be studied, has only recently been reported to have a luteal phase length of 4–5 days in a cycle of 9 days. This finding may need further confirmation, as it would place the timing of preimplantation events in a completely different frame from that reported in all other primates. For a review of comparative reproduction in New World primates, see Hearn (30).

The rather longer time found to embryo attachment in the marmoset may be due to the normal production of twins or triplets and the needs for spacing of embryos in the uterus. While there is no evidence to support this suggestion, it is clear in the marmoset that the first half of embryonic and fetal development during gestation is significantly slower than that reported for the rhesus monkey, baboon, and human (31), perhaps due to the complex intrafetal blood supply set up between the placental disks that is responsible for the blood chimerism seen in this species.

Embryonic loss during periimplantation stages of pregnancy is estimated in women to be up to 50% (32). The causes of this embryonic wastage could be genetic, disorders of differentiation, asynchrony of endocrine events, or of the morphological stages reached by the embryo and the endometrium. In many nonprimate species, including the rodents, ovids, bovids, and the suidae, there is known to be a need for close synchrony between embryonic, tubal, and endometrial developments if implantation is to be successful. In primates, there appears to be more tolerance since transfer of binucleate embryos from the fallopian tube to the uterus will result in successful implantation and live-born young (33, 34). Estimating the *implantation window* from the aspect of first measurement of CG in women and primates, respectively, Lenton et al. (35) and Hearn et al. (26) suggested a 4-day tolerance in these species. While it is obviously possible for embryonic development to continue without passage through the oviduct, thereby calling in question the need for oviduct-embryo interactions in primate embryonic development, this does not rule out the probable facilitatory role of oviduct fluids and proteins in increasing the survival of embryos. There are as yet no clear data in primates that clarify this question.

Chorionic Gonadotropin

There are several factors, mainly proteins or lipoproteins, that have been claimed as embryonic signals during the periimplantation period of pregnancy. Chorionic gonadotropin, perhaps the oldest and best described of the gonadotropins, is still the first clear embryonic signal. There are now data from numerous sources to suggest that this hormone is responsible for the initial revitalization of the CL on which early pregnancy depends. There is also emerging evidence that CG may have a role in intraembryonic differentiation, the attachment process itself, and in the invasion of the endometrium by trophoblast.

The issue is complicated by a number of as yet unknown variables. First, CG is controlled by a multigene family: a single-gene regulation of the α-subunit that is thought to have a largely structural function and a 7-gene regulation of the β-subunit that is thought to carry the biological function of the hormone. Biological activity is exerted only when the hormone is available in dimeric form. While the β-subunit can be transcribed by the 6-cell stage (Table 10.1), there is as yet little evidence to suggest secretion of the biologically active hormone until embryo attachment or immediately before this event. The evidence currently arises from studies in the marmoset monkey, the rhesus monkey, and the human. A further complication in the assessment of the data is that most of the analyses have been carried out by radioimmunoassay—which can show the presence of α- or β-subunits before embryonic attachment—but may require bioassay confirmation in order to show that the hormone is capable of exerting its biological effects. Let us look at the current data from these three species.

Marmoset Monkey (Callithrix jacchus)

There are now several studies that show the profile of CG release in relation to the time of embryonic attachment by the marmoset blastocyst (4, 14, 26, 36). The bioassay used in all of these reports showed exponential rises in CG production by individual embryos from immediately after attachment. Studies of numerous individual embryos cultured for several days before attachment in the presence or absence of human cord serum showed very low to undetectable levels of CG production in culture medium before attachment of the embryo. However, the background threshold of the assay and its interpretation were set high to ensure that only a clear result was counted as positive. It is necessary to confirm these findings with protein-free medium and with CG subunit immunoassays in order to obtain a more comprehensive picture.

Summers (14, 36) bisected marmoset blastocysts and cultured the 2 halves, one containing the *inner cell mass* (ICM) and one without. In each case the half containing the ICM grew and initiated CG production in a manner similar to the whole embryo, but the half without an ICM showed a greatly retarded onset of CG secretion and a far lower rise in its profile after attachment. These studies suggest that the ICM is required for the onset of efficient CG secretion by blastocysts from the marmoset.

Hearn (14, 26) reported the effects of earlier studies in which marmoset monkeys were immunized actively or passively against CG β-subunit during early pregnancy, resulting in the disruption of implantation and the termination of early pregnancy. These findings suggest that CG may be necessary for implantation to occur, but they do not discriminate between a failure of pregnancy due to lack of support of the CL and any direct effects of antisera on the embryo or the implantation site. Preliminary data from a study when marmoset blastocysts were cultured in vitro in the presence of antisera to CG also showed an inhibition of embryo attachment and outgrowth. These results require confirmation and are being further developed at present. If confirmed, they suggest that CG may well be necessary for the late stages of preimplantation embryonic differentiation and at the embryo-maternal interface during implantation, in addition to the accepted role of this hormone in support of the CL.

Rhesus Monkey (Macaca mulatta)

Recently, Seshagiri and Hearn (unpublished observations) carried out a study of the onset of CG secretion in individual rhesus monkey morulae and blastocysts cultured through hatching and attachment stages. The CG production was measured by bioassay. The results to date confirm those seen in the marmoset and human, with an exponential rise in CG production after embryo attachment and the production of low to undetectable levels of CG by the embryo immediately before and after hatching in culture. Our current studies are exploring the manipulation of CG with potential stimulants and the experimental regulation of its release with GnRH.

Human (Homo sapiens)

Recent reports from Lopata (37, 38) and Dokras (39), in which the morphology and production of CG from the human blastocyst were studied, suggest that CG or its subunits, measured by immunoassay, can be secreted from immediately before the time of blastocyst attachment. Once attachment in culture has occurred, there is a rapid rise in the secretion of CG in the culture fluid. The assay kit used by Lopata and Hay measured the hCG dimer, but was not a biologically based assay. The kit used by Dokras et al. was an *immunoradiometric assay* (IRMA) for the intact hCG molecule and any free β-subunit.

Lopata and his colleagues further studied the influence of a number of factors on the secretion of CG by the human blastocyst (38). The addition of 1% human cord serum to the culture medium produced more than a 10-fold-higher increase in CG secretion than was seen with other stimulants, such as insulin, transferrin, and PDGF. EGF failed to provoke hCG secretion until later. The authors suggest that receptors for these factors are present in the preimplantation embryo with the exception of EGF, the latter not becoming effective until 4–5 days after the expanded blastocyst stage. Lopata and Oliva (38) conclude that there are many factors involved in the stimulus of CG production by the blastocyst, several of which may act synergistically. One factor that appeared from their preliminary studies not to be involved at this stage was *platelet activating factor* (PAF) since PAF antagonists did not inhibit the hCG response to serum.

Dokras and her colleagues (39) also showed a release of CG from the late preimplantation blastocyst. In studying 26 blastocysts produced by IVF, hCG secretion was detected by day 8 of culture, increased to a maximum on day 10, and then fell. The CG was not detectable on days 5 or 6 as a diagnostic of viable blastocyst function for selection and transfer to the uterus. The decline in CG seen after day 10 was probably due to embryo degeneration because of the simple culture medium used and, perhaps, the daily replenishment of culture medium that may have removed essential, embryo-derived nutrients. The morphology of these embryos appeared normal, and although they had been cultured for lengthy periods, 10 (38%) hatched, and 5 of these adhered to the culture dish where they were cultured until day 14. No difference in the low levels of CG secretion between hatched or unhatched blastocysts was reported in the immediate preimplantation stages, but those blastocysts that attached showed a significant increase of CG secretion.

The results of these studies show that the human blastocyst appears to be capable of secreting CG β-subunit and probably the dimeric hormone in low quantities immediately before attachment. Once attachment is achieved, there is an exponential rise in the production of CG. While these findings would profit by confirmation with bioassay, it seems likely that low levels of bioactive CG are secreted immediately before attachment.

Conclusions

The regulation of implantation of the primate embryo depends on different mechanisms to those seen in nonprimate species, especially in the embryonic production of CG and the biological effects of this hormone at the embryo maternal interface and at the CL. The sum of results obtained from the three primate species considered in this chapter suggest that CG may be transcribed soon after fertilization, but is not expressed or secreted by the embryo until the blastocyst stage, but before attachment and implantation. In all three species studied once embryonic attachment is achieved, there is a very rapid rise in the embryonic secretion of CG.

There is still a need to clarify the interactions between embryonic secretion of CG subunits and the dimer measured by radioimmunoassay kits in the human studies when compared with the measurement of bioactive CG reported from the studies in the marmoset and rhesus. Studies of the regulation of CG release and the trophic or inhibitory factors that might affect secretion are still at an early stage. There are likely to be a number of factors that can influence release, but their interactions, and possible synergism, are not yet understood. New approaches to fertility regulation, either in enhancing CG production to increase embryonic viability or in inhibiting CG secretion to block this embryonic signal and thereby prevent implantation, require knowledge of the precise time of release of CG and the consequences to embryonic development of its manipulation.

Acknowledgments. We thank our colleagues at the Primate Research Center and at the Institute of Zoology for their expert animal care and assistance with the complex infrastructural requirements for studies in primate embryology, including endocrine monitoring, surgery, tissue culture, and histology. We acknowledge program grant support (J.P.H.

and G.E.W.) from the UK Medical Research and Agriculture and Food Research Councils; program support (J.P.H. and P.B.S.) from NIH Grant RR-00167; and project support from the World Health Organization Special Program of Research, Training, and Research Development in Human Reproduction. We thank Susan Carlson for her help in preparation of the manuscript. This paper is publication number 32-009 of the University of Wisconsin Regional Primate Research Center.

References

1. Enders AC, Hendrickx AG, Schlafke S. Implantation in the rhesus monkey: initial penetration of the endometrium. Am J Anat 1983;167:275–98.
2. Lenton EA, Woodward AJ. The endocrinology of conception cycles and implantation in women. J Reprod Fertil 1988;36:1–15.
3. Hearn JP, Summers PM. Experimental manipulation of embryo implantation in the marmoset monkey and exotic equids. Theriogenology 1986;25:3–11.
4. Hearn JP, Hodges JK, Gems S. Early secretion of chorionic gonadotrophin by marmoset embryos in vivo and in vitro. J Endocrinol 1988a;119:249–55.
5. Dukelow WR. The squirrel monkey. In: Hearn JP, ed. Reproduction in New World primates. MTP Press, 1983:149–80.
6. Enders AC, Lantz KC, Schlafke S. Differentiation of trophoblast of the baboon blastocyst. Anat Record 1989;225:329–40.
7. Tarara R, Enders AC, Gulamhusein N, et al. Early implantation and embryonic development of the baboon: stage 5, 6 and 7. Anat Embryol 1987;175:117–26.
8. Moore HDM, Gems S, Hearn JP. Early implantation stages in the marmoset monkey (*Callithrix jacchus*). Am J Anat 1985;172:265–78.
9. Smith CA, Moore HDM, Hearn JP. The ultrastructure of early implantation in the marmoset monkey (*Callithrix jacchus*). Anat Embryol 1987;175:399–410.
10. Hertig AT, Rock J. Two human ova of the pre-villous stage having a developmental age of about seven and nine days, respectively. Carnegie Contrib Embryol 1945;31:65–84.
11. Stewart HJ, Jones DSC, Pascall JC, Popkin RM, Flint APF. The contribution of recombinant DNA techniques of reproductive biology. J Reprod Fertil 1988;83:1–57.
12. Braude PR, Bolton V, Moore S. Human gene expression first occurs between the four- and eight-cell stages of preimplantation development. Nature 1988;332:459–61.
13. Lopata A, Summers PM, Hearn JP. Births following the transfer of cultured embryos obtained by in vitro fertilization in the marmoset monkey, *Callithrix jacchus*. Fertil Steril 1988;50:503–9.
14. Hearn JP, Gidley-Baird AA, Hodges JK, Summers PM, Webley GE. Embryonic signals during the peri-implantation period in primates. J Reprod Fertil Suppl 1988b;36:49–58.
15. Edwards RG. Current status of human conception in vitro. Proc R Soc Lond [Biol] 1985;B223:417–48.

16. Ziegler TE, Savage A, Scheffler G, Snowdon CT. The endocrinology of puberty and reproductive functioning in female cotton-top tamarins (*Saguinus oedipus*) under varying social conditions. Biol Reprod 1987;37:618–27.
17. Hearn JP. The embryo-maternal dialogue during early pregnancy in primates. J Reprod Fertil 1986;76:809–19.
18. Tesarik J. Involvement of oocyte-coded message in cell differentiation control of early human embryos. Development 1989;105:317–22.
19. Surani MA, Allen ND, Barton SC, et al. Developmental consequences of imprinting of parental chromosomes by DNA methylation. Philos Trans R Soc Lond [Biol] 1990;B326:313–27.
20. Flach G, Johnson MH, Braude PR, Taylor RAS, Bolton VN. The transition from maternal to embryonic control in the 2-cell mouse embryo. EMBO J 1982;1:681–6.
21. Bolton VN, Oades PJ, Johnson MH. The relationship between cleavage, DNA replication and gene expression in the mouse 2-cell embryo. J Embryol Exp Morphol 1984;79:139–63.
22. Manes C. The participation of the embryonic genome during early cleavage in the rabbit. Dev Biol 1973;32:453–9.
23. Crosby IM, Gandolfi F, Moor RM. Control of protein synthesis during early cleavage of sheep embryos. J Reprod Fertil 1988;82:769–75.
24. Frei RE, Schultz GA, Church RB. Qualitative and quantitative changes in protein synthesis occur at the 8–16 cell stage of embryogenesis in the cow. J Reprod Fertil 1989;86:637–41.
25. Bonduelle M-L, Dodd R, Liebaers I, Van Steirteghem A, Williamson R, Akhurst R. Chorionic gonadotrophin-β mRNA, a trophoblast marker, is expressed in human 8-cell embryos derived from tripronucleate zygotes. Hum Reprod 1988;3:909–14.
26. Hearn JP, Webley GE, Gidley-Baird AA. Chorionic gonadotrophin and embryo-maternal recognition during the peri-implantation period in primates. J Reprod Fertil 1991;92:497–509.
27. Reyes FI, Winter JSD, Faiman C, Hobson WC. Serial serum levels of gonadotropins, prolactin and sex steroids in the non pregnant and pregnant chimpanzee. Endocrinology 1975;96:1447–55.
28. Shaikh AA. Animal models for research in human reproduction. NIH Invited Report, Washington, DC, 1978.
29. Atkinson LE, Hotchkiss J, Fritz GR, Surve AH, Neill JD, Knobil E. Circulating levels of steroids and chorionic gonadotropin during pregnancy in the rhesus monkey, with special attention to the rescue of the corpus luteum in early pregnancy. Biol Reprod 1975;12:335–45.
30. Hearn JP, ed. Reproduction in New World primates. Lancaster, UK: MTP Press, 1983.
31. Chambers PL, Hearn JP. Embryonic, foetal and placental development in the common marmoset monkey (*Callithrix jacchus*). J Zool Lond 1985;A207:545–61.
32. Short RV. When a conception fails to become a pregnancy. In: Maternal recognition of pregnancy. Ciba Foundation Symposium 64. Amsterdam: Excerpta Medica, 1979:337–95.
33. Marston JH, Penn R, Sivelle PC. Successful autotransfer of tubal eggs in the rhesus monkey, *Macaca mulatta*. J Reprod Fertil 1977;49:175–6.

34. Summers PM, Shephard AM, Taylor CT, Hearn JP. The effects of cryopreservation and transfer on embryonic development in the common marmoset monkey, *Callithrix jacchus*. J Reprod Fertil 1987;79:241–50.
35. Lenton EA, Hooper M, King H, et al. Normal and abnormal implantation in spontaneous in-vivo and in-vitro human pregnancies. J Reprod Fertil 1991;92(2):555–65.
36. Summers PM, Taylor CT, Miller MW. Inner cell mass is required for efficient chorionic gonadotrophin secretion by common marmoset blastocysts. J Reprod Fertil 1992.
37. Lopata A, Hay DL. The potential of early human embryos to form blastocysts, hatch from their zona and secrete HCG in culture. Hum Reprod 1989;4(suppl):87–94.
38. Lopata A, Oliva K. Regulation of chorionic gonadotropin secretion by cultured human blastocysts. In: Bavister BD, ed. Preimplantation embryo development. New York: Springer-Verlag, 1993.
39. Dokras A, Sargent IL, Ross C, Gardner RL, Barlow DH. The human blastocyst: morphology and human chorionic gonadotrophin secretion in vitro. Hum Reprod 1991;6(8):1143–51.

11
Interactions Between the Embryo and Uterine Endometrium During Implantation and Early Pregnancy in the Baboon (*Papio anubis*)

ASGERALLY T. FAZLEABAS, SHERI HILD-PETITO,
KATHLEEN M. DONNELLY, PATRICIA MAVROGIANIS,
AND HAROLD G. VERHAGE

The establishment of pregnancy in all mammalian species requires a synchronous interaction between the implanting embryo and the maternal endometrium. The mammalian uterus is receptive to the implanting blastocyst for a specific period of time and this *receptive window* appears to be regulated primarily by ovarian steroids. Embryo implantation is the natural culmination of this period, and successful nidation requires the precise preparation of both the blastocyst and endometrium. A remarkable synchrony is achieved by continuous maternal/conceptus interaction even prior to trophoblast invasion. The internal lining of the uterus is a specialized interface where a complex combination of anatomic, biochemical, endocrinologic, and immunologic events occur to ensure successful embryonic development. It is apparent, therefore, that the biological requirements of the early mammalian conceptus must be met by uterine and oviductal secretions since they constitute the primary environmental contact between the developing embryo and its mother prior to implantation.

The concept that uterine secretory materials existed for the nourishment of the developing conceptus was first proposed by Aristotle (384–322 B.C.) and William Harvey in the 17th century. It is now an established fact that the uterine endometrium secretes specific substances (i.e., proteins) and that these secretions are hormonally modulated. For the past several years, the overall goal of our laboratory has been to identify hormonally regulated proteins synthesized and secreted by the nonpregnant and pregnant baboon uterus (1–3). Our studies have demonstrated that the baboon endometrium during explant culture synthesizes and releases at least 15–17 polypeptides (4). These polypeptides can be divided into two groups: *group I proteins* are those that are present throughout the menstrual cycle and show only minor cyclic variation in synthesis, and *group II proteins* are those whose secretion appears to be hormonally modulated (4, 5).

In the first group, proteins with molecular weights of 66,000, 46,000, and 37,000 are electrophoretically similar to previously described human endometrial proteins (6). The second group includes M_r 40,000 and M_r 33,000 proteins that also appear to be electrophoretically similar to ones synthesized by the human endometrium (4, 5). The M_r 33,000 protein appears to be estrogen dependent, whereas the M_r 40,000 protein can be induced in vitro by progesterone in the human (7) and is upregulated by progesterone in vivo in both the pregnant and nonpregnant baboon uterus (4, 5).

Our most recent studies have focused on the regulation of two additional proteins synthesized by the baboon uterus that may be of functional significance. They have been identified as *insulin-like growth factor binding protein 1* (IGFBP-1) and *retinol binding protein* (RBP). This chapter summarizes our findings on IGFBP-1 and RBP regulation in the baboon uterus during pregnancy and discusses their potential functional significance during implantation and early development.

Insulin-Like Growth Factor Binding Proteins

Insulin-like growth factor binding proteins bind to IGF-I and IGF-II with comparable affinity to that of their receptors and have been shown to modulate IGF bioactivity (8, 9). Six IGFBPs have been cloned to date and numerically designated as 1 to 6 (10–20). We have primarily focused our efforts on the expression of IGFBP-1 by the baboon uterus during the menstrual cycle and the first trimester of pregnancy.

Our initial studies (21, 22) demonstrated that although IGFBP-1 in the baboon and IGFBP-1 in the human were immunologically and biochemically similar, their site of synthesis in the nonpregnant uterus was distinctly different. In the human the major site of IGFBP-1 localization is the decidual cell during both the cycle and pregnancy; whereas, in the baboon IGFBP-1 synthesis is confined to the epithelial cells of the deep basal glands during the late luteal stage, but in the third trimester of pregnancy, the decidual cells become the major site of synthesis for this protein. This observation led us to propose that the conceptus may regulate the switch from glandular to decidual synthesis in the baboon. In order to test this hypothesis, uteri from pregnant baboons on days 18, 25, and 32 postovulation were separated into three distinct regions—that is, the implantation site, a region adjacent to the implantation site, and a region opposite to the implantation site—and the tissues were subjected to biochemical and immunological analysis (23).

We observed a close correlation between the establishment of a mature placenta, the decidualization of the stromal cells, and the onset of IGFBP-1 synthesis by these decidual cells.

During the luteal stage IGFBP-1 is localized primarily to the epithelium of the deep basal glands (22). By day 18 of pregnancy, however, IGFBP-1 immunoreactivity was present in the luminal as well as the basal glands, and the most intense staining was observed in the glands just below the placental/endometrial junction. Our studies also showed that as pregnancy progressed, an increase in the stromal cell expression of IGFBP-1 was evident. On day 18 no immunoreactive product was visible in stromal cells, whereas by day 25 groups of stromal cells at the placental/endometrial junction stained positively for IGFBP-1. By day 32 staining was not limited to the stromal cells at the placental/endometrial junction, but extended deeper into the endometrium at the site of implantation. Although stromal staining was observed at adjacent sites beginning at day 32, staining was always most intense in the regions of the endometrium that were in direct contact with the placenta throughout the first trimester.

The ontogeny of glandular to decidual transformation associated with IGFBP-1 synthesis at the implantation site in the baboon uterus during the first trimester of pregnancy is illustrated in Figure 11.1. In addition to the stromal cells at the placental/endometrial junction, the stromal cells adjacent to the spiral arteries in the upper functionalis of the implantation site also showed IGFBP-1 immunoreactivity. As pregnancy progressed, IGFBP-1 was also observed in the stromal cells surrounding the spiral arteries in the functionalis of adjacent sites, but never around the spiral arteries in the basalis region of any endometrial segment. Thus, it appears that IGFBP-1 immunolocalization in stromal cells surrounding the spiral arteries of the functionalis is also correlated with the process of decidualization.

IGFBP-1 synthesis is induced by *progesterone* (P) in the *estrogen*-(E) primed baboon uterus (22). In addition, IGFBP-1 is immunolocalized to the deep glands of the basalis in the nonpregnant baboon uterus (22). These observations, together with the ability of P to induce decidualization and IGFBP-1 synthesis in human stromal cells cultured in vitro (24), led us to examine the distribution of *progesterone recepters* (PR) and *estrogen receptors* (ER) in the baboon uterus during the menstrual cycle and early pregnancy (25). Our findings that E-exposure induced PR and P decreased PR expression in all epithelial cells except for those in the deep glands of the basalis supported the hypothesis that P induces IGFBP-1 via a receptor-mediated mechanism.

During early pregnancy IGFBP-1 production by the deep glands of the basalis was decreased (23), corresponding to an absence of PR in these same cells. Thus, the loss of PR may be the mechanism for reducing IGFBP-1 production in these cells (25). In contrast, during early pregnancy IGFBP-1 production is induced in the glands of the

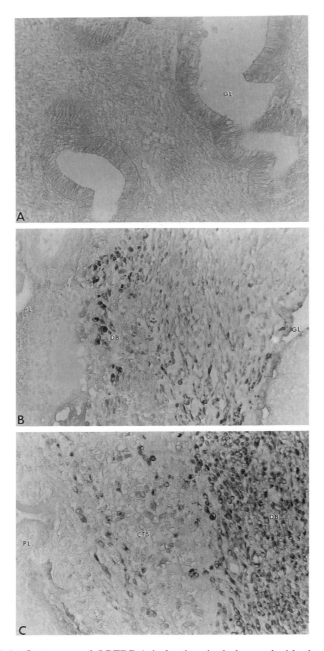

FIGURE 11.1. Ontogeny of IGFBP-1 induction in baboon decidual cells at the implantation site during the first trimester of pregnancy. *A:* Late luteal; *B:* day 32 of pregnancy; and *C:* day 58 of pregnancy. *A* was stained with the monoclonal antibody B2H10, while *B* and *C* were stained with MAb C4H11. (C4H11 does not stain nonpregnant tissues.) Note the limited staining in the deep basal glands during the late luteal stage compared to increasing decidual staining at the implantation site as pregnancy proceeds (205×). (GL = glands; PL = placenta; DB = decidua basalis; CTS = cytotrophoblastic shell.)

functionalis, stromal cells surrounding spiral arteries, and stromal cells directly under the implanting placenta (23). Since PR was absent from all epithelial cells, and although some stromal cells contained PR, there was no direct correlation between cells staining positively for PR and cells producing IGFBP-1 (23, 25).

Thus, IGFBP-1 production does not appear to be regulated solely by P during early pregnancy. Progesterone may serve a permissive role in preparing the uterus for IGFBP-1 production and decidualization; however, other factors, particularly those of placental and/or embryonic origin, appear to regulate IGFBP-1 production and decidualization. Current studies in our laboratory using human chorionic gonadotrophin- (hCG) stimulated, E- and P-treated nonpregnant baboons would support this conclusion. In spite of peripheral concentrations of E and P in these stimulated baboons being comparable to pregnant baboons at 18, 25, and 32 days postovulation, decidual production of IGFBP-1 was not observed, although limited glandular synthesis prevailed (Fazleabas, Verhage, unpublished observations).

In addition to IGFBP-1 the baboon endometrium and decidua also synthesize IGFBP-2 and -3, and the developing placenta produces IGFBP-3 (23). The functionalis region of the baboon endometrium was the major site of synthesis of all of the IGFBPs. Recent studies on the nonpregnant human endometrium indicated that the mRNA for IGFBP-1, -2, and -3 were differentially regulated in the secretory phase compared to the proliferative phase of the menstrual cycle (26). IGFBP-2 and -3 synthesis increased between the proliferative and secretory phases, while IGFBP-1 synthesis was only observed during the secretory phase (13, 26), similar to our observations in the baboon (22, 23).

The developing placenta also secretes IGFBP-3 in culture, and the production of this IGFBP appears to increase with gestational age and placental development. Studies using human syncytiotrophoblasts in monolayer cultures detected the presence of both protein and mRNA for IGFBP-3 and IGF-II (27). These studies led to the suggestion that compartmentalization of IGFBP synthesis within the decidua and placenta (27) may be important for paracrine regulation of physiological functions during pregnancy. Our data in the baboon certainly support this hypothesis since production of the IGFBPs by the pregnant endometrium and placenta appears to be selectively regulated in early pregnancy.

Retinol Binding Protein

Endometrial protein 15, also known as α_2 *pregnancy-associated endometrial globulin* (α_2-PEG), is the major secretory product of the human endometrium during the late luteal phase and early pregnancy (7, 28). This protein is identical to placental protein 14 (29) and has sequence homology with β-lactoglobulin and serum RBPs. The α_2-PEG may be involved in the transport of retinol to the developing embryo during early pregnancy (30). Interestingly, although α_2-PEG constitutes the major endometrial secretory product in the human, it is absent in the baboon (4, 22). Instead, in the baboon RBP is synthesized in the basal glands during the luteal stage, and its synthesis increases markedly in the pregnant uterus through day 32 postovulation and declines thereafter (31).

The pattern of RBP synthesis in the pregnant baboon is very similar to that observed for α_2-PEG in the human. Figure 11.2 compares the message expression for RBP and IGFBP-1 and -2 in the functionalis region of the baboon implantation site. As demonstrated immunocytochemically in Figure 11.1, decidual IGFBP-1 synthesis increases markedly as pregnancy proceeds, and this parallels the expression of IGFBP-1 mRNA. In contrast, RBP mRNA reaches a peak at day 32 coincident with protein synthesis and declines thereafter. This decrease in RBP synthesis is associated with glandular regression in the pregnant uterus. In the nonpregnant baboon, IGFBP-1 and RBP are localized primarily to the epithelium of the deep basal glands (22, 31). In early pregnancy there is marked increase in glandular synthesis of both these proteins, with the exception that IGFBP-1 expression is more pronounced in the upper functionalis in contrast to RBP, whose expression is confined to the glands in the lower functionalis and basalis regions of the endometrium.

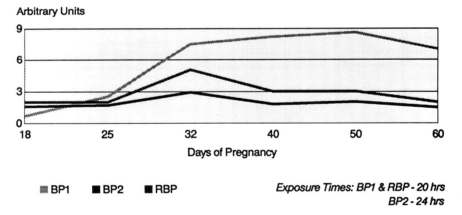

FIGURE 11.2. Densitometric scans of slot blots hybridized with cDNAs to human IGFBP-1 and -2 and porcine RBP. Total RNA (20 µg) from the functionalis region of the baboon implantation site from each time point in pregnancy was analyzed. Note the increase in IGFBP-1 mRNA throughout the first trimester that parallels decidual protein expression (see Fig. 11.1) and the marked increase at day 32 and decline thereafter of RBP message that coincides with glandular regression. IGFBP-2 mRNA expression remains relatively constant throughout this time period.

As pregnancy proceeds and glands regress, a cell-specific change in expression occurs for IGFBP-1, while RBP synthesis remains confined to the glandular epithelium (32). IGFBP-2 mRNA expression is up-regulated by P in both the baboon and human uterus (26, 33) and continues to be expressed at relatively steady state levels.

Potential Functions of IGFBP-1 and RBP in Baboon Pregnancy

Our studies on the regulation of endometrial proteins during pregnancy in the baboon coincide with two distinct periods of placental development in this primate (34). Between days 18 and 25, the placenta undergoes villous branching and angiogenesis that is characterized by (i) lateral growth of the placenta, (ii) formation of the cytotrophoblastic shell, (iii) intermingling of cytotrophoblastic and decidual cells, and (iv) rapid decidualization of the stromal cells. Between days 25 and 40, a definitive embryonic placenta is established, and endometrial differentiation is characterized by (i) deep trophoblastic penetration at the implantation site, (ii) the peripheral movement of trophoblast cells, (iii) thickening of the cytotrophoblastic shell and extracellular matrix deposition, and (iv) abembryonic trophoblast forming direct contact with the endometrium opposite the implantation site.

Alterations in IGFBP-1 and RBP expression in the pregnant uterus parallel these morphological changes. For example, decidualization of stromal cells in the primary phase and trophoblast penetration in the secondary phase is associated with IGFBP-1 synthesis by the decidua at the implantation site and, subsequently, at adjacent sites. Establishment of a definitive embryonic placenta and completion of organogenesis in the fetus, which occurs between days 25 and 40 (34), coincides with the peak of RBP synthesis. The fact that altered changes in expression of both these binding proteins (IGFBP-1 and RBP)—which bind to potentially mitogenic (IGF-I and -II) and differentiation-inducing (retinol) polypeptides—are associated with defined morphological changes in the developing embryo and placenta suggests that these proteins may be of functional significance in the primate.

Production of IGFBP-1 by primate decidual tissue must reflect some unique requirement in pregnancy in species exhibiting hemochorial placentation (35). The function of IGFBP-1 may be associated with the autocrine and/or paracrine regulation of trophoblastic growth by modulating the bioavailability of IGFs. However, the manner in which they act is conflicting at best since numerous studies have shown that IGFBPs could either enhance or inhibit the actions of IGFs or could even act independently of IGFs (reviewed in 36–38).

It has been suggested that IGFBP-1 may play a role in preventing trophoblast invasion into the human endometrium by inhibiting IGF action on placental IGF receptors (9, 39, 40). In the baboon, however, we have proposed that IGFBP-1 may initially play a stimulatory role in stromal cell and trophoblast proliferation since stromal synthesis of IGFBP-1 appears to be regulated by the placenta and/or conceptus at the implantation site (23).

One potential function of IGFBP-1 could be to regulate IGF action on endometrial/decidual cells in an autocrine manner. The endometrium undergoes dramatic growth and differentiation during the first trimester, and if IGFs are required for these processes, the local production of IGFBP-1 in its stimulatory form could provide a mechanism to locally enhance the growth-promoting effects of IGFs. The prerequisite to the hypothesis that IGFs have a direct autocrine effect on decidual cell proliferation is that IGF-I receptors have to be present in the same cell. Preliminary studies in our laboratory (41) show the presence of IGF-I receptors on decidualizing cells at the implantation site on day 18 of pregnancy and a continued increase in receptor expression as the stromal cells continue to decidualize.

Alternatively, a potential paracrine function for IGFBP-1 is suggested when the cellular localization of synthesis and secretion is considered with reference to the behavior of the embryonic trophoblast during implantation and placental development in the primate. IGFBP-1 contains the RGD tripeptide sequence (10–13), a recognition site on extracellular matrix proteins for a class of cell-surface receptors known as *integrins* (42). In addition, extracellular matrix proteins, such as fibronectin, are important for trophoblast/endometrial contact, proliferation, and syncytial formation (43, 44). Thus, it is possible that the induction of IGFBP-1 synthesis in decidual cells at the placental-endometrial interface is a mechanism by which IGFs can be locally concentrated. IGF-I receptors are present in the basal lamina of cytotrophoblastic cells at day 18 in the developing placenta (41) coincident with IGFBP-1 expression in the luminal glandular epithelial cells and decidual cells. These two correlated events may enhance trophoblast proliferation. IGFBP complexed with IGFs and localized in the decidual cells at the placental/endometrial interface and luminal glandular epithelium may act as a mitogen on conceptus tissues and facilitate trophoblastic penetration and contact with the maternal vasculature.

The mechanism by which IGF is released from IGFBP to enable it to bind to its receptor may be regulated locally by plasmin. Campbell et al. (45) have demonstrated that activation of plasminogen to plasmin by plasminogen activators results in the dissociation of bioactive IGF-I from IGFBP-1. The invasive trophoblast produces plasminogen activators (46, 47). Therefore, the potential exists for the penetrating and proliferating trophoblast to locally activate plasminogen to plasmin, which in turn releases IGFs from decidual IGFBP-1, thereby making this mitogenic peptide available for binding to its decidual and placental receptor. Following the establishment of placental contact with the endometrium and the formation of the syncytium, IGFBP-3 synthesis by the placenta increases together with increased decidual synthesis of IGFBP-1. Tissue compartmentalization and co-expression of decidual IGFBP-1 and placental IGFBP-3, both of which can either stimulate or inhibit the mitogenic actions of IGFs (48–50), may be a mechanism to inhibit IGF binding to its receptor following implantation and thereby control additional trophoblastic proliferation and invasion. This may be the process by which the baboon, a superficial implanter, is able to rapidly establish contact with the maternal vasculature and yet control trophoblast invasion.

Retinol binding protein is the carrier protein for retinol (vitamin A). Retinol plays a critical role in normal embryonic development (51), and both deficiency and excess result in congenital defects in the embryo (52, 53). In the baboon an increase in glandular RBP synthesis is associated with the establishment of a definitive embryonic placenta, completion of fetal organogenesis, and stromal cell differentiation (34). Retinol primarily exerts its effects by inducing cellular differentiation and inhibiting cellular proliferation (54). In MCF-7 cells, for example, retinol is capable of inhibiting IGF-I-induced cell proliferation by increasing IGFBP synthesis (55). Thus, we propose that in addition to glandular RBP being able to provide retinol to the developing embryo, it may also act synergistically with IGFs to induce stromal-to-decidual differentiation and cytotrophoblast-to-syncytiotrophoblast transformation and initiate the synthesis of IGFBP-1 by decidual cells and IGFBP-3 by placental tissue.

Summary

Our in vivo studies are suggestive of a close interaction between the developing placenta and endometrial IGFBP-1 and RBP synthesis. However, more definitive in vitro studies using coculture systems will be necessary to better understand the role of IGFBPs, IGFs, and their receptors during the process of implantation and early embryonic development in the primate. These studies are currently underway in our laboratory.

Acknowledgments. These studies were supported by NIH Grants HD-21991 (A.T.F.) and HD-07508 (S.H.P.). The generous gifts of monoclonal antibody to IGFBP-1 by Dr. Stephen C. Bell and cDNAs to

IGFBP-2 and RBP from Drs. J. Schwander and R. Michael Roberts, respectively, are greatly appreciated. We also thank Ms. Esther Vergara for her excellent technical assistance and Ms. Margarita Guerrero for her secretarial skills.

References

1. Fazleabas AT, Verhage HG, Bell SC. Steroid-induced proteins of the primate oviduct and uterus: potential regulators of reproductive function. In: Krey LC, Guylas BJ, McCraken JA, eds. Autocrine and paracrine mechanisms in reproductive endocrinology. New York: Plenum Press, 1989:115–36.
2. Fazleabas AT, Verhage HG, Bell SC. Insulin-like growth factor binding protein and pregnancy: regulation and function in the primate. In: Heyner S, Wiley LM, eds. Early embryo development and paracrine relationships. New York: Alan R. Liss, 1990:137–52.
3. Fazleabas AT, Bell SC, Verhage HG. Insulin-like growth factor binding proteins: a paradigm for conceptus-maternal interactions in the primate. In: Strauss JF III, Lyttle CR, eds. Uterine and embryonic factors in early pregnancy. New York: Plenum Press, 1991:157–66.
4. Fazleabas AT, Verhage HG. Synthesis and release of polypeptides by the baboon (*Papio anubis*) uterine endometrium in culture. Biol Reprod 1987;37:979–88.
5. Fazleabas AT, Miller JB, Verhage HG. Synthesis and release of estrogen- and progesterone-dependent proteins by the baboon (*Papio anubis*) uterine endometrium. Biol Reprod 1988;39:729–36.
6. Bell SC, Hales MW, Patel S, Kirwan PH, Drife JO. Protein synthesis and secretion by the human endometrium and decidua during early pregnancy. Br J Obstet Gynaecol 1985;92:793–803.
7. Bell SC, Patel SR, Kirwan PH, Drife JO. Protein synthesis and secretion by the human endometrium during the menstrual cycle and effect of progesterone in vitro. J Reprod Fertil 1986;77:221–31.
8. Nissley SP, Rechler MM. Insulin-like growth factors: biosynthesis, receptors and carrier proteins. In: Li H, ed. Hormonal proteins and peptides. New York: Academic Press, 1985;XII:128–203.
9. Ritvos O, Ranta T, Jalkanen J, et al. Insulin-like growth factor (IGF) binding protein from human decidua inhibits the binding and biological action of IGF-I in cultured choriocarcinoma cells. Endocrinology 1988;122:2150–7.
10. Lee Y-L, Hintz RL, James PM, Lee PDK, Shively JE, Powell DR. Insulin-like growth factors (IGF) binding protein complementary deoxyribonucleic acid from human Hep G2 hepatoma cells; predicted protein sequence suggests an IGF binding domain different from those of the IGF-I and IGF-II receptors. Mol Endocrinol 1988;2:404–11.
11. Brewer MT, Stetler GL, Squires CH, Thompson RC, Busby WH, Clemmons DR. Cloning, characterization, and expression of a human insulin-like growth factor binding protein. Biochem Biophys Res Commun 1988;152:1289–97.
12. Brinkman A, Groffen C, Kortleve DJ, Geurts Van Kessel A, Drop SLS. Isolation and characterization of a cDNA encoding the low molecular weight insulin-like growth factor binding protein (IBP-I). EMBO J 1988;7:2417–23.

13. Julkunen M, Koistinen R, Aalto-Setalak K. Seppala M, Janne OA, Kontulak K. Primary structure of human insulin-like growth factor binding protein/placental protein 12 and tissue specific expression of its mRNA. FEBS Lett 1988;236:295–300.
14. Binkert C, Landwehr J, Mary J-L, Schwander J, Heinrich G. Cloning sequence analysis and expression of a cDNA encoding a novel insulin-like growth factor binding protein IGFBP-2. EMBO J 1989;8:2497–502.
15. Margot JB, Binkert C, Mary J-L, Landwehr J, Heinrich G, Schwander J. A low molecular weight insulin-like growth factor binding protein from rat; cDNA cloning and tissue distribution of its messenger RNA. Mol Endocrinol 1989;3:1053–60.
16. Wood WI, Cachianes G, Henzel WJ. Cloning and expression of the growth hormone dependent insulin-like growth factor binding protein. Mol Endocrinol 1989;2:1176–85.
17. Shimasaki F, Uchiyama F, Shimonaka M, Ling N. Molecular cloning of the cDNAs encoding a novel insulin-like growth factor-binding protein from rat and human. Mol Endocrinol 1990;4:1451–8.
18. LaTour D, Mohan S, Linkhart TA, Baylink DJ, Strong DD. Inhibitory insulin-like growth factor binding protein (hIGFBP-4): cloning, complete sequence and physiological regulation. Mol Endocrinol 1990;4:1806–14.
19. Shimasaki S, Shimonaka M, Zhang H-P, Ling N. Identification of five different insulin-like growth factor binding proteins (IGFBP's) from adult rat serum and molecular cloning of a novel IGFBP-5 in rat and human. J Biol Chem 1991;266:10646–53.
20. Shimasaki S, Gao L, Shimonaka M, Ling N. Isolation and molecular cloning of insulin-like growth factor binding protein-6. Mol Endocrinol 1991;5:938–48.
21. Fazleabas AT, Verhage HG, Waites G, Bell SC. Characterization of an insulin-like growth factor binding protein analogous to human pregnancy-associated secreted endometrial α_1 globulin, in decidua of the baboon (*Papio anubis*) placenta. Biol Reprod 1989;40:873–85.
22. Fazleabas AT, Jaffe RC, Verhage HG, Waites G, Bell SC. An insulin-like growth factor binding protein (IGFBP) in the baboon (*Papio anubis*) endometrium: synthesis, immunochemical localization and hormonal regulation. Endocrinology 1989;124:2321–9.
23. Tarantino S, Verhage HG, Fazleabas AT. Regulation of insulin-like growth factor binding proteins in the baboon (*Papio anubis*) uterus during early pregnancy. Endocrinology 1992;130:2354–62.
24. Bell SC, Jackson JA, Ashmore J, Zhu HH, Tseng L. Regulation of insulin-like growth factor binding protein-1 synthesis and secretion by progestin and relaxin in long-term cultures of human endometrial stromal cells. J Clin Endocrinol Metab 1991;72:1014–24.
25. Hild-Petito S, Verhage HG, Fazleabas AT. Immunocytochemical localization of estrogen and progestin receptors in the baboon (*Papio anubis*) uterus during implantation and early pregnancy. Endocrinology 1992;130:2343–53.
26. Giudice LC, Milkowski DA, Lamson G, Rosenfeld RG, Irwin JC. Insulin-like growth factor binding proteins in human endometrium: steroid-dependent messenger ribonucleic acid expression and protein synthesis. J Clin Endocrinol Metab 1991;72:779–87.

27. Deal CL, Giudice LC, Lamson G, Rosenfeld RG. Production of insulin-like growth factors (IGFs) and insulin-like growth factor binding proteins (IGFBPs) by human trophoblast [Abstract A54]. 2nd Int Symp on IGFs/Somatomedins, San Francisco, CA, 1991.
28. Bell SC, Smith S. The endometrium as a paracrine organ. In: Chamberland GJP, ed. Contemporary obstetrics and gynecology. London: Butterworths Scientific, 1988;273–98.
29. Bell SC, Bohn H. Immunochemical and biochemical relationship between human pregnancy-associated secreted α_1 and α_2 endometrial globulins and the soluble placental proteins 12 and 14. Placenta 1986;7:282–94.
30. Bell SC. Secretory endometrial/decidual proteins and their function in early pregnancy. J Reprod Fertil 1988;suppl 36:1–17.
31. Donnelly KM, Vergara EF, Mavrogianis PA, Fazleabas AT. Endometrial synthesis of retinol binding protein (RBP) in the non-pregnant and pregnant baboon (*Papio anubis*) [Abstract 20]. Biol Reprod 1991;44(suppl 1).
32. Donnelly KM, Vergara EF, Mavrogianis PA, Fazleabas AT. Insulin-like growth factor binding proteins (IGFBPs) and retinol binding protein (RBP) expression in the baboon (*Papio anubis*) uterus during the first trimester of pregnancy [Abstract 365]. Biol Reprod 1992;46(suppl 1).
33. Giudice LC, Milkowski DA, Fielder PJ, Irwin JC. Characterization of the steroid dependence of insulin-like growth factor binding protein-2 and mRNA expression in cultured human endometrial stromal cells. Hum Reprod 1991;6:632–40.
34. Hendrickx AG. Embryology of the baboon. University of Chicago Press, 1971.
35. Bell SC. Comparative aspects of decidualization in rodents and humans: cell types, secreted products and associated function. In: Edwards RG, Purdy J, Steptoe PC, eds. Implantation of the human embryo. London: Academic Press, 1985:71–122.
36. Ooi GT. Insulin-like growth factor binding proteins (IGFBPs): more than just 1, 2, 3. Mol Cell Endocrinol 1990;71:C39–43.
37. Sara VR, Hall K. Insulin-like growth factors and their binding proteins: Physiol Rev 1990;70:591–614.
38. Rosenfeld RG, Lamson G, Pham H, et al. Insulin-like growth factor binding proteins. Recent Prog Horm Res 1990;46:99–159.
39. Rutanen E-M, Pekonen F, Makinen T. Soluble 34K binding protein inhibits the binding of insulin-like growth factor I to its receptors in human secretory phase endometrium: evidence for autocrine/paracrine regulation of growth factor action. J Clin Endocrinol Metab 1988;66:173–80.
40. Pekonen F, Suikkari A-M, Makinen T, Rutanen E-M. Different insulin-like growth factor binding species in human placenta and decidua. J Clin Endocrinol Metab 1988;67:1250–7.
41. Hild-Petito S, Fazleabas AT. Characterization of receptors for insulin-like growth factor I in the baboon uterus during the cycle and pregnancy [Abstract 360]. Biol Reprod 1992;46(suppl 1).
42. Ruoslahti E, Pierschbacher MD. New perspectives in cell adhesion: RGD and antigens. Science 1987;238:491–7.
43. Kao L-C, Caltabiano S, Wu S, Strauss JF III, Kliman HJ. The human villous cytotrophoblast: interaction with extracellular matrix proteins, endocrine

function and cytoplasmic differentiation in the absence of syncytium formation. Dev Biol 1988;130:693–702.
44. Kliman HJ, Feinberg RF. Human trophoblast-extracellular matrix (ECM) interactions in vitro: ECM thickness modulates morphology and proteolytic activity. Proc Natl Acad Sci USA 1990;87:3057–61.
45. Campbell PG, Novak JF, Yanosick TB, McMaster JH. Involvement of the plasmin system in dissociation of the insulin-like growth factor binding protein complex. Endocrinology 1992;130:1401–12.
46. Strickland S, Reich E, Sherman MI. Plasminogen activator in early embryogenesis: enzyme production by trophoblast and parietal endoderm. Cell 1976:231–40.
47. Fisher SJ, Cui T-Y, Zhang L, et al. Adhesive and degradative properties of human placental cytotrophoblast cells in vitro. J Cell Biol 1989;891–902.
48. Elgin RG, Busby WH, Clemmons DR. An insulin-like growth factor (IGF) binding protein enhances the biologic response to IGF I. Proc Natl Acad Sci USA 1987;84:3254–8.
49. Blum WF, Jenne EW, Reppin F, Kietzmann K, Ranke MB, Bierich JR. Insulin-like growth factor I binding protein complex is a better mitogen than free IGF I. Endocrinology 1989;125:766–72.
50. Conover CA, Powell DR. Insulin-like growth factor (IGF)-binding protein-3 blocks IGF I induced receptor down-regulation and cell desensitization in cultured bovine fibroblasts. Endocrinology 1991;129:710–9.
51. Goodman DS. Vitamin A and retinoids in health and disease. N Engl J Med 1984;310:1023–31.
52. Warkany J, Roth CP. Congenital malformations induced in rats by maternal vitamin A deficiency, II. Effects of varying the preparatory diet in the yield of abnormal young. J Nutr 1948;35:1–11.
53. Cohlan SQ. Excessive vitamin A uptake as a cause of congenital anomalies in rat. Science 1953;117:535–6.
54. Sporn MB, Roberts AB, Goodman DS. The retinoids. New York: Academic Press, 1984.
55. Fontana JA, Mezu-Burrows A, Clemmons DR, LeRoith D. Retinoid modulation of insulin-like growth factor binding proteins and inhibition of breast carcinoma proliferation. Endocrinology 1991;128:1115–22.

12

Regulation of Human Cytotrophoblast Invasion

KATHRYN E. BASS, IRIS ROTH, CAROLINE H. DAMSKY, AND SUSAN J. FISHER

Human cytotrophoblast invasion of the uterus is the result of an unusual differentiation process in which polarized epithelial cells, anchored to the chorionic villus basement membrane, become detached, aggregate into multilayered columns of nonpolarized cells, and rapidly penetrate the endometrium, the first third of the myometrium, and the associated spiral arterioles (Fig. 12.1). This process continues through the first trimester, peaks during the 12th week of pregnancy, and declines rapidly thereafter (1–4). The result is formation of the hemochorial placenta in which the fetal trophoblast cells are constantly bathed by maternal blood.

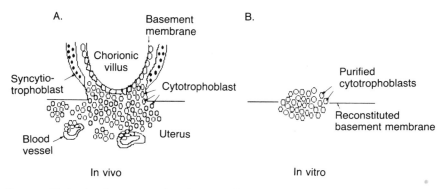

FIGURE 12.1. A: Anatomy of a human chorionic villus. In vivo, noninvasive cytotrophoblasts that are anchored to the chorionic villus basement membrane become detached and aggregate into multilayered columns of nonpolarized cells that rapidly penetrate the endometrium. A multinucleated syncytium composed of syncytiotrophoblasts surrounds the villus. B: In vitro invasion assay depicting highly purified cytotrophoblasts penetrating a reconstituted basement membrane.

Using a combination of experimental strategies, we have been studying the role of proteinases, adhesion molecules, and histocompatability antigens in the invasion process. We have applied immunolocalization techniques to placental bed biopsy specimens containing cytotrophoblasts that have invaded the uterus to demonstrate which molecules are differentially regulated at the protein level during invasion in vivo. However, this approach does not allow access to the dynamic process, a prerequisite for studies designed to understand the functional consequences of the differential expression of certain molecules during the invasion process. For this purpose we have also developed in vitro models in which early gestation human cytotrophoblasts plated as a monolayer aggregate on basement membrane-like matrices, then rapidly invade these substrates (Fig. 12.1).

As described in the following sections, we have used both immunolocalization and biochemical techniques to show that cytotrophoblast differentiation and invasion are accompanied by dramatic changes in the expression of matrix-degrading metalloproteinases, extracellular matrix ligands and their integrin receptors, and the trophoblast-specific class III molecule, HLA-G (Table 12.1). Using our in vitro models of the invasion process, we have uncovered functional roles for many of these molecules. Growth and differentiation of many cell types are regulated by a complex milieu of growth factors and cytokines. Data from our laboratory suggest that early gestation cytotrophoblasts that are differentiating along the invasive pathway express a particular repertoire of growth factors and cytokines. We postulate that these molecules play an important role in the regulation of trophoblast invasion and differentiation.

Role of Proteinases in Cytotrophoblast Invasion

Many elements of trophoblast invasion are similar to events that occur during tumor cell invasion. The phenotypic change from carcinoma in situ

TABLE 12.1. Modulation of key molecules by cytotrophoblasts.

		Cytotrophoblasts	
Class of molecules		Noninvasive	Invasive
I.	92-kd type IV collagenase	−	+
II.	Integrins		
	α6	+	−
	α5	−	+
	α1	−	+
	β1	−	+
	β4	+	−
III.	HLA-G	−	+

to invasive carcinoma occurs when tumor cells acquire the ability to penetrate an epithelial basement membrane and invade the underlying stroma (5, 6). Likewise, after a brief adherent stage, cytotrophoblasts penetrate the basement membrane underlying the uterine epithelial cells and invade the stroma and its associated arterioles. These similarities suggest that the two invasive processes may share certain common mechanisms. However, unlike tumor invasion, trophoblast invasion is precisely regulated, confined spatially to the uterus and temporally to early pregnancy.

Breaching of the *extracellular matrix* (ECM) barrier by cells involves matrix-degrading metalloproteinases (7–9) and plasminogen activators, members of the serine proteinase family. Metalloproteinases constitute a multigene family that includes PUMP-1, interstitial collagenase, and the stromelysins, as well as the 72- and 92-kd type IV collagenases. The wide specificities that metalloproteinases exhibit may be important in invasion. For example, 92-kd type IV collagenase degrades collagen types IV, V, and VII, gelatin, fibronectin, proteoglycans, and elastin. *Plasminogen activators* include the *tissue-type* (tPA) and the *urokinase-type* (uPA), which can cleave fibronectin and other extracellular matrix components.

To determine which proteinases are made by cytotrophoblasts in a developmentally regulated manner, it was necessary to obtain highly purified preparations of these cells from all three trimesters of pregnancy. Removal of bone marrow-derived cells is particularly important since they synthesize high levels of both metallo- and serine proteinases. In the purification process dissociated villus cytotrophoblasts are layered over a Percoll gradient (10). The gradient is centrifuged, and a broad band in the middle of the tube containing the cytotrophoblasts is removed. Any remaining leukocytes among the cytotrophoblasts are removed by using an antibody to CD-45, a protein tyrosine phosphatase found on bone marrow-derived cells (11), but not on cytotrophoblasts. Briefly, the antibody (HLe, Becton Dickinson and Co., Mountain View, CA, or IgG affinity-purified from the GAP 8.3 hybridoma, American Type Culture Collection, Rockville, MD) is coupled to magnetic beads (Advanced Magnetics Inc., Cambridge, MA) and mixed with the cytotrophoblast-enriched Percoll gradient fraction at a density of 25 particles/cell. After incubation for 20 min at 4°C with occasional gentle mixing, the CD-45-positive cells are removed by means of a Bio-Mag Separator (Advanced Magnetics).

We used zymography on substrate gels to profile proteinases of highly purified cytotrophoblasts isolated from first-, second-, and third-trimester human placentas, as well as first-trimester human placental fibroblasts (12). Metalloproteinases were demonstrated in gelatin- or casein-containing gels. Their identity was confirmed by using specific inhibitors and antibodies. Cytotrophoblasts synthesized a variety of metalloproteinases; of these, the 92-kd type IV collagenase was regulated in accordance with the invasive properties of the cells; that is, expression was highest in early gestation and greatly reduced at term. Human placental fibroblasts did not synthesize this proteinase. We also examined the expression of PAs. Like Queenan et al. (13), we found that cytotrophoblast cells from all three trimesters express only uPA in vitro.

To study the function of the cytotrophoblast proteinases, we used the in vitro model shown in Figure 12.1B. Specific inhibitors, including protein inhibitors (tissue inhibitor of metalloproteinase 1 and 2, PA inhibitor 1, and PA inhibitor 2) and function-perturbing antibodies, were copolymerized with the reconstituted basement membrane. Metalloproteinase inhibitors and function-perturbing antibody specific for the 92-kd type IV collagen-degrading metalloproteinase completely inhibited cytotrophoblast invasion in vitro, whereas inhibitors of the PA system had only a partial (20%–40%) inhibitory effect. These experiments demonstrated that the 92-kd type IV collagenase is rate limiting for human cytotrophoblast invasion in vitro (14) and, therefore, is likely to be an important component of the invasion apparatus used by these cells in vivo.

Role of Adhesion Molecules in Cytotrophoblast Invasion

We also studied the expression of adhesion receptors and their extracellular matrix ligands during human trophoblast invasion in vivo. The results showed that the distribution pattern of integrin cell-matrix receptors is modulated extensively as cells differentiate along the invasive pathway, suggesting that development of the appropriate adhesion phenotype is also critical during the invasion process (15). Villus cytotrophoblasts, anchored to basement membrane, stained for α6 and β4 integrin subunits (the combination of which is thought to act as a laminin receptor) and for multiple forms of laminin. In contrast, nonpolarized cytotrophoblasts in columns expressed primarily α5 and β1 integrin subunits (when complexed, the major fibronectin receptor) and a fibronectin-rich matrix. Cytotrophoblast clusters in the uterine wall stained for α1 (α1/β1 integrin is a collagen and laminin receptor) as well as for α5 and β1 integrins, but not for most extracellular matrix antigens, suggesting that these cells interact primarily with surrounding maternal cells and matrices. Thus, the down-regulation of α6/β4 and the sequential up-regulation of α5/β1 and α1/β1 are characteristic of the invasive differentiation pathway in situ. The intricate way in which trophoblasts modulate their expression of adhesion molecules during uterine invasion suggests that these changes have important functional significance in this process.

This hypothesis is further supported by our recent observation that cytotrophoblast expression of a subset of integrin receptors is disregulated in preeclampsia. This disorder (reviewed in 16) affects 4%–7% of all pregnancies, most commonly develops in primagravidas, and is

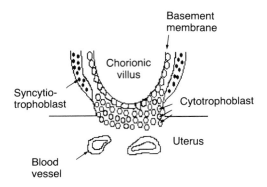

FIGURE 12.2. Characterization of preeclampsia. The disorder is distinguished by abnormally shallow invasion of the uterus by cytotrophoblasts, as well as failure of these cells to invade the uterine vasculature.

characterized by abnormally shallow trophoblast invasion of the uterus (Fig. 12.2). The signs, including increased maternal blood pressure and renal dysfunction evidenced by proteinuria and edema, usually develop in the late second or early third trimester. In addition to these effects on the mother, there is a profound impact on the fetus, resulting in increased perinatal mortality and frequent intrauterine growth retardation. There is no cure for this disease other than termination of pregnancy. When the maternal or fetal condition mandates expeditious delivery, iatrogenic prematurity is often the result.

We used placental bed biopsy specimens to show that cytotrophoblast expression of a subset of integrin cell-extracellular matrix receptors is abnormal in preeclampsia. Staining for the $\alpha 5$ and $\beta 1$ subunits was similar to that observed in samples from age-matched control patients. In contrast, the pattern of expression of the other cell adhesion molecules was very different. The $\alpha 6$ and $\beta 4$ subunits continued to be strongly expressed within the uterine wall, and $\alpha 1/\beta 1$ was either not up-regulated or only weakly expressed by a small proportion ($<5\%$) of cytotrophoblasts within the uterine wall. These observations suggest that in preeclampsia, cytotrophoblasts that invade the uterus retain cell adhesion molecules that under normal conditions are expressed only by villus stem cells and cytotrophoblasts in the proximal region of the cell columns. In addition, they fail to up-regulate the expression of one integrin receptor that in normal pregnancy is expressed only by cytotrophoblasts that have invaded the uterine wall. We hypothesize that the failure of cytotrophoblasts to switch their adhesion molecule phenotype in preeclampsia could tip the delicate balance of molecules that normally permit cytotrophoblast invasion in favor of those that restrain this process, with the net effect of shallow uterine invasion.

Role of Class I Molecules in Maternal Tolerance of Cytotrophoblast Invasion

The exact mechanisms that protect the genetically different fetus from maternal rejection remain obscure. However, important insights have recently been gained into molecules that could play interesting roles. For example, it is likely that a subset of the growth factors and cytokines produced by cytotrophoblasts can regulate some aspects of maternal immune function. For example, as discussed in the following section, the immunosuppressive cytokine IL-10 (17, 18) is produced by cytotrophoblasts.

Also interesting in this regard is the recent discovery that the trophoblast component of the placenta, although apparently devoid of class II antigens, does express a highly unusual class I molecule. We (19) and Ellis et al. (20) showed that the nonclassical class I molecule HLA-G is expressed by human cytotrophoblasts. This was the first evidence that the HLA-G gene is actually transcribed. The same antigen is also present on some clonal lines of choriocarcinoma cells, including BeWo (21) and JEG (19), but has not been detected on any other normal human cells, except possibly those in the anterior chamber of the eye (22), another "immunologically privileged" site.

HLA-G, like the other classical (A, B, and C) and nonclassical (E and F) HLA class I molecules, is located on the short arm of chromosome 6 and is telomeric to the classical HLA-A locus (23). The protein encoded by this gene is similar in organization to HLA-A, -B, and -C except that an in-frame termination codon prevents transcription of most of the cytoplasmic domain (24). Like these other molecules, HLA-G associates with β-2 microglobulin. Comparison of the amino acid sequences shows that certain regions of the extracellular domain are very similar (α3), while others show extensive differences (α1 and α2). By two-dimensional gel electrophoresis we detected no allelic variation in samples analyzed from 6 individuals, suggesting that this molecule is not obviously polymorphic (19). A similar conclusion was reached by Ellis et al. (21). Interestingly, these investigators did note some nucleic acid sequence polymorphisms, but none produced alterations in the protein sequence of the molecule. Our laboratory is currently investigating whether this molecule is in fact nonpolymorphic by cloning and sequencing HLA-G from individuals of various genetic backgrounds.

The role of HLA-G in producing maternal tolerance of the fetus has still not been clearly defined. Several hypotheses concerning the function of this unusual molecule have been proposed. The lack of obvious polymorphisms suggests that cytotrophoblasts would not stimulate MHC-restricted rejection by maternal effector cells and might protect the fetal tissues from non-MHC-restricted cytolytic lymphoid cells present in the decidua (25). In support of this hypothesis, Kovats et al. (26) have recently shown that the expression of HLA-G α-chain on an HLA-A,B,C-null lymphocyte cell line reduced its susceptibility to lysis by non-MHC-restricted NK and γΔ T cell clones.

Cytotrophoblast Expression of Growth Factors, Cytokines, and Their Receptors

Since growth factors and cytokines can have autocrine and/or paracrine actions, a comprehensive experimental strategy is required to identify with certainty the cells that synthesize and respond to these factors. The first step in understanding the role of these molecules in placental development is to describe the repertoire of trophoblast growth factors, cytokines, and their related receptors. Placental expression of the EGF receptor (27, 28), EGF (29), TGFβ (30, 31), IGF-I (32), IGF-II (33), CSF-I (34), M-CSF (35), IL-1β (30), IL-2 (36), IL-6 (37), and the IL-6 receptor (38) by the placenta has been described. In some cases localization approaches have been used. However, in situ hybridization, which detects cells producing mRNA, does not by itself actually prove protein synthesis. Conversely, detection of protein by immunolocalization techniques does not allow unequivocal identification of cells that produce growth factors and cytokines, as opposed to ones that express the appropriate receptors.

Other approaches that have been used, such as detection of secreted protein (ELISA, radioimmunoassay) or mRNA (Northern blot analyses), are often confounded by the fact that few of these studies have used purified populations of cytotrophoblasts. This is of particular concern since the widely used isolation procedures do not include specific steps to remove bone marrow-derived cells, which produce a wide spectrum of growth factors and cytokines, as well as their receptors. In these circumstances the identity of the cells that are actually producing and responding to the various factors is ambiguous. Nevertheless, some interesting correlations have been noted. For example, the CSF-I receptor is present on human intermediate trophoblasts and syncytiotrophoblasts (39). Since CSF-I is expressed by the glandular epithelial and endothelial cells of the uterus (40), this growth factor could act in a paracrine mechanism.

Beyond these largely descriptive data, limited evidence supports a role for growth factors in regulating human trophoblast differentiation. EGF induces the differentiation of isolated first-trimester cytotrophoblasts into syncytiotrophoblasts and increases the secretion of human chorionic gonadotropin by these cells in vitro (41, 42). Morrish et al. (43) induced the differentiation of term cytotrophoblasts into syncytiotrophoblasts with EGF and inhibited it by the addition of TGFβ1. Thus, it is possible to hypothesize that the balanced action of several growth factors and cytokines on cytotrophoblasts during differentiation is important in producing the spatially and temporally regulated cytotrophoblast invasion required for normal pregnancy.

Our laboratory is currently investigating this hypothesis. Our first step has been to correlate the expression of a particular repertoire of these factors and their receptors with specific stages of trophoblast development along this pathway. To do so, we used the reverse *transcriptase-polymerase chain reaction method* (RT-PCR) to analyze mRNA from highly purified cytotrophoblasts at precise stages of differentiation. These cells were isolated by positive selection on the basis of their stage-specific expression of cell adhesion and class I molecules (described in detail in the previous sections; see also, Table 12.1).

For this purpose we used two antibodies that were produced in this laboratory. One specifically recognizes integrin α6 that is expressed by cytotrophoblast stem cells and column cytotrophoblasts, but not by the cytotrophoblasts within the uterine wall. The other reacts with HLA-G, expressed only by cytotrophoblasts that have differentiated along the invasive pathway; that is, those in the distal column and uterine wall. These antibodies were used for immunoaffinity cell chromatography. Briefly, each antibody was coupled to a cell separation column (CellPro, Inc., Seattle, WA). First-trimester cytotrophoblasts, isolated and depleted of bone marrow-derived cells as described above, were applied to the columns in Dulbecco's Modified Eagle's H21 MEM containing 2% Nutridoma (Boehringer Mannhein, Indianapolis, IN) (SFM). Cells that bound to the columns were released by a change in pressure. In this manner we isolated 4 cell fractions: α6-positive, α6-negative, HLA-G-positive, and HLA-G-negative. The cells were washed once in SFM, and total RNA was collected (44), reverse transcribed, and subjected to PCR reactions according to the method of Rappolee et al. (45–47). As a negative control, reactions were carried out in the absence of reverse-transcribed RNA. The PCR products were electrophoresed in 2% agarose gels, stained with ethidium bromide, and photographed.

The resulting data (Table 12.2) indicate that the cytotrophoblast population selected by use of the integrin α6 monoclonal antibody expresses message for the α6, α5, and β1 subunits, but not the α1 subunit. This is consistent with what we know about the expression of these integrins by cytotrophoblasts in vivo. The cytotrophoblast stem cells and those in the proximal portions of columns express α6, but not α1, which is expressed only by cytotrophoblasts that have entered the uterine wall (15).

We also found evidence for both the autocrine and the paracrine action of growth factors and cytokines. Autocrine processes are exemplified by interactions involving IL-1β and its receptor. The IL-1β mRNA is expressed by highly purified populations of cytotrophoblasts prior to

TABLE 12.2. RT-PCR products.

Targets of PCR primers	α6	(−)α6	HLA-G	(−)HLA-G
I. Integrins				
α6	+	+	+	+
α1	−	−	−	−
α5	+	+	+	+
β1	+	+	+	+
II. Growth factors				
EGF	−	−	−	−
EGF receptor	+	+	+	+
IGF-II	+	+	+	+
III. Cytokines				
GM-CSF	−	−	−	−
IL-1α	−	−	−	−
IL-1β	+	+	+	+
IL-10	+	+	+	+
IV. Control				
GAPDH	+	+	+	+

Note: Cells were chromatographed on columns derivatized with either anti-α6 or anti-HLA-G. Those that did not bind to the column are denoted by (−) in the column headings. (GAPDH = glyceraldehyde phosphate dehydrogenase; + = PCR product of the expected size; − = no PCR product seen.)

culture. We have recently shown by immunoprecipitation that they also have the receptor for this cytokine (Librach et al., manuscript in preparation). These results suggest that cytotrophoblasts can both produce and respond to IL-1β. Paracrine interactions involving maternally derived factors are exemplified by our finding that cytotrophoblasts express the mRNA for the EGF receptor, but not for EGF itself. Additional experiments using a radioreceptor assay demonstrated that cytotrophoblasts produce neither EGF nor TGFα (Roth, Taylor, Fisher, unpublished observations). Thus, it is likely that the EGF receptor ligand is maternally produced. We also have evidence that fetally derived factors can influence the mother. Cytotrophoblasts express the mRNA for the newly described cytokine IL-10. Since this cytokine is strongly immunosuppresive (17, 48), it is tempting to speculate that IL-10 may down-regulate the activity of the maternal immune system.

Conclusions

Normal development of the human placenta requires differentiation of cytotrophoblasts along the invasive pathway, a process that is closely regulated both spatially and temporally. Expression of molecules that mediate important functions relative to the invasion process, including the 92-kd type IV collagenase, a number of adhesion molecules, and HLA-G, is also tightly regulated. Growth factors and cytokines are known to influence the expression of many of these molecules in other cells. Thus, we are determining whether a particular repertoire of these factors and their receptors can be correlated with specific stages of trophoblast differentiation.

Our preliminary studies suggest that both autocrine and paracrine regulatory pathways exist at the maternal-fetal interface. Our next step will be a detailed analysis of the functional consequences of the timed expression of growth factors, cytokines, and their related receptors during cytotrophoblast differentiation along the invasive pathway in vitro. The results will contribute substantially to our understanding of factors controlling normal trophoblast invasion, as well as the etiology of disorders, such as preeclampsia, that are related to defects in this process.

References

1. Brosens I, Dixon HG. Anatomy of the maternal side of the placenta. Br J Obstet Gynaecol 1966;73:357–63.
2. Boyd JD, Hamilton WJ. The human placenta. Cambridge: Heffer and Sons, 1970.
3. Ramsey EM, Houston ML, Harris JWS. Interactions of the trophoblast and maternal tissues in three closely related primate species. Am J Obstet Gynecol 1976;124:647–52.
4. Tuttle SE, O'Toole RV, O'Shaughnessy RW, Zuspan FP. Immunochemical evaluation of human placental implantation: an initial study. Am J Obstet Gynecol 1985;153:239–44.
5. Liotta LA, Rao CN, Wewer UM. Biochemical interactions of tumor cells with the basement membrane. Annu Rev Biochem 1986;55:1037–58.
6. Liotta LA, Stetler-Stevenson WG. Tumor invasion and metastasis: an imbalance of positive and negative regulation. Cancer Res 1991;51(18 suppl):5054s-9s.
7. Alexander CM, Werb Z. Proteinases and extracellular matrix remodeling. Curr Opin Cell Biol 1989;1:974–82.
8. Matrisian L. Metalloproteinases and their inhibitors in matrix remodeling. Trends Genet 1990;6:121–5.
9. Hendrix MJ, Seftor EA, Grogan TM, et al. Expression of type IV collagenase correlates with the invasion of human lymphoblastoid cell lines and pathogenesis in SCID mice. Mol Cell Probes 1992;6:59–65.
10. Kliman HJ, Nestler JE, Sermasi E, Sanger JM, Strauss JF. Purification, characterization and in vitro differentiation of cytotrophoblast from human term placentae. Endocrinology 1986;118:1567–82.
11. Charbonneau H, Tonks NK, Kumar S, et al. Human placenta protein-tyrosine-phosphatase: amino acid sequence and relationship to a family of receptor-like proteins. Proc Natl Acad Sci USA 1989;86:5252–6.
12. Fisher SJ, Cui T, Zhang L, et al. Adhesive and degradative properties of human placental cytotrophoblast cells in vitro. J Cell Biol 1989;109:891–902.

13. Queenan J, Kao L, Arboleda CE, et al. Regulation of urokinase-type plasminogen activator production by cultured human cytotrophoblasts. J Biol Chem 1987;262:10903-6.
14. Librach C, Werb Z, Fitzgerald ML, et al. 92-kd type IV collagenase mediates invasion of human cytotrophoblasts. J Cell Biol 1991;112:437-49.
15. Damsky CH, Fitzgerald ML, Fisher SJ. Distribution patterns of extracellular matrix components and adhesion receptors are intricately modulated during first trimester cytotrophoblast differentiation along the invasive pathway, in vivo. J Clin Invest 1992;89:210-22.
16. Roberts J, Taylor RN, Friedman SA, Goldfien A. New developments in pre-eclampsia. In: Dunlop W, ed. Fetal medical review. London: Edward Arnold, 1990.
17. de Waal Malefyt R, Abrams J, Bennett B, Figdor CG, de Vries JE. Interleukin 10 (IL-10) inhibits cytokine synthesis by human monocytes: an autoregulatory role of IL-10 produced by monocytes. J Exp Med 1991;174:1209-20.
18. Zlotnik A, Moore KW. Interleukin 10. Cytokine 1991;3:366-71.
19. Kovats S, Main EK, Librach C, Stubblebine M, Fisher SJ, DeMars R. A class I antigen, HLA-G, expressed in human trophoblasts. Science 1990;248:220-3.
20. Ellis SA, Sargent IL, Redman CW, McMichael AJ. Evidence for a novel HLA antigen found on human extravillous trophoblast and a choriocarcinoma cell line. Immunology 1986;59:595-601.
21. Ellis SA, Palmer MS, McMichael AJ. Human trophoblast and the choriocarcinoma cell line BeWo express a truncated HLA class I molecule. J Immunol 1990;144:731-5.
22. Ishitani A, Geraghty DE. Alternative splicing of HLA-G transcripts yields proteins with primary structures resembling both class I and class II antigens. Proc Natl Acad Sci USA 1992;89:3947-51.
23. Koller BH, Geraghty DE, DeMars R, Duvick L, Rich SS, Orr HT. Chromosomal organization of the human major histocompatibility complex class I gene family. J Exp Med 1989;169:469-80.
24. Geraghty DE, Koller BH, Orr HT. A human major histocompatibility complex class I gene that encodes a protein with a shortened cytoplasmic segment. Proc Natl Acad Sci USA 1987:84:9145-55.
25. Starkey PM, Sargent IL, Redman CW. Cell populations in human early pregnancy decidua: characterization and isolation of large granular lymphocytes by flow cytometry. Immunology 1988;65:129-34.
26. Kovats S, Librach C, Fisch P, et al. Expression and possible function of the HLA-G α chain in human cytotrophoblasts. In: Chaouat J, Mowbray J, eds. Cellular and molecular biology of the materno-fetal relationship; vol 212. New York: John Libbey Eurotext Ltd, 1991:21-9.
27. Ladines-Llave CA, Maruo T, Manalo AS, Mochizuki M. Cytologic localization of epidermal growth factor and its receptor in developing human placenta varies over the course of pregnancy. Am J Obstet Gynecol 1991;165:1377-82.
28. Mirlesse V, Alsat E, Fondacci C, Evain-Brion D. Epidermal growth factor receptors in cultured human trophoblast cells from first- and third-trimester placentas. Horm Res 1990;34:234-9.

29. Hofmann GE, Scott RT, Bergh PA, Deligdisch L. Immunohistochemical localization of epidermal growth factor in human endometrium, decidua, and placenta. J Clin Endocrinol Metab 1991;73:882–7.
30. Kauma S, Matt D, Strom S, Eierman D, Turner T. Interleukin-1β (IL-1β), HLA-DRα and transforming growth factor-β (TGF-β) expression in endometrium, placenta, and placental membranes. Am J Obstet Gynecol 1990;163:1430–7.
31. Dungy LJ, Siddiqui TA, Khan S. Transforming growth factor-β1 expression during placental development. Am J Obstet Gynecol 1991;165:853–7.
32. Wang C, Daimon M, Shen S, Engelmann GL, Ilan J. Insulin-like growth factor-1 messenger ribonucleic acid in the developing human placenta and in term placenta of diabetics. Mol Endocrinol 1988;2:217–29.
33. Ohlsson R, Larsson E, Nilsson O, Wahlstrom T, Sundstrom P. Blastocyst implantation precedes induction of insulin-like growth factor II gene expression in human trophoblasts. Development 1989;106:555–9.
34. Kauma SW, Aukerman SL, Eierman D, Turner T. Colony-stimulating factor-1 and c-fms expression in human endometrial tissues and placenta during the menstrual cycle and early pregnancy. J Clin Endocrinol Metabol 1991:73:746–51.
35. Saji F, Azuma C, Kimura T, Koyama M, Ohashi K, Tanizawa O. Gene expression of macrophage colony-stimulating factor and its receptor in human placenta and decidua. Am J Reprod Immunol 1990;24:99–104.
36. Boehm KD, Kelley MF, Ilan J, Ilan J. The interleukin 2 gene is expressed in the syncytiotrophoblast of the human placenta. Proc Natl Acad Sci USA 1989;86:656–60.
37. Kameda T, Noboru M, Sawai K, et al. Production of interleukin-6 by normal human trophoblast. Placenta 1990;11:205–13.
38. Nishino E, Matsuzaki N, Masuhiro K, et al. Trophoblast-derived interleukin-6 (IL-6) regulates human chorionic gonadotropin release through IL-6 receptor on human trophoblasts. J Clin Endocrinol Metab 1990;71:436–41.
39. Pampfer S, Daiter E, Barad D, Pollard JW. Expression of the colony-stimulating factor-1 receptor (c-fms proto-oncogene product) in the human uterus and placenta. Biol Reprod 1992;46:48–57.
40. Daiter E, Pampfer S, Yeung YG, Barad D, Stanley ER, Pollard JW. Expression of colony-stimulating factor-1 in the human uterus and placenta. J Clin Endocrinol Metab 1992;74:850–8.
41. Wilson EA, Jawad MJ, Vernon MW. Effect of epidermal growth factor on hormone secretion by term placenta in organ culture. Am J Obstet Gynecol 1984;579–80.
42. Maruo T, Matsuo H, Oishi T, Hayashi M, Nishino R, Mochizuki M. Induction of differentiated trophoblast function by epidermal growth factor: relation of immunohistochemically detected cellular epidermal growth factor receptor levels. J Clin Endocrinol Metab 1987;64:744–50.
43. Morrish DW, Bhardwaj D, Paras MT. Transforming growth factor β1 inhibits placental differentiation and human chorionic gonadotropin and human placental lactogen secretion. Endocrinology 1991;129:22–6.
44. Chirgwin J, Przybyla A, MacDonald R, Rutter W. Isolation of biologically active ribonucleic acid from sources enriched in ribonuclease. Biochemistry 1979;18:5294–9.

45. Rappolee DA, Mark D, Banda MJ, Werb Z. Wound macrophages express TGF-α and other growth factors in vivo. Analysis by mRNA phenotyping. Science 1988;24:708–12.
46. Rappolee DA, Brenner CA, Schultz R, Mark D, Werb Z. Developmental expression of PDGF, TGF-α, and TGF-β genes in preimplantation mouse embryos. Science 1988;241:1823–5.
47. Rappolee DA, Mark D, Werb Z. A novel method for studying mRNA phenotypes in single or small numbers of cells. J Cell Biochem 1988;39:1–11.
48. Fiorentino DF, Zlotnik A, Vieira P, et al. IL-10 acts on the antigen-presenting cell to inhibit cytokine production by Th1 cells. J Immunol 1991;146:3444–51.

13

Expression and Binding of Transforming Growth Factor βs in the Mouse Embryo and Uterus During the Periimplantation Period

S.K. Dey, S.K. Das, B.C. Paria, K.C. Flanders, and G.K. Andrews

The implantation process involves complex interactions between embryonic and uterine cells. The major events of the implantation process are (i) synchronized development of the preimplantation embryo into a blastocyst and establishment of the receptive uterus (1); (ii) escape of the embryo from immunological responses of the mother (2); (iii) increased endometrial capillary permeability at the site of the blastocyst apposition (1); (iv) localized decidualization of the endometrial stroma following blastocyst attachment (1, 3); and (v) controlled uterine invasion by trophoblasts (4). These events are a conglomerate of temporally and spatially regulated proliferation, differentiation, migration, and remodeling of heterogeneous cell types of both the embryonic and uterine tissues.

Activation and cleavage of the egg after fertilization, as well as differentiation of embryonic cells into *inner cell mass* (ICM) and *trophectoderm* (Tr) at the blastocyst stage, are critical features of the preimplantation embryo development. Furthermore, a landmark event in the embryo's life is its implantation into the receptive uterus at the blastocyst stage. Although both *progesterone* (P_4) and *estrogen* (E) are necessary for implantation in the mouse, the mechanism by which E initiates blastocyst implantation in the P_4-primed uterus is not yet completely clear.

In this species the first detectable sign of initiation of implantation (increased localized endometrial vascular permeability at the site of blastocyst apposition) occurs in the evening of day 4 of pregnancy (day 1 = vaginal plug). However, removal of gonadal steroids by ovariectomy before midmorning on day 4 results in blastocyst dormancy and delayed implantation. While delayed implantation can be maintained by P_4 treatment, an injection of E in the P_4-primed animal will reactivate the blastocyst and initiate its implantation (5, 6). The mechanisms that direct proliferation and differentiation of the preimplantation embryo and its implantation are poorly understood. Since the documentation of expression of several polypeptide growth factors and their receptors in the preimplantation embryo and uterus, the participation of growth factors as autocrine/paracrine mediators in preimplantation embryo development and embryo-uterine interactions during implantation is being investigated.

Although several of the polypeptide growth factors are likely to be involved in embryonic development and implantation (7–10), *transforming growth factor βs* (TGFβs), a group of structurally homologous dimeric proteins, may play critical roles in these processes because of their diverse regulatory functions (11–13). Five isoforms of TGFβ (TGFβ 1–5) have been identified, of which TGFβ 1–3 are abundant in mammals. While studies of TGFβs on postimplantation embryogenesis are abundant (14–24), only a few studies on preimplantation embryo development and implantation have been reported (25–28). In order to gain more insight regarding the roles of TGFβs in preimplantation embryo development and implantation, we studied expression of mammalian isoforms of TGFβ as well as TGFβ binding in the embryo and uterus during the periimplantation period.

Expression of TGFβ Isoforms in the Preimplantation Embryo

Preimplantation mouse embryos have been shown to express TGFβ1 and TGFβ2 genes (25–29). However, detailed studies regarding the stage and

---

FIGURE 13.1. Immunocytochemical localization of TGFβ isoforms in preimplantation embryos developed in vivo. Embryos were freed of zonae by a brief exposure to 0.5% pronase solution and cytospun onto poly-L-lysine-coated slides. They were washed in phosphate-buffered saline (PBS) following fixation in Bouin's solution and incubated with primary antibodies to TGFβ1, TGFβ2, and TGFβ3 for 20 h at 4°C. Immunostaining employed the avidin-biotin-peroxidase complex (ABC) system. Dark deposits indicate sites of immunoreactive TGFβs. Immunostaining of TGFβ 1–3 is represented by photomicrographs in the left, middle, and right columns, respectively (400×). Rows from top to bottom depict embryos at 1-cell, 2-cell, 4-cell, 8-cell, morula, and day 4 blastocyst stages, respectively. Greatly reduced or no immunostaining was observed when embryos were incubated in specific primary antibodies preneutralized with excess specific antigenic peptides (data not shown). Reprinted with permission from Paria, Jones, Flanders, and Dey (37).

13. Expression and Binding of Transforming Growth Factor βs 197

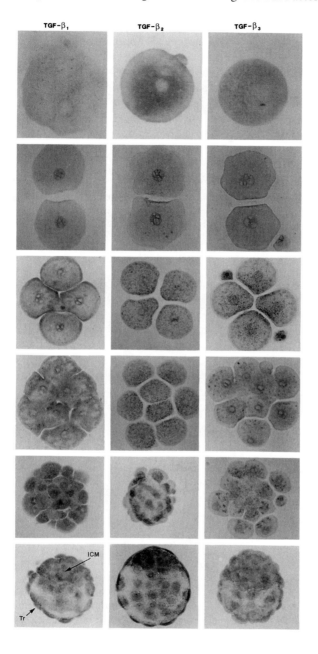

cell-specific distribution and production of all 3 mammalian isoforms of TGFβ in the embryo throughout the preimplantation period had been limited. Thus, the stage and cell-specific accumulation of mammalian isoforms of TGFβ (TGFβ 1–3) was examined in the preimplantation embryo, as well as in P_4-treated delayed or P_4 plus *estradiol-17β* (E_2) treated activated blastocysts in the mouse (5). Immunocytochemical studies that used rabbit polyclonal antipeptide antibodies specific to TGFβ 1–3 (30–32) revealed that while all 3 immunoreactive TGFβ isoforms were present in 1-cell embryos, very little or no immunostaining was observed in 2-cell embryos (Fig. 13.1). However, distinct immunostaining of these isoforms was again observed in 4-cell embryos and persisted through the blastocyst stage. TGFβ2 immunostaining showed a unique pattern in morulae; the staining was primarily observed

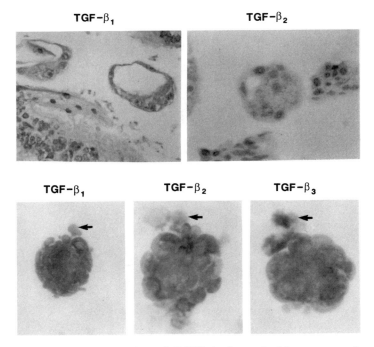

FIGURE 13.2. Immunolocalization of TGFβ isoforms in blastocyst sections and immunosurgically isolated ICMs. Day 4 blastocysts were transferred into oviducts, fixed in Bouin's solution, paraffin-embedded, sectioned, and mounted onto poly-L-lysine-coated slides. Immunostaining was performed as described in Figure 13.1. ICMs from blastocysts were isolated by immunosurgery, as described in reference 37. ICMs were fixed in Bouin's solution, washed in PBS, and cytospun onto slides for immunostaining. The top and bottom rows are microphotographs of blastocyst and ICM sections, respectively (400×). Arrows indicate remnants of Tr cells that adhered to the ICM. Reprinted with permission from Paria, Jones, Flanders, and Dey (37).

| TGF-β_1 | TGF-β_2 | TGF-β_3 |

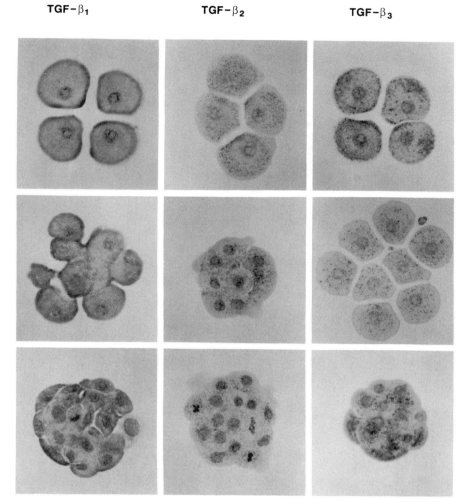

FIGURE 13.3. Immunolocalization of TGFβ isoforms in preimplantation embryos grown in vitro. Two-cell embryos (10–15/group) were cultured in 25-μL medium (37). Zona-free embryos at different developmental stages were processed for immunostaining. Immunostaining of TGFβ 1–3 is represented by photomicrographs in the left, middle, and right columns, respectively (400×). Rows from top to bottom represent embryos at 4-cell, 8-cell, and morula stages, respectively. Immunostaining of TGFβ isoforms in cultured blastocysts was comparable to that in day 4 blastocysts developed in vivo (data not shown). Reprinted with permission from Paria, Jones, Flanders, and Dey (37).

in outside cells (Fig. 13.1). However, in blastocysts immunostaining for all 3 isoforms was present both in the ICM and Tr, as confirmed by immunostaining in sectioned blastocysts and immunosurgically isolated ICMs (Fig. 13.2).

To ascertain whether preimplantation embryos can produce TGFβs, immunostaining was performed in embryos grown in vitro from the 2-cell

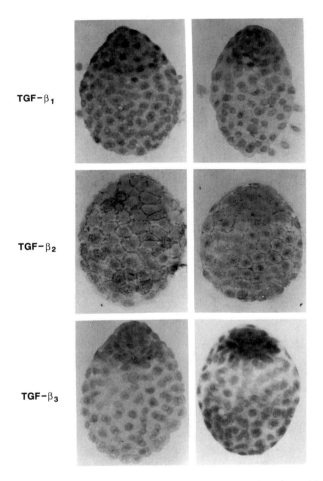

FIGURE 13.4. Immunolocalization of TGFβs in delayed and activated blastocysts. Delayed blastocysts were obtained from mice ovariectomized on day 4 of pregnancy and treated with P for 4 days, while activated blastocysts were obtained from P-primed mice 24 h after an injection of E_2. Immunostaining in delayed and activated blastocysts is shown in photomicrographs in the left and right columns, respectively (400×). Rows from top to bottom represent immunostaining of TGFβ1, TGFβ2, and TGFβ3, respectively. Reprinted with permission from Paria, Jones, Flanders, and Dey (37).

stage in Whitten's medium supplemented with 0.3% bovine serum albumin (33). Immunoreactive TGFβ 1–3 were present in embryos at all stages of development examined (4-cell embryos through blastocysts) (Fig. 13.3). The absence of immunoreactive TGFβs in 2-cell embryos, but their accumulation in embryos at later stages of development in vitro suggests that these growth factors are produced by embryos. The status of immunoreactive TGFβ isoforms was examined in delayed and activated blastocysts. Although immunostaining of TGFβ1 and TGFβ3 was virtually absent in delayed blastocysts, staining of TGFβ3, but not of TGFβ1, reappeared primarily in cells at the abembryonic pole of the

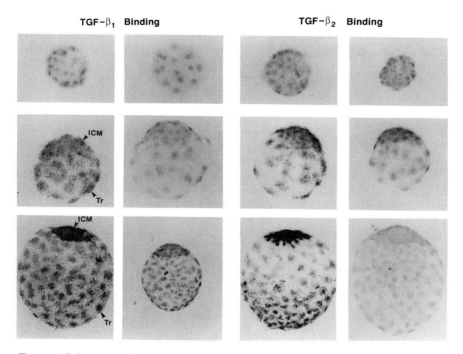

FIGURE 13.5. Autoradiographic localization of TGFβ1 and TGFβ2 binding in the preimplantation embryo. Zona-free embryos were incubated with 100 pM ^{125}I-TGFβ1 or 100 pM ^{125}I-TGFβ2 for 2 h in the presence or absence of a 500-fold molar excess of unlabeled TGFβ1 or TGFβ2, respectively. They were washed, cytospun onto poly-L-lysine-coated slides, fixed in paraformaldehyde, processed for autoradiography, and exposed for 7 days. After development embryos were poststained in hematoxylin. Autoradiographic signals are represented by black grains under bright field. Photomicrographs of TGFβ1 and TGFβ2 binding are shown in the left and right panels, respectively (400×). Rows from top to bottom represent morula, day 4 blastocyst, and P-treated delayed blastocyst stages, respectively. The right column of each panel represents nonspecific binding. (ICM = inner cell mass; Tr = trophectoderm.) Reprinted with permission from Paria, Jones, Flanders, and Dey (37).

blastocyst following E_2 activation. On the other hand, distinct TGFβ2 immunostaining was evident in the delayed, and with a slightly lower intensity, in the activated blastocyst (Fig. 13.4).

TGFβ Binding to the Preimplantation Embryo

No information was available regarding TGFβ binding (receptor) to the preimplantation embryo. In order to assess at which stages of development preimplantation mouse embryos could be responsive to TGFβs, specific binding of ^{125}I-TGFβ1 and ^{125}I-TGFβ2 was performed in embryos and examined by autoradiography. Low levels of binding were first detected in 8-cell embryos. The binding increased in morulae, followed by a further increase in blastocysts (Fig. 13.5). Analysis of

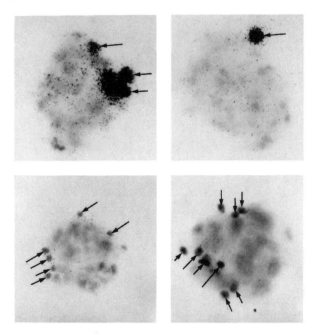

FIGURE 13.6. Autoradiographic localization of TGFβ2 binding in the ICM. ICMs were isolated by immunosurgery of blastocysts. Isolated ICMs were incubated with 100 pM ^{125}I-TGFβ2 in the presence or absence of a 500-fold molar excess of unlabeled TGFβ2 and processed for autoradiography. Autoradiographic exposure was for 8 days. Bright-field photomicrographs of binding in 2 representative ICMs are shown in the top row, while those of nonspecific binding are depicted in the bottom row (400×). Arrows indicate remnants of Tr cells that adhered to isolated ICMs. Note intense labeling of these areas as compared to ICM cells. Darkly stained bodies appear to be pycnotic nuclei of damaged Tr cells. Reprinted with permission from Paria, Jones, Flanders, and Dey (37).

binding of ^{125}I-TGFβ2 in immunosurgically isolated ICMs indicated that binding was primarily abundant in Tr cells (Fig. 13.6). Crosslinking experiments using ^{125}I-TGFβ1 (data not shown) or ^{125}I-TGFβ2 in day 4 blastocysts revealed 3 size classes of binding proteins with approximate molecular sizes of 65 kd (type I), 90 kd (type II), and >250 kd (type III), in addition to a doublet of 130- and 140-kd proteins (Fig. 13.7). This observation is similar to those reported for other cell types (34–36). The specific binding of ^{125}I-TGFβ1 and ^{125}I-TGFβ2, as determined by autoradiography, persisted in delayed and activated blastocysts (Fig. 13.5). The data suggest that embryos are likely to be responsive to TGFβs after the third cleavage and that specific isoforms of TGFβ may have distinct functions in blastocysts at their different states of activity.

Expression of TGFβ Isoforms in the Periimplantation Uterus

Although TGFβ has major roles in cell proliferation and differentiation, limited information was available regarding the expression of TGFβ genes in the uterus around the time of implantation. Immunohistochemistry and in situ and Northern blot hybridization were employed to determine

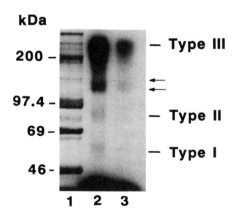

FIGURE 13.7. Crosslinking of TGFβ2 to day 4 blastocysts. Blastocysts (80 embryos/group) were incubated with 80 pM of ^{125}I-TGFβ2 in the absence (lane 2) or presence (lane 3) of a 500-fold molar excess of unlabeled TGFβ2 followed by crosslinking with disuccinimidyl suberate. The proteins were separated on 7.5% polyacrylamide gel that was dried and exposed for autoradiography. Molecular weight markers are shown in lane 1. Approximate molecular sizes of the binding proteins are: 65 kd (type 1), 90 kd (type II), and >250 kd (type III). Arrows indicate a doublet of 130- and 140-kd proteins. Reprinted with permission from Paria, Jones, Flanders, and Dey (37).

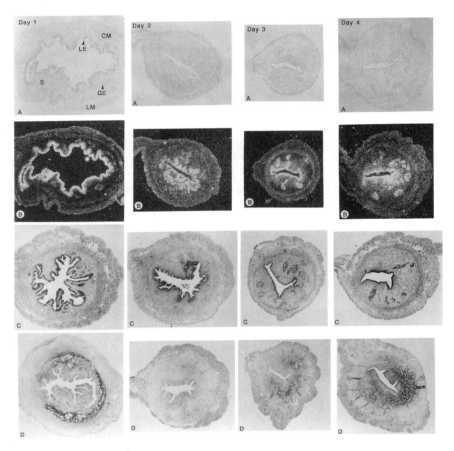

FIGURE 13.8. In situ hybridization of TGFβ1 mRNA and immunolocalization of TGFβ1 in the mouse uterus on days 1–4 of pregnancy. In situ hybridization: Uteri on indicated days of pregnancy were perfusion fixed with 4% paraformaldehyde in PBS. Paraffin-embedded tissues were sectioned at 7 μm and placed onto poly-L-lysine-coated slides. TGFβ1 mRNA was hybridized in situ for 5 h at 42°C with a ^{35}S-labeled human TGFβ1 cRNA probe, and RNase A-resistant hybrids were detected after 4 days of autoradiography using Kodak NTB-2 liquid emulsion. Slides were poststained lightly in hematoxylin. Shown are bright-field (A) and dark-field (B) photomicrographs (40×) of uterine sections on specific days of pregnancy. (LE = luminal epithelium; GE = glandular epithelium; S = stroma; LM = longitudinal muscle; CM = circular muscle.) Immunohistochemistry: Bouin's-fixed paraffin-embedded uterine sections (7 μm) from specific days of pregnancy were mounted onto slides. After deparaffinization and hydration, sections were incubated in primary antibodies for 24 h at 4°C. Immunostaining was performed by employing the ABC technique. Photomicrographs (40×) of immunostaining by anti-LC (intracellular staining) and anti-CC (extracellular staining) are represented by C and D, respectively, for each day of pregnancy. Reprinted with permission from Tamada, McMaster, Flanders, Andrews, and Dey (9).

temporal and cell type-specific expression of TGFβ 1–3 in the periimplantation mouse uterus (9, 38). Two rabbit polyclonal antipeptide antibodies were used to detect intracellular (anti-LC) and extracellular (anti-CC) TGFβ1 (30).

The co-localization of intracellular TGFβ1 protein with the mRNA in the luminal and glandular epithelia on days 1–4 of pregnancy indicates that the epithelial cells are the major sites of TGFβ1 synthesis in the uterus during the preimplantation period (Fig. 13.8A). In contrast, staining of the extracellular matrix of the stroma by anti-CC during this period suggests an active accumulation of TGFβ1 that is produced and secreted by the epithelia (Fig. 13.8). The extracellular staining of the stroma noted on day 1 was absent on day 2, but was again observed in the stroma on day 3, and further increased on day 4 (Fig. 13.8). On day 5 (early postimplantation period), the intracellular staining was limited to the luminal epithelium and *primary decidual zone* (PDZ), whereas an intense extracellular staining was observed in the decidualizing stroma around the PDZ (Fig. 13.9). On days 6 and 7, the PDZ still showed intracellular staining, but the extracellular staining was present in the *secondary decidual zone* (SDZ) on day 6 and in the decidua capsularis on day 7. The extracellular staining persisted in the decidua capsularis on day 8, while the intracellular staining became diffuse in the regressing decidua (Figs. 13.9 and 13.10). TGFβ1 mRNA was diffusely distributed throughout the decidua on days 5–8 (Figs. 13.8 through 13.10). However, the presence of the intracellular immunoreactive TGFβ1 in the PDZ suggests that this zone is the major site of TGFβ1 synthesis, while extracellular immunostaining in the SDZ and decidua capsularis are indicative of the sites of accumulation. Northern blot hybridization of total RNA confirmed the authenticity of the 2.4-kb TGFβ1 transcript in uteri and decidua on days 1–8. The results suggest that TGFβ1 could be important in regulating epithelial-stromal interactions during the preimplantation period and PDZ-SDZ interactions during the early postimplantation period.

In the preimplantation uterus (days 1–4 of pregnancy), TGFβ2 immunostaining was present in the luminal and glandular epithelia, as well as in the myometrium and vascular smooth muscle. During the early postimplantation period (days 5–8), immunoreactive TGFβ2 was also detected in decidual cells (Fig. 13.11). In contrast, TGFβ3 immunostaining was restricted to the myometrium and vascular smooth muscle throughout the periimplantation period (days 1–8) (Fig. 13.12). Northern blot hybridization showed 4 TGFβ2 transcripts ($\simeq$6.0, 5.0, 4.0, and 3.5 kb) in total uterine poly(A)$^+$ RNA on days 1–6 and in poly(A)$^+$ RNA from the deciduum and myometrium collected on days 7 and 8 of pregnancy (Fig. 13.13). These TGFβ2 transcripts were also detected in isolated samples of deciduomata or myometrium obtained from day 8 pseudopregnant mice in which the decidual cell reaction was induced by intraluminal oil injection on day 4 (Fig. 13.13).

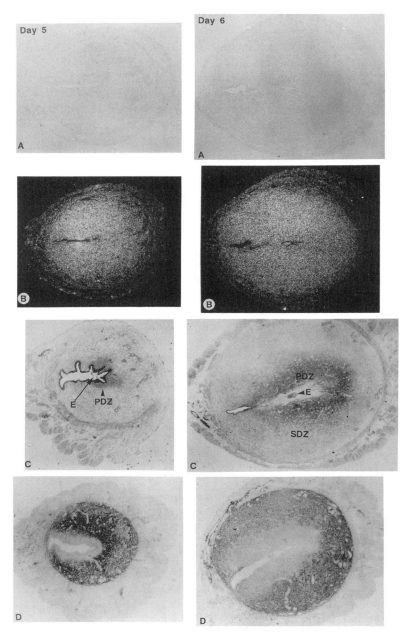

FIGURE 13.9. In situ hybridization of TGFβ1 mRNA and immunolocalization of TGFβ1 in sections of implantation sites on days 5 and 6 of pregnancy. In situ hybridization: Shown are bright-field (A) and dark-field (B) photomicrographs (40×) of sections of implantation sites. Autoradiographic exposure was for 8 days. Immunohistochemistry: C and D represent photomicrographs (40×) of immunostaining in sections of implantation sites by anti-LC and anti-CC, respectively. (E = embryo; PDZ = primary decidual zone; SDZ = secondary decidual zone.) The left and right sides of each photomicrograph represent mesometrial and antimesometrial poles, respectively. Reprinted with permission from Tamada, McMaster, Flanders, Andrews, and Dey (9).

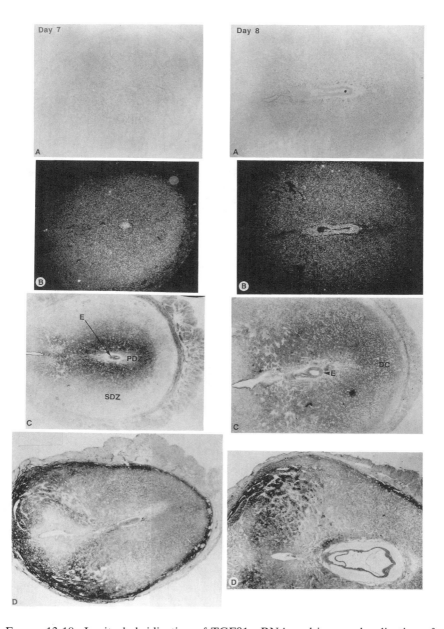

FIGURE 13.10. In situ hybridization of TGFβ1 mRNA and immunolocalization of TGFβ1 in sections of implantation sites on days 7 and 8 of pregnancy. In situ hybridization: Shown are bright-field (*A*) and dark-field (*B*) photomicrographs (40×) of sections of implantation sites. Autoradiographic exposure was for 8 days. Immunohistochemistry: *C* and *D* represent photomicrographs (40×) of immunostaining in sections of implantation sites by anti-LC and anti-CC, respectively. (E = embryo; PDZ = primary decidual zone; SDZ = secondary decidual zones; DC = decidua capsularis.) The orientation of mesometrial and antimesometrial poles are the same as in Figure 13.9. Reprinted with permission from Tamada, McMaster, Flanders; Andrews, and Dey (9).

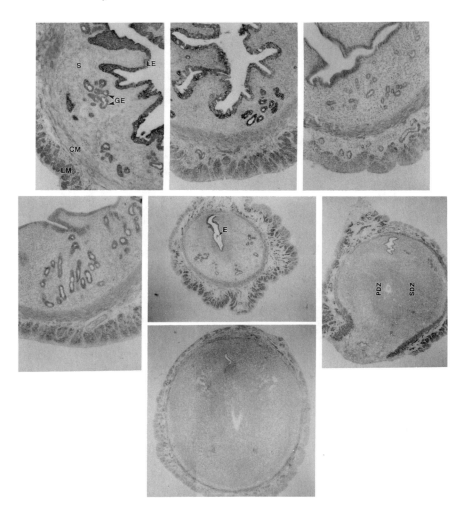

FIGURE 13.11. Immunolocalization of TGFβ2 in the mouse uterus on days 1–8 of pregnancy. Bouin's-fixed paraffin-embedded sections (7 μm) were mounted onto poly-L-lysine-coated slides. After deparaffinization and hydration, sections were incubated with primary antibody. Immunostaining employed the ABC technique in which red deposits indicate positive immunostaining. The top row from left to right represents photomicrographs of uterine sections on days 1–3, respectively. The middle row from left to right represents those on days 4–6, respectively. The photomicrograph in the bottom row at center represents a uterine section on day 8 of pregnancy. Photomicrographs of days 1–4 uterine sections were at 100× and of days 5, 6, and 8 were at 40×. Although not shown here, the immunostaining pattern for TGFβ2 in day 7 uterine sections was similar to that in day 6 or day 8 sections. Reprinted with permission from Das, Flanders, Andrews, and Dey (38), © by The Endocrine Society, 1992.

13. Expression and Binding of Transforming Growth Factor βs 209

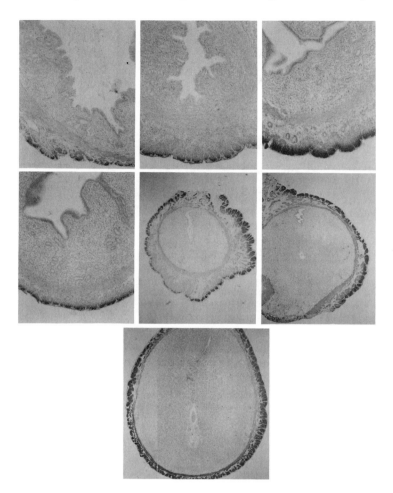

FIGURE 13.12. Immunocytochemical localization of TGFβ3 in the mouse uterus on days 1–8 of pregnancy. Immunolocalization of TGFβ3 was performed as described in Figure 13.9. Reprinted with permission from Das, Flanders, Andrews, and Dey (38), © by The Endocrine Society, 1992.

The levels of these transcripts remained relatively constant during the periimplantation period. Northern blot analysis detected a 3.8-kb TGFβ3 transcript in total uterine poly(A)$^+$ RNA on days 1–6 (Fig. 13.13). This transcript was detected in myometrial RNA samples on days 7 and 8 of pregnancy or day 8 of pseudopregnancy, but was not detected in RNA from the deciduum on these days or in deciduomata on day 8 (Fig. 13.13). The results establish that TGFβ2 and TGFβ3 genes are expressed in the periimplantation uterus. Epithelial, myometrial, and decidual cells

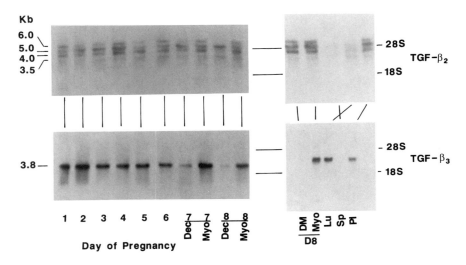

FIGURE 13.13. Northern blot detection of TGFβ2 and TGFβ3 mRNAs in the mouse uterus during the periimplantation period (days 1–8). Whole uterine poly(A)$^+$ RNA from days 1–6 or deciduum (Dec) and myometrium (Myo) on days 7 or 8 of pregnancy was analyzed (left panels). Poly(A)$^+$ RNA was also obtained from deciduomata (DM) and myometrium (Myo) on day 8 of pseudopregnancy (right panels). Deciduomata was induced experimentally by injection of 50-μL sesame oil into the uterine lumen on day 4 of pseudopregnancy. Poly(A)$^+$ RNA (2.0 μg) was separated by formaldehyde-agarose gel electrophoresis, transferred to nylon membrane, and hybridized to ^{32}P-labeled RNA probes complementary to mouse TGFβ2 and TGFβ3 mRNAs. RNA samples in duplicate gels were stained with acridine orange, and the mobilities of the small amounts of 28S and 18S ribosomal RNAs remaining in the poly(A)$^+$ RNA samples are indicated. Four major TGFβ2 transcripts (approximate sizes of 6.0, 5.0, 4.0, and 3.5 kb) and a single 3.8-kb TGFβ3 transcript were detected. Poly(A)$^+$ RNA samples from mouse lung (Lu) and placenta (Pl) served as positive controls and from spleen (Sp) as a negative control. Reprinted with permission from Das, Flanders, Andrews, and Dey (38), © by The Endocrine Society, 1992.

are primary sites of TGFβ2 synthesis, whereas myometrial cells are the primary site of synthesis of TGFβ3 during the periimplantation period. The distinct uterine cell type-specific expression of TGFβ 1–3 suggests that these growth factors may have different functional roles during the periimplantation period.

TGFβ Binding to Uterine Membranes

Specific binding of TGFβ isoforms to uterine membrane preparations was examined using ^{125}I-TGFβ1, ^{125}I-TGFβ2, or ^{125}I-TGFβ3 as ligands. Binding was performed on isolated decidual and myometrial membrane preparations on day 8 following intraluminal oil injection on day 4 of pseudopregnancy (34, 37). Scatchard analysis showed a single class of binding sites for TGFβ1 in myometrial membrane preparation (Kd = 3.6 nM) as compared to 2 classes of binding sites in decidual membranes (Kd_1 = 0.9 nM; Kd_2 = 5.6 nM). In contrast, a single class of binding sites for TGFβ2 was noted in decidual membranes (Kd = 1.02 nM) as compared to 2 classes of binding sites in myometrial preparations (Kd_1 = 0.32 nM; Kd_2 = 4.7 nM). Interestingly, 2 classes of binding sites were noted for TGFβ3 in both myometrial (Kd_1 = 2.6 nM; Kd_2 = 7.9 nM) and decidual (Kd_1 = 0.8 nM; Kd_2 = 7.6 nM) membranes. Crosslinking of TGFβ binding revealed 3 size classes of binding proteins (types I, II, and III) in the uterus during the periimplantation period.

Comments

The importance of TGFβ isoforms in proliferation and differentiation, as well as in the formation of extracellular matrix and cell-surface molecules, is well established. Thus, these growth factors are likely to participate in various events of preimplantation embryo development, implantation, decidualization, placentation, and embryogenesis. The findings from our laboratory and others (25, 27, 28) provide evidence that mouse preimplantation embryos produce all 3 mammalian isoforms of TGFβ (38). Maternal to zygotic transition occurs in the mouse at the 2-cell stage (39, 40). The presence of immunoreactive TGFβ isoforms in 1-cell embryos and their absence in 2-cell embryos suggest that in 1-cell embryos these growth factors are the products of maternal transcription, whereas their accumulation in 4-cell and later stages of development apparently results from activation of the embryonic genome. Although TGFβ1 affects preimplantation embryo development in vitro beneficially (7), whether TGFβ2 or TGFβ3 has similar and/or different functions will require further examination. The absence of detectable TGFβ binding before the embryo reaches the 8-cell stage suggests that embryos are not responsive to TGFβs prior to the third cleavage. This raises questions regarding the implications of the accumulation of these growth factors during early cleavages. If secreted and processed to mature forms, they may, in a paracrine fashion, influence the oviduct where embryos at these stages of development are being nurtured.

As TGFβ binding was most abundant in the morula and blastocyst, the functions of embryonic TGFβs at these stages of development may be to influence morula to blastocyst transformation (compaction and blastocoel formation) and maturation of the blastocyst (blastocoel expansion) in an autocrine manner. Because of TGFβ's diverse roles (13, 41, 42), these growth factors alone or in combination with other growth factors may direct and coordinate the proliferation and differentiation in the embryo in creating the differentiated blastocyst. This is consistent with the findings that TGFβ1 plus fibroblast growth factor, or TGFβ2 alone, are important morphogens for mesoderm induction in frog blastulae (43, 44). Localization of TGFβ2 in outside cells of the morula may have a special connotation in morula to blastocyst transformation since outside cells give rise to the Tr, while the inside cells contribute to the ICM in the process of blastocyst formation (45). Another function of embryonic TGFβs could be to participate in embryo-uterine interactions during implantation in a paracrine fashion.

Because of the overlapping actions of TGFβ isoforms (13, 46), neither distinctive functions nor the relative importance of specific isoforms in preimplantation embryo development can be easily speculated. However, the model of delayed implantation provides some clues as to the regulation and specific functions of TGFβ isoforms in the blastocyst. The disappearance of TGFβ1 from the blastocyst during delayed implantation and its failure to return in the activated blastocyst suggest that this isoform may be involved in the induction of blastocyst dormancy but not in its activation.

On the other hand, the presence of immunoreactive TGFβ2 in both dormant and activated blastocysts suggests that this isoform could be important for maintaining the blastocyst in a differentiated and responsive state so that its activation and subsequent implantation can occur in the uterus following E-treatment. Persistence of TGFβ binding to blastocysts under these conditions is consistent with this speculation. This is in contrast to EGF binding to the blastocyst, which disappears rapidly following the induction of delay, but reappears in the activated blastocyst (to be published). The absence of immunoreactive TGFβ3 in delayed blastocysts and its reappearance in activated blastocysts are suggestive of a role for this isoform in the induction of blastocyst dormancy and activation. Since the abembryonic cells of the Tr make the first association with the uterine luminal epithelium and initiate the process of implantation, the accumulation of immunoreactive TGFβ3 at this region of the activated blastocyst after E-treatment may play a role in embryo-uterine interactions during initiation of implantation.

In regard to expression of TGFβs in the uterus, our studies demonstrate that all 3 mammalian TGFβ isoforms (TGFβ 1–3) are expressed in the periimplantation mouse uterus in a temporal and cell type-specific manner. Expression of TGFβ1 is restricted mostly to the luminal and glandular epithelia and to the deciduum during the periimplantation period (9). The pattern of TGFβ2 gene expression is similar to that of TGFβ1, except that the TGFβ2 gene is also expressed in the uterine myometrium. However, the TGFβ3 gene is expressed exclusively in the myometrium. The significance of multiple TGFβ2 transcripts in the uterus is not yet clearly understood. The observed absence of 6.0- and 4.0-kb TGFβ2 transcripts in decidua but not in deciduomata suggests that the presence of the embryo may influence the levels of these transcripts.

The differential expression of TGFβ isoforms in various uterine cell types during early pregnancy suggests that in addition to overlapping functions (47), each TGFβ isoform may have a unique role in embryo-uterine interactions during the process of implantation. Differential binding affinities of TGFβ isoforms to various uterine cell types are consistent with this suggestion. Preparation of the uterus to the receptive state for implantation, as well as decidualization of the stroma after the initiation of implantation, involve well-orchestrated proliferation and differentiation of various uterine cell types, tissue remodeling, and extracellular matrix formation (9, 10, 48, 49). Thus, TGFβs are likely to play important roles in these events of the implantation process.

These growth factors may also be important in neutralizing immunorejection of the *allogenic* embryo by the maternal immune system and modulating the inflammatory responses to mating or the distribution and activation of inflammatory cells in the uterus and/or deciduum (50, 51). Furthermore, because of their role in mesoderm induction in *Xenopus* embryos (43, 44), uterine-derived TGFβs could be important in early mammalian embryogenesis (17, 20–23). The expression of TGFβ1 and TGFβ2 in the uterine epithelium and deciduum is consistent with this suggestion.

In contrast, the selective expression of TGFβ3 in the uterine myometrium suggests more restricted functions for this TGFβ isoform. Although participation of TGFβs in angiogenesis, inhibition of vascular smooth muscle cell proliferation, and stimulation of cellular hypertrophy has been suggested (52–54), expression of TGFβ isoforms in smooth muscle cells has not been reported previously. Several lines of evidence suggest a role for TGFβ3 in myogenesis. Levels of TGFβ3 mRNA are high in embryonic and adult cardiac muscles (15, 55, 56), skeletal muscle, and skeletal myoblast cell lines (56). Furthermore, TGFβ3 mRNA levels rise following myoblast fusion. However, all 3 isotypes of this growth factor can inhibit myoblast fusion, and TGFβ1 can inhibit expression of myogenin, a muscle-specific gene (56–59). Interestingly, loss of TGFβ receptors occurs with muscle cell differentiation, which suggests that differentiated muscle cells are not responsive to TGFβ (58).

The 3 different size classes of TGFβ binding proteins in the blastocyst and uterus are similar to those identified in other cell types (34, 60–65). Therefore, the biological effects of TGFβ in the preimplantation embryo and uterus are likely to be mediated by these binding proteins (receptors). That type I TGFβ binding protein appears to mediate TGFβ actions is evident from the unresponsiveness to TGFβ of mutant cell lines lacking this binding protein (66, 67). It is also known from other studies that type I and type II binding proteins are not disulfide linked to other binding subunits and have higher affinity for TGFβ1 than for TGFβ2 (61,

62). On the other hand, type III binding protein, known as TGFβ proteoglycans, binds TGFβ1 and TGFβ2 with equal affinity (61, 62) and may serve as storage for extracellular TGFβ (68). Furthermore, recent studies suggest that TGFβ3 also associates with these 3 types of cell-surface binding proteins (69, 70). Although the signal transduction system(s) that mediates TGFβ functions has not yet been fully explored, the recent cloning of type II and type III receptor cDNAs is likely to generate important information in this regard (71–73).

A comprehensive review, as described above, regarding the expression of TGFβ isoforms and their binding in the embryo and uterus during the periimplantation period suggests that these growth factors play unique autocrine/paracrine roles in preimplantation embryo development and in the implantation process. However, further consideration of the autocrine/paracrine roles of embryonic and/or uterine TGFβs will require evidence that these growth factors are, in fact, secreted and processed to mature forms by embryonic and/or uterine tissues.

Acknowledgments. This work was supported in part by grants from NICHD (HD-12304 to S.K.D.) and NIEHS (ES-04725 to G.K.A.). Thanks are due to Crystal Gore and Susan Strong for technical assistance and Sharon McCray for preparation of the manuscript.

References

1. Psychoyos A. Endocrine control of egg implantation. In: Greep RO, Astwood EG, Geiger SR, eds. Handbook of physiology. Washington, DC: American Physiological Soc, 1973:187–215.
2. Beer AE, Billingham RE. Immunoregulatory aspects of pregnancy. Fed Proc 1978;37:2374–8.
3. DeFeo VJ. Decidualization. In: Wynn RM, ed. Cellular biology of the uterus. Amsterdam: North Holland, 1967:191–290.
4. Kirby DRS, Cowell TP. Trophoblast-host interactions. In: Fleischmeyer R, Billingham RE, eds. Epithelial-mesenchymal interactions. Baltimore, MD: Williams and Wilkins, 1968:64–77.
5. Huet-Hudson YM, Dey SK. Requirement for progesterone priming and its long-term effects on implantation in the mouse. Proc Soc Exp Biol Med 1990;193:259–63.
6. Yoshinaga K, Adams CE. Delayed implantation in the spayed progesterone treated adult mouse. J Reprod Fertil 1966;12:593–5.
7. Paria BC, Dey SK. Preimplantation embryo development in vitro: cooperative interactions among embryos and role of growth factors. Proc Natl Acad Sci USA 1990;87:4756–60.
8. Paria BC, Tsukamura H, Dey SK. Epidermal growth factor-specific protein tyrosine phosphorylation in preimplantation embryo development. Biol Reprod 1991;45:711–8.

9. Tamada H, McMaster MT, Flanders KC, Andrews GK, Dey SK. Cell type-specific expression of transforming growth factor-β1 in the mouse uterus during the periimplantation period. Mol Endocrinol 1990;4:965–72.
10. Tamada H, Das SK, Andrews GK, Dey SK. Cell type-specific expression of transforming growth factor-α in the mouse uterus during the periimplantation period. Biol Reprod 1991;45:365–72.
11. Massagué J. The transforming growth factor-β family. Annu Rev Cell Biol 1990;6:597–641.
12. Moses HL, Tucker RF, Leof EB, Coffey RJ, Halper J, Shipley GD. Type-beta transforming growth factor is growth stimulator and growth inhibitor. In: Feramisco J, Ozanne B, Stiles C, eds. Cancer cells; vol 3. Cold Spring Harbor, 1985:65–71.
13. Roberts AB, Sporn MB. Transforming growth factor-betas. In: Sporn MB, Roberts AB, eds. Handbook of experimental pharmacology; vol 95. Heidelberg: Springer, 1990: 419–72.
14. Akhurst RJ, Lehnert SA, Gatherer D, Duffie E. The role of TGF beta in mouse development. Ann NY Acad Sci 1990;593:259–71.
15. Denhez F, Lafayatis R, Kondaiah P, Roberts AB, Sporn MB. Cloning by polymerase chain reaction of a new mouse TGF-beta, mTGF-beta 3. Growth Factors 1990;13:139–46.
16. Fitzpatrick DR, Denhez F, Kondaiah P, Akhurst RJ. Differential expression of TGF beta isoforms in murine paletogenesis. Development 1990;109:585–95.
17. Gatherer D, ten Dijke P, Baird DT, Akhurst RJ. Expression of TGF-β isoforms during first trimester human embryogenesis. Development 1990;110:445–60.
18. Heine UI, Flanders KC, Roberts AB, Minoz EF, Sporn MB. Role of transforming growth factor-β in the development of the mouse embryo. J Cell Biol 1987;105:2861–76.
19. Lehnert SA, Akhurst RJ. Embryonic expression pattern of TGF beta type-1 RNA suggests both paracrine and autocrine mechanisms of action. Development 1988;104:263–72.
20. Millan FA, Denhez F, Kondaiah P, Akhurst RJ. Embryonic gene expression patterns of TGF-β1, β2 and β3 suggest different developmental functions in vivo. Development 1991;111:131–44.
21. Miller DA, Lee A, Matsui Y, Chen EY, Moses HL, Derynck R. Complementary DNA cloning of murine transforming growth factor beta 3 (TGF-β3) precursor and the comparative expression of TGF-β1 and TGF-β3 and TGF-β1 in murine embryos and adult tissues. Mol Endocrinol 1989;3:1926–34.
22. Pelton RW, Nomura S, Moses HL, Hogan BLM. Expression of transforming growth factor beta-2 RNA during murine embryogenesis. Development 1989;106:759–67.
23. Pelton RW, Dickinson ME, Moses HL, Hogan BLM. In situ hybridization analysis of TGF β3 RNA expression during mouse development: comparative studies with TGF β1 and β2. Development 1990;110:609–20.
24. Wilcox JN, Derynck R. Developmental expression of transforming growth factors alpha and beta in the mouse fetus. Mol Cell Biol 1988;8:3415–22.

25. Kelly D, Campbell J, Tiesman I, Rizzino A. Regulation and expression of transforming growth factor type β during early mammalian development. Cytotechnology 1990;4:227–42.
26. Mummery CL, Slager HG, Kruijer W, et al. Expression of transforming growth factor-β2 during the differentiation of murine embryonal carcinoma and embryonic stem cells. Dev Biol 1990;137:161–70.
27. Rappolee DA, Brenner CA, Schultz R, Mark D, Werb Z. Developmental expression of PDGF, TGFα and TGFβ genes in preimplantation mouse embryos. Science 1988;241:1823–5.
28. Slager HG, Lawson KA, Van Den Eijnden-Van Raaij AJM, DeLaat SW, Mummery CL. Differential localization of TGF-β2 in mouse preimplantation and early postimplantation development. Dev Biol 1991;145:205–18.
29. Rappolee DA, Sturm KS, Schultz GA, Pedersen RA, Werb Z. The expression of growth factor ligands and receptors in preimplantation mouse embryos. In: Heyner S, Wiley LM, eds. Early embryo development and paracrine relationships. New York: Alan R. Liss, 1990:11–25.
30. Flanders KC, Thompson NL, Cissel DS, et al. Transforming growth factor-β1: histochemical localization with antibodies to different epitopes. J Cell Biol 1989;108:653–60.
31. Flanders KC, Cissel DS, Mullan LT, Danielpour D, Sporn MB, Roberts AB. Antibodies to transforming growth factor-β2 peptides: specific detection of TGF-β2 in immunoassays. Growth Factors 1990;3:45–52.
32. Flanders KC, Ludencke G, Engels S, et al. Localization and actions of transforming growth factors-βs in the embryonic nervous system. Development 1991;113:183–91.
33. Whitten WK. Nutritional requirements for the culture of preimplantation embryos in vitro. Adv Biosci 1971;6:129–39.
34. Massagué J, Like B. Cellular receptor for type β transforming factor. Ligand binding and affinity labeling in human and rodent cell lines. J Biol Chem 1985;260:2636–45.
35. MacKay K, Danielpour D. Novel 150- and 180-kDa glycoproteins that bind transforming growth factor (TGF)-β1 but not TGF-β2 are present in several cell lines. J Biol Chem 1991;266:9907–11.
36. O'Grady P, Huang SS, Huang JS. Expression of a new type high molecular weight receptor (type V receptor) of transforming growth factor β in normal and transformed cells. Biochem Biophys Res Comm 1991;179:378–85.
37. Paria BC, Jones KL, Flanders KC, Dey SK. Localization and binding of transforming growth factor-β isoforms in mouse preimplantation embryos, and in delayed and activated blastocysts. Dev Biol 1992;151:91–104.
38. Das S, Flanders KC, Andrews GK, Dey SK. Expression of transforming growth factor-β isoforms (β2 and β3) in the mouse uterus: analysis of the periimplantation period and effects of ovarian steroids. Endocrinology 1992;130:3459–66.
39. Bolton VN, Oades PJ, Johnson MH. The relationship between cleavage, DNA replication and gene expression in the mouse 2-cell embryo. J Embryol Exp Morphol 1984;79:139–63.
40. Flach G, Johnson MH, Braude PR, Taylor RAS, Bolton VN. The transition from maternal to embryonic control in the 2-cell mouse embryo. EMBO J 1982;1:681–6.

41. Heino J, Ignotz RA, Hemler ME, Crouse C, Massagué J. Regulation of cell adhesion receptors by transforming growth factor-β: concomitant regulation of integrins that share a β1 subunit. J Biol Chem 1989;264:380–8.
42. Ignotz RA, Heino J, Massagué J. Regulation of cell adhesion receptors by transforming growth factor-β: regulation of vitronectin receptor and LFA-1. J Biol Chem 1989;264:389–92.
43. Kimelman D, Kirschner M. Synergistic induction of mesoderm by FGF and TGF-β and the identification of an mRNA coding for FGF in early *Xenopus* embryo. Cell 1987;51:869–77.
44. Rosa F, Roberts AB, Danielpour D, Dart LL, Sporn MB, Dawid IB. Mesoderm induction in amphibians: the role of TGF-β2–like factors. Science 1988;239:783–6.
45. Balakier H, Pederson RA. Allocation of cells to inner cell mass and trophectoderm lineages in preimplantation mouse embryos. Dev Biol 1982;90:352–62.
46. Miller DA, Pelton RW, Derynck R, Moses RL. Transforming growth factor-β: a family of growth regulatory peptides. Ann NY Acad Sci 1990;593:208–17.
47. Grayar JL, Miller DA, Arrick BA, Lyons RM, Moses HL, Derynck R. Human transforming growth factor-β3: recombinant expression, purification and biological activities in comparison with transforming growth factors-β1 and β2. Mol Endocrinol 1989;7:1977–86.
48. Aplin JD, Charlton AK, Ayad S. An immunohistochemical study of human endometrial extracellular matrix during the menstrual cycle and first trimester of pregnancy. Cell Tissue Res 1988;253:231–40.
49. Wewer UM, Damjanov A, Weiss J, Liotta LA, Damjanov I. Mouse endometrial stromal cells produce basement-membrane components. Differentiation 1986;32:49–58.
50. Clark DA, Flanders KC, Banwatt D, et al. Murine pregnancy decidua produces a unique immunosuppressive molecule related to transforming growth factor β2. J Immunol 1990;144:3008–14.
51. McMaster MT, Newton RC, Dey SK, Andrews GK. Activation and distribution of inflammatory cells in the mouse uterus during the preimplantation period. J Immunol 1992;148:1699–705.
52. Assoian RK, Sporn MB. Type β transforming growth factor in human platelets: release during platelet degranulation and action on vascular smooth muscle cells. J Cell Biol 1986;102:1217–23.
53. Majack RA. Beta type transforming growth factor specifies organizational behavior in vascular smooth muscle cell cultures. J Cell Biol 1987;105:465–71.
54. Owens GK, Geisterfer AAT, Yang YW, Komoriya A. Transforming growth factor-β-induced growth inhibition and cellular hypertrophy in cultured vascular smooth muscle cells. J Cell Biol 1988;107:771–80.
55. Potts JD, Dagle JM, Walder JA, Weeks DL, Runyan RB. Epithelial-mesenchymal transformation of embryonic cardiac endothelial cells is inhibited by a modified antisense oligodeoxynucleotide to transforming growth factor-β3. Proc Natl Acad Sci USA 1991;88:1516–20.
56. Lafyatis R, Lechleider R, Roberts AB, Sporn MB. Secretion and transcriptional regulation of transforming growth factor-β3 during myogenesis. Mol Cell Biol 1991;11:3795–803.

57. Florini JR, Roberts AB, Ewton DZ, Falen SL, Flanders KC, Sporn MB. Transforming growth factor-β: a very potent inhibitor of myoblast differentiation, identical to the differentiation inhibitor secreted by buffalo rat liver cells. J Biol Chem 1986;261:16509–13.
58. Ewaton DZ, Spizz G, Olson EN, Florini JR. Decrease in transforming growth factor-β binding and action during differentiation in muscle cells. J Biol Chem 1988;263:4029–32.
59. Massagué J, Cheifetz S, Endo T, Nadal-Ginard B. Type-β transforming growth factor is an inhibitor of myogenic differentiation. Proc Natl Acad Sci USA 1986;83:8206–10.
60. Cheifetz S, Like B, Massagué J. Cellular distribution of type 1 and type II receptors for transforming growth factor-β. J Biol Chem 1986;261:9972–8.
61. Cheifetz S, Weatherbee JA, Tsang MLS, et al. The transforming growth factor beta system, a complex pattern of cross-reactive ligands and receptors. Cell 1987;48:409–15.
62. Cheifetz S, Bassols A, Stanley K, Ohta M, Greenberger J, Massagué J. Heterodimeric transforming growth factor B: biological properties and interaction with three types of cell surface receptors. J Biol Chem 1988;263:10783–9.
63. MacKay K, Robbins AR, Bruce MD, Danielpour D. Identification of disulfide-linked transforming growth factor-β1–specific binding proteins in rat glomeruli. J Biol Chem 1990;265:9351–6.
64. Mitchell EJ, O'Connor-McCourt MD. A transforming growth factor β (TGF-β) receptor from human placenta exhibits a greater affinity for TGF-β2 than for TGF-β1. Biochemistry 1991;30:4350–6.
65. Wakefield LM, Smith DM, Masui T, Harris CC, Sporn MB. Distribution and modulation of the cellular receptor for transforming growth factor-beta. J Cell Biol 1987;105:965–75.
66. Boyd FT, Massagué J. Transforming growth factor-β inhibition of epithelial cell proliferation linked to expression of a 53-kDa membrane receptor. J Biol Chem 1989;264:2272–8.
67. Laiho M, Weis FMB, Massagué J. Concomitant loss of transforming growth factor (TGF)-β receptor types I and II in TGF-β-resistant cell mutants implicates both receptor types in signal transduction. J Biol Chem 1990;265:18518–24.
68. Cantrella M, McCarthy TL, Canalis E. Glucocorticoid regulation of transforming growth factor β1 activity and binding in osteoblast-enriched cultures from fetal rat bone. Mol Cell Biol 1991;11:4490–6.
69. Graycar JL, Miller DA, Arrcik BA, Lyons RM, Moses HL, Derynck R. Human transforming growth factor β3: recombinant expression, purification, and biological activities in comparison with transforming growth factors β1 and β2. Mol Endocrinol 1989;3:1977–86.
70. ten Dijke P, Iwata KK, Goddard C, et al. Recombinant transforming growth factor type β3: biological activities and receptor binding properties in isolated bone cells. Mol Cell Biol 1990;10:4473–9.
71. Wang X-F, Lin HY, Ng-Eaton E, Downward J, Lodish HF, Weinberg RA. Expression cloning and characterization of the TGF-β type III receptor. Cell 1991;67:797–805.

72. Lopez-Casillas F, Chiefetz S, Doody J, Andres JL, Lane WS, Massagué J. Structure and expression of the membrane proteoglycan betaglycan, a component of the TGF-β receptor system. Cell 1991;67:785–95.
73. Lin HY, Wang X-F, Ng-Eaton E, Weinberg RA, Lodish HF. Expression cloning of the TGF-β type II receptor, a functional transmembrane serine/threonine kinase. Cell 1992;68:775–85.

Part IV

Gamete/Embryo Manipulation

14

Cryobiology of Gametes and Embryos from Nonhuman Primates

W.F. RALL

Modern cryobiology and its application to gamete and embryo preservation began in earnest in 1949 when Chris Polge and his colleagues reported on the protective action of glycerol during the freezing of spermatozoa (1). In the ensuing four decades, the driving force for further progress in this area has been the need for information and cryopreservation technology for practical application to (i) the genetic improvement of domestic livestock, (ii) the treatment of infertility in humans, and (iii) the banking of rare or unique genetic stocks for research and species conservation purposes. This has given this subspecialty a uniquely practical and often empirical bias unlike that associated with other reproductive biotechnologies.

Continued interest in the cryopreservation and banking of gametes and embryos from mammals and other animals originates from its ability to arrest all biological processes and place cells into a state of suspended animation. This unique ability to interrupt the life cycle of animals for any desired period of time provides a useful tool for basic studies of reproductive and developmental processes and for strategies to assist the propagation of rare and endangered species. The successful application of gamete and embryo cryopreservation requires extensive knowledge of the underlying reproductive physiology of each target species. This is because reproductive cyclicity of mammals is characterized by a complex temporal sequence of physiological processes with specific evolutionary variations to suit the life-history strategies of each species (2). The most important challenge to the application of gamete/embryo cryopreservation biotechniques is the ability to determine and/or control the appropriate time for oocyte and semen collection, *embryo transfer* (ET), and *artificial insemination* (AI) of semen.

This chapter (i) reviews basic aspects of the cryobiology of spermatozoa, oocytes, and embryos; (ii) describes current progress in applying gamete and embryo cryopreservation to primates; and (iii) identifies areas of research that may improve basic and applied applications of germ plasm cryopreservation to primates.

Three Phases of Cryopreservation

Although the details of successful cryopreservation procedures vary enormously, the basic steps of cryopreservation share common features that can be divided into three phases (Table 14.1). The first phase is related to the preparation and cooling of the biological materials. This requires the collection of the biological material (i.e., spermatozoa, oocytes, or embryos) and use of appropriate tests to ensure that only high-quality material is cryopreserved. Once selected, cells are suspended and equilibrated in a solution containing a cryoprotective solute, such as glycerol, propylene glycol, or *dimethyl sulfoxide* (DMSO). Then, the cell suspension is cooled to liquid nitrogen temperature ($-196°C$) using carefully controlled conditions.

The second phase of cryopreservation concerns low-temperature storage. The viability of cryopreserved cells is unaffected by storage period provided that the temperature of the samples remains below $-130°C$ *at all times*. This is because cryopreserved samples completely solidify into stable crystalline (i.e., ice) and/or glassy solids at temperatures below about $-130°C$ (3, 4). It is the lack of molecular translocation during storage and low levels of thermal energy (5) that prevent normal chemical reactions from degrading cryopreserved materials. The only known damaging chemical reactions during storage are photophysical events related to the accumulation of DNA breaks and damage from direct "hits" of cosmic rays and ionizing background radiation (6, 7). Recent reports suggest that normal levels of environmental irradiation do not represent a major problem. Experiments with cryopreserved mouse embryos indicate that exposure to the equivalent of 2000 years of background radiation produced no measurable genetic changes or decrease in viability (8).

TABLE 14.1. Three phases of cryopreservation.

Phase 1—preparation and cooling
  1. Collect and test biological material (e.g., semen, oocytes, or embryos).
  2. Equilibrate cells in solution containing a cryoprotective solute (e.g., glycerol, DMSO, propylene glycol, etc.).
  3. Cool cell suspension using carefully controlled conditions to liquid nitrogen temperatures ($-196°C$).

Phase 2—low-temperature storage

Phase 3—warming and processing
  1. Warm and thaw cell suspension using carefully controlled conditions.
  2. Remove cryoprotective solute from the cell suspension.
  3. Return cells to normal physiological conditions.

The third phase of cryopreservation is related to the warming and processing of the biological material at the end of the storage period. First, cryopreserved suspensions must be warmed and thawed using carefully controlled conditions. Then, the cryoprotectant is removed from the cell suspension using a procedure that prevents excessive osmotic swelling. Finally, the cells are returned to normal physiological conditions (e.g., AI of spermatozoa, *in vitro fertilization* (IVF) of oocytes, and ET to recipients).

The appropriate conditions for each step of cryopreservation are determined by a large number of interacting factors (reviewed in 7, 9). These factors can be divided into two groups. The first group of factors are related to the specific intrinsic properties of the cells in question. These include the permeability characteristics of the cells to water and cryoprotectants, the size of the cells, their water content, and potential heterogeneity of these properties within the cell population. Some cells have special requirements for one or more components of the cryopreservation process as a result of unusual physiological or morphological features. The most important include a high sensitivity to chilling, cryoprotectant toxicity, or cold shock. These intrinsic cellular properties determine the basic requirements for the steps of successful cryopreservation.

The second group of factors is related to the procedural steps or conditions that have been selected to prepare and cryopreserve the cell suspension. Examples of these factors include the choice of freezing container, type and concentration of cryoprotectant, cryoprotectant equilibration conditions, and rates of cooling and warming. These choices are often influenced by practical considerations, such as the type of equipment available for controlling the rate of cooling, the volume and concentration of sperm suspension needed for one insemination, or the need to eliminate or simplify the dilution procedure following thawing.

The complex interactions between cryobiological factors usually necessitate empirical optimization of the steps of cryopreservation to minimize sources of injury and maximize survival. A final complication is related to differences in the intrinsic properties of the same cells isolated from different species and changes in the size and permeability properties of embryos during preimplantation development. One important consequence of these complications is that successful cryopreservation procedures developed for animal germ plasm from one species and particular developmental stage may be inappropriate for germ plasm from another species.

Methods of Cryopreservation

Many different procedures have been developed during the past 40 years for the successful cryopreservation of animal germ plasm. These can be divided into three approaches: (i) controlled slow freezing, (ii) rapid freezing, and (iii) vitrification. Each approach results in a characteristic sequence of changes in the osmotic volume of cells during the steps of cryopreservation. The procedural steps and resultant osmotic consequences of each procedure are compared and contrasted below.

Controlled Slow Freezing

Controlled slow freezing procedures allowed the first successful cryopreservation of animal germ plasm and remain the most widely used method. Successful controlled slow freezing has two distinctive characteristics: (i) A low concentration of cryoprotectants (<2 molar) is added to the cell suspension, and (ii) the suspension is frozen using a controlled, usually low, rate of cooling.

Comprehensive reviews appear elsewhere on the basic mechanisms of cryoprotection and cryoinjury to cells during controlled freezing procedures (7). The primary physical, chemical, and osmotic events during the first phase of controlled freezing are illustrated in Figure 14.1. When cells suspended in a molar concentration are cooled to $-5°C$, the suspension usually supercools and remains unfrozen due, in part, to the depression of the equilibrium freezing point by cryoprotective solutes. However, ice forms in the suspension at temperatures between -5 and $-15°C$ as a result of either spontaneous nucleation or seeding with an ice crystal. Ice crystals grow throughout the solution surrounding the cells, but not into the supercooled cytoplasm, presumably because the cell membrane acts as barrier to ice crystal growth. One consequence of such compartmentalization of freezing is the formation of a concentration gradient between the partially frozen extracellular solution and the supercooled cytoplasm. Cells respond osmotically to this gradient by the movement of cellular water across the plasma membrane. The rate of water loss is determined by the membrane's permeability properties, which vary enormously for different types of cells. Some cells exhibit high permeability to water and lose water rapidly, while other types of cells exhibit low permeability and lose water slowly.

The rate at which the cell suspension is cooled once ice forms determines the subsequent osmotic behavior of the cells (Fig. 14.1, right side). Two extreme conditions need to be considered. In the first, the suspension is cooled slowly (Fig. 14.1, lower right). Under these conditions sufficient time is available for water to leave the cell; as a result, the cell dehydrates and is exposed to freeze-concentrated solutions for prolonged periods of time. The second extreme occurs when the suspension is cooled rapidly (Fig. 14.1, upper right). Rapid cooling does not provide sufficient time for water to leave the cell; as a result, the cytoplasm becomes increasingly supercooled and eventually spontaneously freezes. Suspensions cooled at either extreme usually lead to cell death. The optimum levels of survival are obtained at an

14. Cryobiology of Gametes and Embryos from Nonhuman Primates 227

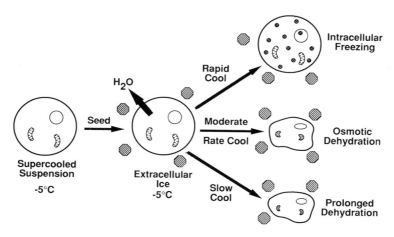

FIGURE 14.1. Diagrammatic representation of physical events during cell freezing. See text for details. Modified with permission from Mazur (10).

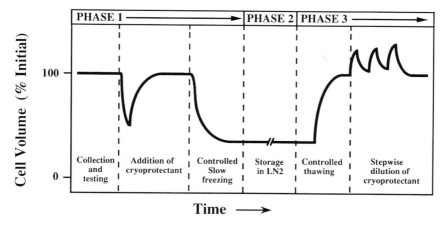

FIGURE 14.2. Diagrammatic representation of typical cell volume changes during "successful" controlled rate freezing. See text for details. Modified with permission from Rall (17).

intermediate cooling rate (Fig. 14.1, middle right), a rate that is low enough to allow the cells to dehydrate and avoid the deleterious effects of intracellular freezing (10), but high enough to prevent prolonged periods of dehydration.

Analysis of the physical, chemical, and osmotic consequences of controlled freezing procedures has yielded a thermodynamic model that predicts the appropriate rate of cooling for successful cryopreservation (7, 11). Such predictions assisted in the first reports of successful cryopreservation of mammalian embryos (12) and oocytes (13). The typical steps of controlled slow freezing procedures and resultant osmotic consequences are shown in Figure 14.2.

The first phase of cryopreservation by controlled slow freezing results in two changes in the osmotic volume of the cells. First, cells undergo a transient shrink-swell change in volume when placed in a saline containing 1-2 molar concentration of a permeable cryoprotectant, such as glycerol, DMSO, or propylene glycol. Cells initially shrink due to the exosmosis of cell water when placed in the hypertonic cryoprotectant solution. Once osmotic equilibrium is restored, cells gradually swell as the cryoprotectant permeates into the cytoplasm. The extent and kinetics of cell shrinkage and swelling are determined by the permeability properties of the cell membrane to water and cryoprotectants (9). The second change in cell volume results from the osmotic dehydration of cells when the suspension is frozen slowly to temperatures below about $-40°C$, as described above. At temperatures below about $-120°C$, the freeze-concentrated cytoplasm and residual extracellular liquid solidify (vitrify) into a glass (3).

During the second phase of cryopreservation—storage in liquid nitrogen (LN_2, $-196°C$)—cells remain shrunken because the suspension has completely solidified. Two changes in the osmotic volume of cells occur during the third phase of cryopreservation. First, controlled warming and thawing results in a gradual reduction in the concentration of the extracellular solution as the ice melts. Cells gradually rehydrate and swell during thawing as they restore osmotic equilibrium. The rate of cellular rehydration depends on the rate at which the suspensions are thawed. Rapid warming may result in the complete thawing of the suspension before appreciable rehydration has occurred. The final change in osmotic volume occurs when the cryoprotectant is removed from the cell suspension. Figure 14.2 illustrates the osmotic consequences of a three-step dilution of the cell suspension with an isotonic saline. Each dilution results in a transient osmotic swelling of the cells. The extent and length of swelling depend on the amount of cryoprotectant in the cytoplasm and the permeability of the cell membrane to cryoprotectant and water (14).

Vitrification

The vitrification approach was originally proposed by Luyet (15) and has been successfully applied recently (16). Vitrification procedures have two distinctive features: (i) Cells are osmotically dehydrated by brief exposure to concentrated solutions of cryoprotectants (>6 molar) *prior to* cooling to $-196°C$, and (ii) no ice forms in the cell suspension during cooling, storage, or warming. Vitrification has been used to cryopreserve embryos and cells from a variety of mammalian, plant, and insect species, but not, as yet, to cryopreserve spermatozoa (17).

14. Cryobiology of Gametes and Embryos from Nonhuman Primates 229

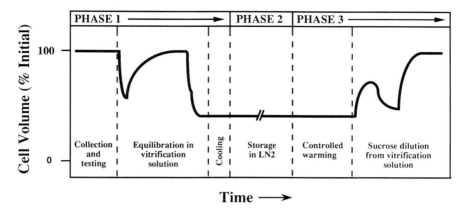

FIGURE 14.3. Diagrammatic representation of typical cell volume changes during "successful" vitrification of mammalian embryos using VS3a vitrification solution. See text for details. Modified with permission from Rall (17).

Reviews of the mechanisms of cryoprotection and cryoinjury during vitrification appear elsewhere (4, 18). The steps for vitrification of embryos using a glycerol-based solution (VS3a: 6.5 molar glycerol and 6% BSA in saline) (19) and resultant osmotic consequences are shown in Figure 14.3.

The first phase of vitrification results in three changes in the osmotic volume of cells (Fig. 14.3). First, embryos undergo a shrink-swell change in volume when placed in saline containing 1.6 molar glycerol at room temperature for 20 min. (Note that this step is similar to the equilibration step of controlled slow freezing.) Then, the embryos are osmotically dehydrated by transfer into the vitrification solution in two short steps (4.2 and 6.5 molar glycerol for about 1 min at room temperature). This period of exposure to each of the concentrated solutions ensures osmotic dehydration without extensive additional permeation of glycerol. Once embryos dehydrate, the suspension is then vitrified by cooling at rates of 20°C–2500°C/min to the storage temperature. Many alternative equilibration protocols and vitrification solutions have been developed to yield a similar extent of osmotic dehydration prior to the vitrification of mammalian embryo suspensions (20, 21).

During the second phase of vitrification—storage at −196°C—cells remain shrunken because the suspension has completely solidified into a glass (Fig. 14.3). At the end of the storage period, vitrified suspensions are usually warmed rapidly to reduce the likelihood of devitrification (crystallization) of the glassy suspension and injury associated with intracellular freezing (4). However, unlike controlled slow freezing procedures, embryos remain shrunken during warming because the concentration of solute in the suspension does not change. Embryos must be diluted out of the vitrification solution immediately after warming to prevent further permeation of cryoprotectants into the cells. This is usually accomplished using a sucrose dilution procedure (9, 22). First, embryos undergo a transient swell-shrink change in volume when placed in saline containing 1.0 molar sucrose (Fig. 14.3). Embryos initially swell as water flows into the cells to restore osmotic equilibrium with the lower osmolality of the sucrose diluent. Then, embryos progressively shrink as the glycerol leaves the cytoplasm. Finally, the embryos swell back to their normal volume when transferred into isotonic saline.

Rapid Freezing

Rapid freezing procedures provide an alternative method for eliminating controlled slow freezing during embryo cryopreservation. In this approach embryos are dehydrated prior to cooling in moderately concentrated solutions of cryoprotectants prior to rapid cooling to −196°C. High levels of survival (65%–98%) have been reported for mouse embryos equilibrated in a wide range of mixtures of sucrose (0.25–0.5 molar) plus either glycerol (3–4 molar) (23, 24), DMSO (2–4.5 molar) (25, 26), or ethylene glycol (3 molar) (27, 28). Unfortunately, rapid freezing has been less effective for embryos from other mammalian species (29).

Reviews of the mechanisms of cryoprotection and cryoinjury during rapid freezing appear elsewhere (30, 31). Successful rapid freezing has three characteristic features: (i) controlled osmotic dehydration of cells prior to cooling by equilibration in a mixture of permeating and

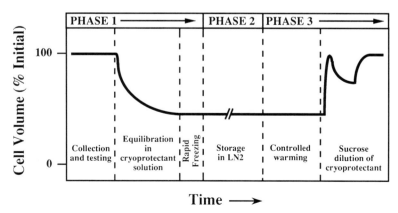

FIGURE 14.4. Diagrammatic representation of typical cell volume changes during "successful" rapid freezing of mouse embryos. See text for details.

nonpermeating cryoprotectants, (ii) rapid cooling of the suspension to the storage temperature, and (iii) rapid warming at the end of the storage period. The steps for rapid freezing of mouse embryos in saline containing DMSO (4.5) and sucrose (0.25) by the method of Shaw et al. (26) and the approximate osmotic consequences are shown in Figure 14.4.

The first phase of rapid freezing produces a single change in the osmotic volume of the embryos (Fig. 14.4). Embryos shrink rapidly when placed into saline containing DMSO and sucrose for 10 min at 0°C due to the osmotic loss of cell water. Although the low exposure temperature, 0°C, limits the amount of DMSO permeating into the cells (32), calculations using Mazur's solute permeability equations (31) indicate that the intracellular DMSO concentration is about 1/3 of that outside. It is likely that the intracellular solute concentration is high enough to yield a vitreous cytoplasm during ultrarapid cooling to the storage temperature. Alternative rapid freezing protocols presumably yield similar levels of osmotic dehydration and intracellular cryoprotectant concentrations after equilibration in the cryoprotectant solution (23, 30, 33).

During the second phase of rapid freezing—storage at $-196°C$—cells remain shrunken because the suspension has completely solidified into a mixture of metastable glass and possibly small ice crystals. At the end of the storage period, rapidly frozen suspensions are warmed rapidly to reduce the likelihood of injury associated with the inevitable devitrification (crystallization) of the embryo suspension. Despite the transient increase in extracellular solute concentration due to freezing during warming, insufficient time is available for the embryos to respond osmotically. The only change in the volume of embryos occurs when they are diluted from the cryoprotectant solution. As in the case of vitrification, rapidly frozen embryos are usually diluted by the sucrose method. In this case, embryos are transferred into a saline containing 0.25 molar sucrose. Embryos initially swell as water flows into the cells to restore osmotic equilibrium with the lower osmolality of the sucrose diluent (Fig. 14.4). Then, embryos progressively shrink as the DMSO leaves the cytoplasm. Finally, the embryos swell back to their normal volume when transferred into isotonic saline.

Recent studies indicate that a high frequency of structural chromosomal abnormalities are associated with some rapid freezing procedures (34, 35). The incidence and severity of chromosomal damage varied depending on the concentration of cryoprotectants. Embryos frozen rapidly in low concentrations of DMSO (1.5–3.0 molar) yielded the highest incidence of chromosomal damage (22%–77%). Little or no damage was observed when embryos were frozen rapidly in 4.5 DMSO using conditions that ensure vitrification during cooling, but allow devitrification during warming (35). Further research is needed to determine the mechanisms of chromosome damage during rapid freezing procedures.

Current Status of Mammalian Sperm Cryopreservation

Considerable progress has been made since the first report in 1951 of normal progeny following AI with cryopreserved spermatozoa in cattle in applying sperm cryopreservation to other mammals. Excellent general reviews are available on the cryopreservation of mammalian spermatozoa (36–38).

Mammalian spermatozoa provide the best example of the limits to predictive cryobiology. From a cryobiological point of view, the general properties of spermatozoa are very favorable for successful cryopreservation. They are small, exhibit a high permeability to water and cryoprotectant, and have a low water content. Analysis of the osmotic behavior of spermatozoa during freezing using Mazur's thermodynamic model predicts no special precautions except possibly for a requirement of moderately high rates of cooling (39).

However, practical experience indicates that special properties of sperm and other factors complicate the application of sperm cryopreservation to a wide variety of mammalian species. First, sperm are highly compartmentalized in terms of structure and function (40, 41), and there is some evidence of differential sensitivity of these compartments to osmotic and toxic stresses (42). Second, sperm membranes can exhibit a high sensitivity to cold shock, which can be reduced by slow cooling and equilibration at 0°C in the presence of special additives, such as egg yolk (43). Other complicating factors include heterogeneity in the properties of individual sperm in the population, the presence of complex seminal plasma components, and the need to preserve motility and associated metabolism. Finally, perhaps the most important complication is that the relative importance of these special properties often varies between species and even between individuals within a species.

The complexity of species-specific variations in the spermatozoal properties and a poor understanding of the cryobiological consequences of these factors limit quantitative prediction of the appropriate cryopreservation conditions for each new species. Additional difficulties in developing and using sperm cryopreservation procedures for new species are related to a general lack of information concerning their basic reproductive physiology. Efficient application of sperm cryopreservation usually requires information concerning (i) normal reproductive cyclicity and seasonality, (ii) appropriate semen collection and AI procedures, (iii) characteristics of normal fertile ejaculates, and (iv) in vitro tests of sperm structure and function that correlate with fertility. The usual strategy for developing a cryopreservation procedure for a new species is to select a successful protocol for a taxonomically related species and empirically modify it until acceptable postthaw fertility is obtained.

Despite these difficulties, empirical approaches have been successfully applied to at least 34 mammalian species, where success is defined as

TABLE 14.2. Mammalian species yielding offspring following AI with cryopreserved sperm.

Domesticated
  Cattle, sheep, horse, pig, goat, water buffalo, red deer, rabbit, dog, cat, domestic ferret, laboratory mouse

Nondomesticated
  Addax, bighorn sheep, bison, black buck, scimitar-horned oryx, black-footed ferret, Siberian ferret, leopard cat, giant panda, gaur, Eld's deer, fallow deer, wapiti, reindeer, white-tailed deer, chital deer, fox, wolf

Primates
  Human, chimpanzee, cynomolgus monkey, gorilla

---

normal offspring following AI with cryopreserved spermatozoa (Table 14.2). High levels of success have been obtained for the human, most economically important domesticated livestock, and some laboratory and companion species. The level of success for other species listed in Table 14.2 is lower or unknown due to a small number of inseminations and/or lack of independent confirmation. Indeed, for some species only a single offspring has been produced using cryopreserved spermatozoa.

Human Spermatozoa

Reviews are available on basic and applied aspects of the cryopreservation of spermatozoa from the human (44, 45). Considerable progress has been made since the first report of births in 1954 (46) in applying sperm cryopreservation to the treatment of infertility in the human. A wide range of cryoprotectant protocols yield acceptable levels of fertility postthaw (44). Nearly all human sperm cryopreservation protocols currently employ (i) rigid quality control criteria to select donors and ejaculates for cryopreservation, (ii) glycerol (7%–15%) as the cryoprotectant, (iii) controlled cooling to the storage temperature, and (iv) rapid warming. The chief difference is the composition of glycerol solution used to dilute raw semen (ranging from neat glycerol to various mixtures of glycerol, egg yolk, glucose, water, zwitterionic buffers, and/or citrate). Artificial insemination with cryopreserved sperm from either a donor or the husband has become an accepted procedure in the treatment of infertility. Although there are no authoritative figures of the number of births annually, one survey conservatively estimated 30,000 in the United States during 1987 (47). Human sperm cryopreservation and banking has become increasingly important because it provides the only method of quarantining semen for the recommended HIV testing of semen donors (48, 49).

The fertility of cryopreserved human sperm following AI is generally thought to be 50%–70% of that using fresh sperm (45). This conclusion is based on the fact that cryopreservation decreases the quality of sperm suspensions—with regard to motility, speed of forward progressive motion, and longevity of motility after thawing (45)—and on the results of retrospective studies and a limited number of prospective studies (50, 51). However, a decrease in the quality of cryopreserved sperm suspensions is not incompatible with high fertility after AI if the conditions of cryopreservation and methods of AI are adjusted to minimize or compensate for cryoinjury. Examples of factors reported to contribute to high rates of fertility following AI with frozen sperm include (i) insemination of high numbers of spermatozoa (e.g., 40×10^6) (52); (ii) careful timing of AI to ovulation (51, 53); (iii) intrauterine insemination of thawed, washed sperm (54, 55); (iv) diagnosis and treatment of female factors of infertility, especially ovulatory dysfunction (51); and (v) cryopreservation of semen within 1 h of delivery (56).

Nonhuman Primates

In contrast to the human, only limited progress has been made in applying sperm cryopreservation to the breeding of nonhuman primates. Semen has been collected from at least 18 species of nonhuman primates, and the effects of cryopreservation have been examined in vitro for at least 10 species (reviewed in 57–60). Generally speaking, empirical optimization of sperm cryopreservation procedures developed for the human or bull yielded postthaw motility from about 45% to 85% of prefreeze motility. Despite this apparent cryopreservation success by in vitro assays of viability, a total of at least 80 reported inseminations with frozen sperm since 1972 have yielded only 3 offspring (Table 14.3): one each in the gorilla (61), chimpanzee (62), and cynomolgus macaque (63). The low fertility of cryopreserved spermatozoa contrasts sharply with reports of high fertility of fresh spermatozoa. For example, reported conception rates (pregnancies per insemination) following AI with fresh spermatozoa in the rhesus macaque vary from 4% (n = 124) for vaginal inseminations (64) to 40% (n = 218) for intrauterine inseminations (65).

TABLE 14.3. Cryopreservation of primate spermatozoa.

Species	Freezing extender	Cooling method	Number of inseminations	Number of offspring	Reference
Human	Many	Many	Many	>50,000	44
Cynomolgus macaque	TES-TRIS, egg yolk, milk, 3% glycerol	LN_2 vapor	3	1	63
Chimpanzee	Ham's F10, serum, 7.8% glycerol	Controlled rate freezer	11	1	62
Gorilla	?	Dry ice	1	1	61

A 21% conception rate (n = 29) was reported following intrauterine insemination in the chimpanzee (66).

Further basic research is needed to determine those factors that reduce the fertility of cryopreserved nonhuman primate spermatozoa and then to modify the appropriate conditions of cryopreservation and/or AI to compensate for the deficiency. One hypothesis is that the fertile longevity of cryopreserved spermatozoa in the reproductive tract of the female is lower than that of fresh spermatozoa, as suggested in the human (51). If so, then research to achieve closer synchrony between AI with cryopreserved spermatozoa and ovulation may improve conception rates. This may take several forms, such as improvements in cryopreservation methods to increase the period of postthaw fertility; elimination or reduction of complicating factors that disrupt the normal timing of ovulation, such as anesthetic (67), stress, or exogenous hormone treatment; and modifications of AI procedures to place spermatozoa at the optimum site of the reproductive tract (intrauterine or intraoviductal) and/or to enhance natural sperm transport in the female reproductive tract (68).

Current Status of Mammalian Oocyte Cryopreservation

The cryopreservation of mammalian oocytes would greatly assist the application of several reproductive biotechnologies, such as IVF, cloning, and gene banking. From a cryobiological point of view, several special characteristics of mature oocytes may present problems for successful cryopreservation. First, mature oocytes are among the largest cells in mammals, and their spherical geometry yields a low surface area: volume ratio. Both of these properties act to limit the rate of permeation of cryoprotectants and water into and out of the oocyte. Second, mature oocytes of most mammals arrest with chromosomes aligned on the metaphase II spindle, cortical granules positioned immediately adjacent to the plasma membrane, and cytoskeletal elements associated with the spindle and cortical regions. Maintenance of these cytological features until the time of fertilization is required for normal developmental processes. Unfortunately, many reports indicate that these and other features are modified by cooling to subambient temperatures or equilibration with cryoprotective solutes (19, 69, 70).

Despite these difficulties, considerable progress has been made in the past 15 years in the cryopreservation of mouse oocytes. Over 100 late-stage fetuses or normal offspring have been produced from cryopreserved mouse oocytes following IVF and ET. However, the overall efficiency of the process is low: Only 6%–14% of oocytes cryopreserved by controlled slow freezing (13, 71, 72) and 20%–38% of oocytes cryopreserved by vitrification (73, 74) produced fetuses or offspring. The application of oocyte cryopreservation to other mammals has yielded limited success. Cattle oocytes exhibit low fertilization rates in vitro after slow freezing, and few (0%–13%) develop to the 2-cell stage (75, 76). Live late-stage fetuses and young have been reported from cryopreserved rabbit oocytes, but the overall survival rate was low (2%–6%) (77, 78). Finally, 2 births have been reported following IVF of cryopreserved human oocytes (79, 80), but the current overall efficiency is too low for practical applications (81).

The reasons for the difficulty in applying cryopreservation methods for mouse oocytes to other mammalian species are not fully understood. Evidence from the mouse (82), hamster (83), cattle (70, 84), and human (85) suggests that the difficulty is not related to unusual osmotic behavior during cryoprotectant permeation or dilution or osmotic dehydration during controlled slow freezing. Further research is needed to characterize the cytological and physiological features of oocytes that are sensitive to cooling and/or cryoprotectants; then, strategies must be developed to prevent or minimize deleterious changes.

Nonhuman Primates

Only one report has been made on the cryobiology of nonhuman primate embryos (86). In that study squirrel monkey oocytes were cryopreserved by slow freezing in DMSO using procedures similar to those developed for mouse oocytes. Approximately 50% (3/6) and 28% (8/29) of cryopreserved squirrel monkey oocytes fertilized following, respectively, insemination with spermatozoa in vitro or xenogenously in a rabbit oviduct. No differences in maturation or fertilization rates were observed between control or cryopreserved oocytes. Although these preliminary data are promising, insufficient information is available to predict the general applicability of oocyte cryopreservation to other nonhuman primates.

Current Status of Mammalian Embryo Cryopreservation

Since the first report of live-born young from cryopreserved mouse embryos (12), similar success has been reported for 14 additional mammalian species (Table 14.4). Excellent general reviews are available on basic and applied aspects of the cryobiology of mammalian embryos (4, 9, 69, 81).

In general, the optimum conditions for embryo cryopreservation exhibit much lower species variation than those for mammalian spermatozoa. However, two important developmental/species variations in the intrinsic cryobiological properties of mammalian embryos have been reported. First, the permeability of embryos to cryoprotectants

TABLE 14.4. Mammalian species yielding offspring following transfer of cryopreserved embryos.

Domesticated
  Laboratory mouse, cattle, sheep, horse, pig, goat, rabbit, rat, cat

Nondomesticated
  Eland

Primates
  Baboon, cynomolgus macaque, marmoset, rhesus macaque, macaque hybrid, human

---

often increases during the course of embryonic development from zygote to blastocyst (87) and may vary for embryos from different species at the same embryonic stage. One consequence of these differences is the need to adjust the type of cryoprotectant and the conditions for equilibration and dilution to be consistent with the embryo's permeability properties. Second, embryos from at least two species, cattle and pig, are reported to exhibit a high sensitivity to cooling to 0°C during development, respectively, from zygote to early morulae or blastocysts (88). Such chilling injury is thought to be associated with lipid phase transitions in membranes and/or the dark cytoplasmic lipid droplets in these embryos.

Most research and application of embryo cryopreservation have concentrated on three species: mice, cattle, and humans. Mouse and cattle receive the greatest attention and have yielded tens of thousands of live offspring. The number of offspring produced from cryopreserved embryos of other mammalian species is much lower. Optimization of cryopreservation and ET procedures for mice and cattle currently permits remarkably high overall rates of development (9, 89). For example, in a direct comparison of controlled slow freezing and vitrification procedures using 8-cell mouse embryos, 75% of frozen and 64% of vitrified embryos developed into normal late-stage fetuses and live-born pups (19).

Nonhuman Primates

Considerable progress has been made in applying embryo cryopreservation to nonhuman primates (Table 14.5). Normal live-born young have been reported in four species—baboon (90, 91), cynomolgus macaque (92, 93), marmoset (94), and rhesus macaque (95, 96)—and one macaque hybrid (pig-tailed × lion-tailed embryo transferred into a pig-tailed recipient) (97). The cryopreservation procedures employed in these studies were based on those developed for cattle or humans. Modifications were required in some cases to account for the specific cryoprotectant permeation properties of the embryos. It is interesting to note that DMSO yielded more pregnancies than glycerol when 6- to 10-cell marmoset embryos were cryopreserved. This result presumably reflects the lower permeability

TABLE 14.5. Successful implantation and production of normal offspring following transfer of cryopreserved primate embryos.

Species	Stage frozen	Cryoprotectant	Postthaw survival[a] (%)	Implantation rate[b] (%)	Estimated overall survival[c] (%)	Reference
Baboon (*Papio sp*)	6-cell to blastocyst	Glycerol	6/6 (100)	2/6 (33)	33	90, 91
Cynomolgus macaque (*Macaca fascicularis*)	2- to 8-cell	DMSO	39/56 (70)	3/25 (12)	8	92, 93
Marmoset (*Callithrix jacchus*)	4- to 10-cell	DMSO	16/18 (89)	6/16 (38)	28	94
		Glycerol	12/15 (80)	1/12 (8)	7	
	Morula	DMSO	15/15 (100)	6/15 (40)	33	94
Rhesus macaque (*M. mulatta*)	3- to 6-cell	Propylene glycol	11/11 (100)	1/6 (17)	9	95
	Pronuclear to 6-cell	Propylene glycol	10/15 (67)	6/8 (75)	25	96
Hybrid macaque (*M. nemestrina* × *silenus*)	2-cell	Propylene glycol	9/16 (56)	1/9 (11)	6	97

[a] Defined as embryos with ≥50% intact blastomeres.
[b] Number of offspring/number of embryos transferred.
[c] Estimate of number of offspring/number of embryos frozen (corrected for embryos not transferred).

of this stage of embryo to glycerol that may lead to excessive swelling and injury during the dilution step. Morulae-stage marmoset embryos exhibited high survival when cryopreserved using DMSO.

The high rates of development of cryopreserved nonhuman primate embryos are remarkable considering that 10 offspring resulted from embryos produced by IVF. This result demonstrates that the use of multiple reproductive biotechnologies can result in high overall pregnancy rates provided that the efficiency of each procedure is high. One implication of this finding is that current nonhuman primate embryo cryopreservation procedures could be used as a tool for research, germ plasm banking, and captive breeding purposes with little or no further research or development.

Summary

The cryopreservation and banking of gametes and embryos provides a powerful method to control animal reproduction. The benefits of germ plasm cryopreservation originate from its ability to arrest all biological processes and place cells into a state of suspended animation for any desired period of time. Most work has focused on male gametes due to the relative ease of collecting and using spermatozoa. "Brilliant empiricism" led to the first report of successful sperm cryopreservation in the bull (1951). This approach has been extended to the cryopreservation of spermatozoa from a wide variety of mammalian, avian, and some fish species. For example, normal offspring from at least 30 mammalian species have resulted from the AI of cryopreserved sperm.

Unfortunately, empirical approaches have yielded only modest success with nonhuman primates. Three live offspring have been reported from the AI of cryopreserved nonhuman primate spermatozoa, one each in the gorilla (*Gorilla gorilla*), chimpanzee (*Pan troglodytes*), and cynomolgus macaque monkey (*Macaca fascicularis*). The difficulty in applying sperm cryopreservation to nonhuman primates likely results from an inadequate understanding of the intrinsic cryobiological properties of primate spermatozoa and the uncertainty of when and where to deposit the spermatozoa during AI.

Although cryopreserved human oocytes have produced several births following IVF and ET, only one report has been published on the cryobiology of nonhuman primate oocytes. Current data from the human and other mammals indicate that the overall efficiency of oocyte cryopreservation is too low for germ plasm banking. The susceptibility of the meiotic spindle, cortical granules, and cytoskeleton of mature oocytes to damage during cooling and exposure to cryoprotectants is likely the limiting factor.

Considerably greater success has been reported for the cryopreservation of nonhuman primate embryos. A total of at least 25 normal offspring have resulted from the transfer of cryopreserved embryos to foster mothers. Success has been achieved in four species—baboon (*Papio sp.*), cynomolgus macaque (*M. fascicularis*), rhesus macaque (*M. mulatta*), and marmoset (*Callithrix jacchus*)—and one macaque hybrid (*M. nemestrina* × *M. silenus* embryo transferred into a *M. nemestrina* foster mother). The reported efficacy of embryo cryopreservation is high. For example, immediate postthaw survival ranges from 56% to 100%, the implantation rate varied from 8% to 75%, and the estimated overall efficiency of cryopreservation ranged from 6% to 33%. These results are remarkable when one considers that (i) no data exist on the cryobiological characteristics of nonhuman primate embryos (e.g., water and cryoprotectant permeabilities); (ii) cryopreservation protocols developed for human, cattle or mouse embryos were used with little or no modification; and (iii) 10 offspring were produced using embryos produced by IVF.

Further progress in developing effective cryopreservation procedures for the gametes and embryos of nonhuman primates requires systematic characterization of their intrinsic cellular properties that determine susceptibility to cryoinjury. Specifically, these include (i) sensitivity of cytological structures to chilling and cold shock injury, and (ii) osmometric properties, such as measurement of osmotically inactive volume, permeability of cell membranes to cryoprotectants and water, and the effect of temperature on permeability. Current levels of success with nonhuman primate embryos indicate that a modest amount of research will yield efficient embryo cryopreservation methods for conservation and research applications.

Acknowledgments. I thank Dr. D.E. Wildt and P.M. Schmidt for discussions and comments on the manuscript. Research was supported by NIH (National Institute on Aging, AG-10164-01) and Friends of the National Zoo.

References

1. Polge C, Smith AU, Parkes AS. Revival of spermatozoa after vitrification and dehydration at low temperatures. Nature 1949;164:666.
2. Bronson FH. Mammalian reproductive biology. University of Chicago Press, 1989.
3. Rall WF, Reid DS, Polge C. Analysis of slow-warming injury of mouse embryos by cryomicroscopy and physiochemical methods. Cryobiology 1984;21:106–21.
4. Rall WF. Factors affecting the survival of mouse embryos cryopreserved by vitrification. Cryobiology 1987;24:387–402.

5. McGee HA, Martin WJ. Cryochemistry. Cryogenics 1962;2:1–11.
6. Mazur P. Freezing and low-temperature storage of living cells. In: Mühlbock O, ed. Basic aspects of freeze preservation of mouse strains. Stuttgart: Gustav Fischer Verlag, 1976:1–12.
7. Mazur P. Freezing of living cells: mechanisms and implications. Am J Physiol 1984;247C:125–42.
8. Glenister PH, Whittingham DG, Lyon MF. Further studies on the effect of radiation during the storage of frozen 8-cell mouse embryos at −196°C. J Reprod Fertil 1984;70:229–34.
9. Leibo SP. Cryobiology: preservation of mammalian embryos. In: Evans JW, Hollander A, eds. Genetic engineering of animals. New York: Plenum Press, 1986:251–72.
10. Mazur P. The role of intracellular freezing in the death of cells cooled at supraoptimal rates. Cryobiology 1977;14:251–72.
11. Mazur P. Kinetics of water loss from cells at subzero temperatures and the likelihood of intracellular freezing. J Gen Physiol 1963;47:347–69.
12. Whittingham DG, Leibo SP, Mazur P. Survival of mouse embryos frozen to −196 and −269°C. Science 1972;178:411–4.
13. Whittingham DG. Fertilization in vitro and development to term of unfertilized mouse oocytes previously stored at −196°C. J Reprod Fertil 1977;49:89–94.
14. Levin RL, Miller TW. An optimum method for the introduction and removal of permeable cryoprotectants: isolated cells. Cryobiology 1981;18:32–48.
15. Luyet BJ. The vitrification of organic colloids and of protoplasm. Biodynamica 1937;1(39):1–14.
16. Rall WF, Fahy GM. Ice-free cryopreservation of mouse embryos at −196°C by vitrification. Nature 1985;313:573–5.
17. Rall WF. Prospects for the cryopreservation of mammalian spermatozoa by vitrification. In: Johnson LA, Rath D, eds. Reproduction in domestic animals, supplement 1, (boar semen preservation II). Hamburg: Paul Parey Scientific, 1991:65–80.
18. Fahy GM, MacFarlane DR, Angell CA, Meryman HT. Vitrification as an approach to cryopreservation. Cryobiology 1984;21:407–26.
19. Rall WF. Cryopreservation of oocytes and embryos: methods and applications. Anim Reprod Sci 1992;28:237–45.
20. Massip A, Van Der Zwalmen P, Ectors F. Recent progress in cryopreservation of cattle embryos. Theriogenology 1987;27:69–79.
21. Kasai M, Komi JH, Takakamo A, Tsudera H, Sakurai T, Machida T. A simple method for mouse embryo cryopreservation in a low toxicity vitrification solution, without appreciable loss of viability. J Reprod Fertil 1990;89:91–7.
22. Leibo SP, Mazur P. Methods for the preservation of mammalian embryos by freezing. In: Daniel JC Jr, ed. Methods in mammalian reproduction. New York: Academic Press, 1978:179–201.
23. Takeda T, Elsden RP, Seidel GE Jr. Cryopreservation of mouse embryos by direct plunging into liquid nitrogen. Theriogenology 1984;21:266.
24. Széll A, Shelton JN. Sucrose dilution of glycerol from mouse embryos frozen rapidly in liquid nitrogen vapour. J Reprod Fertil 1986;76:401–8.

25. Trounson A, Peura A, Kirby C. Ultrarapid freezing: a new low-cost and effective method of embryo cryopreservation. Fertil Steril 1987;48:843–50.
26. Shaw JM, Diotallevi L, Trounson AO. A simple rapid 4.5 M dimethyl-sulfoxide freezing technique for the cryopreservation of one-cell to blastocyst stage preimplantation mouse embryos. Reprod Fertil Dev 1991;3:621–6.
27. Abas-Mazni O, Valdez CA, Takahashi Y, Hishinuma M, Kanagawa H. Quick freezing of mouse embryos using ethylene glycol with lactose or sucrose. Anim Reprod Sci 1990;22:161–9.
28. Oda K, Leibo SP. Volumetric responses of mouse zygotes in ethylene glycol solutions and their survival after rapid cooling. Hum Reprod 1992.
29. Gordts S, Roziers P, Campo R, Noto V. Survival and pregnancy outcome after ultrarapid freezing of human embryos. Fertil Steril 1990;53:469–72.
30. Széll A, Shelton JN. Osmotic and cryoprotective effects of glycerol-sucrose solutions on day-3 mouse embryos. J Reprod Fertil 1987;80:309–16.
31. Mazur P. Equilibrium, quasi-equilibrium and nonequilibrium freezing of mammalian embryos. Cell Biophys 1990;17:53–92.
32. Leibo SP, Mazur P, Jackowski SC. Factors affecting survival of mouse embryos during freezing and thawing. Exp Cell Res 1974;89:79–88.
33. Miyamoto H, Ishibashi T. Liquid nitrogen vapor freezing of mouse embryos. J Reprod Fertil 1986;78:471–8.
34. Bongso DVM, Mok H, Ng SC, et al. Chromosomal analysis of two cell embryos frozen by slow and ultrarapid methods using two different cryoprotectants. Fertil Steril 1988;49:908–12.
35. Shaw JM, Kola I, MacFarlane DR, Trounson AO. An association between chromosomal abnormalities in rapidly frozen 2-cell mouse embryos and the ice-forming properties of the cryoprotectant solution. J Reprod Fertil 1991;91:9–18.
36. Graham EF, Schmehl MKL, Evensen BK, Nelson DS. Semen preservation in nondomestic mammals. Symp Zool Soc London 1978;43:153–73.
37. Watson PF. The preservation of semen in mammals. In: Finn CA, ed. Oxford reviews of reproductive biology. New York: Oxford University Press, 1979: 283–350.
38. Watson PF. Recent advances in sperm freezing. In: Thompson W, Joyce DN, Newton JR, eds. In vitro fertilization and donor insemination. London: Perinatology Press, 1985:261–7.
39. Duncan AE, Watson PF. Predictive water loss curves for ram spermatozoa during cryopreservation: comparison with experimental observations. Cryobiology 1992;29:95–105.
40. Holt WV. Membrane heterogeneity in the mammalian spermatozoon. Int Rev Cytol 1984;87:159–94.
41. Koehler JK. Sperm membranes: segregated domains of structure and function. In: Johnson LA, Larsson K, eds. Deep freezing of boar semen. Upsala: Swedish University of Agric Sci, 1985:37–60.
42. Watson PF, Martin ICA. Effects of egg-yolk, glycerol and the freezing rate on the viability and acrosomal structures of frozen ram spermatozoa. Aust J Biol Sci 1975;28:153–9.
43. Watson PF. The effects of cold shock on sperm cell membranes. In: Morris GJ, Clarke A, eds. Effects of low temperatures on biological membranes. New York: Academic Press, 1982:189–218.

44. Sherman JK. Cryopreservation of human semen. In: Keel BA, Webster BW, eds. CRC handbook of the laboratory diagnosis and treatment of infertility. Boca Raton, FL: CRC Press, 1990:229–59.
45. Brotherton J. Cryopreservation of human semen. Arch Androl 1990;25:181–95.
46. Bunge RG, Keettel WC, Sherman JK. Clinical use of frozen semen. Fertil Steril 1954;5:520–9.
47. Office of Technology Assessment. Artificial insemination: practice in the United States. Washington, DC: Superintendent of Documents, 1988:3–12.
48. American Fertility Society. Revised new guidelines for the use of semen-donor insemination. Fertil Steril 1988;49:211–4.
49. Centers for Disease Control. Semen banking, organ and tissue transplantation and HIV antibody testing. MMWR 1988;28:1314–7.
50. Richter MA, Haning RV Jr, Shapiro SS. Artificial donor insemination: fresh versus frozen semen; the patient as her own control. Fertil Steril 1984;41:277–80.
51. Smith KD, Rodriguez-Rigau LJ, Steinberger E. The influence of ovulatory dysfunction and timing of insemination on the success of artificial insemination donor (AID) with fresh or cryopreserved semen. Fertil Steril 1981;36:496–502.
52. Brown CA, Boone WR, Shapiro SS. Improved cryopreserved semen fecudability in an alternating fresh-frozen artificial insemination program. Fertil Steril 1988;50:825–7.
53. Bordsonn BL, Ricci E, Dickey RP, Dunaway H, Taylor SN, Curole DN. Comparison of fecundability with fresh and frozen semen in therapeutic donor insemination. Fertil Steril 1986;46:466–9.
54. Byrd W, Bradshaw K, Carr B, Edman C, Odom J, Ackerman G. A prospective, randomized study of pregnancy rates following intrauterine and intracervical insemination using frozen donor semen. Fertil Steril 1990;53:521–7.
55. Patton PE, Burry KA, Thurmond A, Novy MJ, Wolf DP. Intrauterine insemination outperforms intracervical insemination in a randomized, controlled study with frozen, donor semen. Fertil Steril 1992;57:559–64.
56. Yavetz H, Yogev L, Homonnai Z, Paz G. Prerequisites for successful human sperm cryobanking: sperm quality and prefreezing holding time. Fertil Steril 1991;55:812–6.
57. Hendrickx AG, Thompson RS, Hess DL, Prahalada S. Artificial insemination and a note on pregnancy detection in the non-human primate. Symp Zool Soc London 1978;43:219–40.
58. Holt WV. Collection, assessment and storage of sperm. In: Benirschke K, ed. Primates: the road to self-sustaining populations. New York: Springer-Verlag, 1986:413–24.
59. Wildt DE. Spermatozoa: collection, evaluation, metabolism, freezing and artificial insemination. In: Dukelow WR, Erwin J, eds. Comparative primate biology; vol 3. New York: A.R. Liss, 1986:171–93.
60. Loskutoff NM, Kraemer DC, Raphael BL, Huntress SL, Wildt DE. Advances in reproduction in captive, female great apes: value of biotechniques. Am J Primatol 1991;24:151–66.

61. Douglass EM, Gould KG. Artificial insemination in a gorilla. In: Proc 1981 meet Am Assoc Zoo Vet, Seattle, WA, Oct 3–9, 1981, 1981:128–30.
62. Gould KG, Styperek RP. Improved methods for freeze preservation of chimpanzee sperm. Am J Primatol 1989;18:275–84.
63. Tollner TL, VandeVoort CA, Overstreet JW, Drobnis EZ. Cryopreservation of spermatozoa from cynomolgus monkeys (*Macaca fascicularis*). J Reprod Fertil 1990;90:347–52.
64. Valerio DA, Leverage WE, Bensenheaver JC, Thornett HD. The analysis of male fertility, artificial insemination, and natural matings in the laboratory breeding of macaques. In: Goldsmith EI, Moor-Jankowski J, eds. Medical primatology. Basel: Karger, 1971:515–26.
65. Czaja JA, Eisele SG, Goy RW. Cyclical changes in the sexual skin of female rhesus: relationships to mating behavior and successful artificial insemination. Fed Proc 1975;34:1680–4.
66. Gould KG, Martin DE, Warner H. Improved method for artificial insemination in the great apes. Am J Primatol 1985;8:61–7.
67. Lasley BL, Czekala NM, Presley S. A practical approach to evaluation of fertility in the female gorilla. Am J Primatol Suppl 1982;1:45–50.
68. Gould KG, Martin DE. Artificial insemination of nonhuman primates. In: Benirschke K, ed. Primates: the road to self-sustaining populations. New York: Springer-Verlag, 1986:425–43.
69. Niemann H. Cryopreservation of ova and embryos from livestock: current status and research needs. Theriogenology 1991;35:109–24.
70. Parks JE, Ruffing NA. Factors affecting low temperature survival of mammalian oocytes. Theriogenology 1992;37:59–73.
71. Glenister PH, Wood MJ, Kirby C, Whittingham DG. Incidence of chromosomal anomalies in first-cleavage mouse embryos obtained from frozen-thawed oocytes fertilized in vitro. Gamete Res 1987;16:205–16.
72. Schroeder AC, Champlin AK, Mobraaten LE, Eppig JJ. Developmental capacity of mouse oocytes cryopreserved before and after maturation in vitro. J Reprod Fertil 1990;89:43–50.
73. Nakagata N. High survival rate of unfertilized mouse oocytes after vitrification. J Reprod Fertil 1989;87:479–83.
74. Kono T, Kwon OY, Nakahara T. Development of vitrified mouse oocytes after in vitro fertilization. Cryobiology 1991;28:50–54.
75. Schellander K, Brackett BG, Führer F, Schleger W. In vitro fertilization of frozen-thawed cattle oocytes. In: Proc 11th Int Con on Animal Reproduction and Artificial Insemination, Dublin, Ireland, June 26–30, 1988; vol 3.
76. Lim JM, Fukui Y, Ono H. The post-thaw developmental capacity of frozen bovine oocytes following in vitro maturation and fertilization. Theriogenology 1991;35:1225–35.
77. Vincent C, Garnier V, Heyman Y, Renard JP. Solvent effects on cytoskeletal organization and in vivo survival after freezing of rabbit oocytes. J Reprod Fertil 1989;87:809–20.
78. Al-Hasani S, Kirsch J, Diedrich K, Blanke S, Van Der Ven H, Krebs D. Successful embryo transfer of cryopreserved and in vitro fertilized rabbit oocytes. Hum Reprod 1989;4:77–9.
79. Chen C. Pregnancy after human oocyte cryopreservation. Lancet 1986;1:884–6.

80. van Uem JFHM, Siebzehnrubl ER, Schuh B, Koch R, Trotnow S, Lang N. Birth after cryopreservation of unfertilized oocytes. Lancet 1987;1:752–3.
81. Trounson A. Cryopreservation. Br Med Bull 1990;46:695–708.
82. Leibo SP. Water permeability and its activation energy of fertilized and unfertilized mouse ova. J Membr Biol 1980;53:179–88.
83. Shabana M, McGrath JJ. Cryomicroscope investigation and thermodynamic modeling of the freezing of unfertilized hamster ova. Cryobiology 1988;25:338–54.
84. Meyers SP, Lin TT, Pitt RE, Steponkus PL. Cryobehavior of immature bovine oocytes. Cryobiol-Lett 1987;8:260–75.
85. Hunter J, Bernard A, Fuller B, McGrath JJ, Shaw RW. Plasma membrane water permeabilities of human oocytes: the temperature dependence of water movement in individual cells. J Cell Physiol 1992;150:175–9.
86. DeMayo FJ, Rawlins RG, Dukelow WR. Xenogenous and in vitro fertilization of frozen/thawed primate oocytes and separation of blastomeres. Fertil Steril 1985;43:295–300.
87. Mazur P, Rigopoulos N, Jackowski SC, Leibo SP. Preliminary estimates of the permeability of mouse ova and early embryos to glycerol. Biophys J 1976;16:232a.
88. Polge C, Willadsen SM. Freezing eggs and embryos of farm animals. Cryobiology 1978;15:370–3.
89. Glenister PH, Whittingham DG, Wood MJ. Genome cryopreservation: a valuable contribution to mammalian genetic research. Genet Res 1990;56:253–8.
90. Pope CE, Pope VZ, Beck LR. Live birth following cryopreservation and transfer of a baboon embryo. Fertil Steril 1984;42:143–5.
91. Pope CE, Pope VZ, Beck LR. Cryopreservation and transfer of baboon embryos. J In Vitro Fertil Embryo Transfer 1986;3:33–9.
92. Balmaceda JP, Heitman TO, Garcia MR, Pauerstein CJ, Pool TB. Embryo cryopreservation in cynomolgus monkeys. Fertil Steril 1986;45:403–6.
93. Balmaceda JP, Gastaldi C, Ord T, Borrero C, Asch RH. Tubal embryo transfer in cynomolgus monkeys: effects of hyperstimulation and synchrony. Hum Reprod 1988;3:441–3.
94. Summers PM, Shepard AM, Taylor CT, Hearn JP. The effects of cryopreservation and transfer on embryonic development in the common marmoset monkey, *Callithrix jacchus*. J Reprod Fertil 1987;79:241–50.
95. Wolf DP, VandeVoort CA, Meyer-Haas GR, et al. In vitro fertilization and embryo transfer in the rhesus monkey. Biol Reprod 1989;41:335–46.
96. Lanzendorf SE, Zelinski-Wooten MB, Stouffer RL, Wolf DP. Maturity at collection and the developmental potential of rhesus monkey oocytes. Biol Reprod 1990;42:703–11.
97. Cranfield MR, Berger NG, Kempske S, Bavister BD, Boatman DE, Ialeggio DM. Macaque monkey birth following transfer of in vitro fertilized, frozen-thawed embryos to a surrogate mother. Theriogenology 1992;37:197.

15
Application of Micromanipulation in the Human

JACQUES COHEN, MINA ALIKANI, ALEXIS ADLER, ADRIENNE REING,
TONI A. FERRARA, ELENA KISSIN, AND CINDY ANDERSON

New micromanipulation techniques in human *in vitro fertilization* (IVF) have generated means by which to study gender, gene abnormalities, dysfunctions of fertilization and embryonic implantation (1, 2). Techniques have been mainly restricted to relatively simple mechanical procedures for enhancing fertilization, such as zona drilling (3) or subzonal insertion (4), techniques for restoring diploidy, such as pronuclear extraction (5), methods for facilitating embryonic implantation, such as assisted hatching (6), and procedures for performing preimplantation genetic diagnosis, such as polar body or blastomere biopsy (1, 7). The latter is rapidly emerging as an autonomous field, bridging the gap between molecular genetics and embryology, and is therefore described in a separate chapter of this book by Dr. Handyside (Chapter 21). Such subcellular modifications as chromosome ablation (8), removal of refractile bodies, and microinjection of DNA-constructs or metabolic substances may soon be clinically feasible, but should be further refined in rodent or primate models. Micromanipulation of preimplantation embryos may be novel to those involved in clinical assisted reproduction, but it is an established discipline in experimental and veterinarian embryology (2). Introduction of this technological field to the human IVF laboratory has been slow due to the ethical and legal restrictions put forth by the scientific community, as well as by society and national governments.

There are a number of important reasons for recent changes in the philosophical approach of IVF specialists toward micromanipulation. First, precise control of the fertilization process is needed to treat patients for whom conventional IVF fails and for others who are excluded from alternative assisted reproduction technologies based on suboptimal semen profiles (9). Second, the advent of safe preconception and preimplantation genetic diagnosis methods that pose little or no risk to embryonic viability may now allow for replacement of genetically fit embryos in couples at risk for genetic disease (1, 7). Third, IVF technology is still considered controversial due to its poor statistical performance. A review of the recent literature shows that there have been only two promising proposals for enhancing embryonic viability. The first of these involves the simultaneous culture (coculture) of embryos on helper cells (10, 11). The second method has been broadly termed *assisted hatching* and is based on the hypothesis that manipulation of the zona pellucida either by drilling a hole through it, by thinning it, or by altering its structure may promote hatching of embryos that are otherwise unable to escape intact from the zona (2, 6, 12, 13).

The purpose of this chapter is to present recent developments in the field of gamete and embryo micromanipulation in the human. For methodology and historical perspectives, the reader is referred to a recent book in which these and other topics are described in detail (2). This chapter is divided into four sections: (i) gamete micromanipulation for assisted fertilization, (ii) zygote micromanipulation for repair of polyspermy, (iii) embryo micromanipulation for assisted hatching, and (iv) future applications of micromanipulation.

Assisted Fertilization

Gamete micromanipulation involves a set of techniques that can be applied to specific patients in whom the fertilization process in vitro is inhibited. It enables the reproductive scientist to circumvent specific stages in the fertilization process without correcting the actual basis for the disorder. Three classes of assisted or microsurgical fertilization by gamete micromanipulation have been explored in several mammalian species (Fig. 15.1). The first involves the creation of an artificial opening in the zona pellucida in order to allow spermatozoa to interact with the oocyte independent from the zona following insemination, a procedure that was originally termed *zona drilling* (3, 14). A second category of assisted fertilization techniques directed at facilitating sperm-egg interaction is the *subzonal placement or insertion* (SZI) of sperm (also referred to as SUZI or MIST) (4, 15–17). Possibly more invasive than zona drilling, SZI completely bypasses the zona and involves direct placement of the live sperm cell(s) into the perivitelline space. Finally, the third mode of microsurgical fertilization is the direct microinjection of a single sperm into the cytoplasm of the oocyte.

Both zona drilling and SZI methods have yielded offspring in the human. Direct cytoplasmic microinjection is now successfully employed in one program in Belgium (Palermo, van Steirtechem, unpublished results) and has already led to live offspring involving studies in rabbits and cows (18, 19). However, since published data regarding the clinical

248   J. Cohen et al.

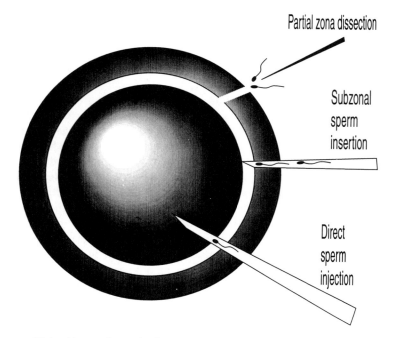

FIGURE 15.1. Alternatives of microsurgically assisted fertilization.

application of this technique are lacking at the present, only the first two procedures are discussed below.

Assisted Fertilization in the Human

Zona Drilling and Partial Zona Dissection

In the zona drilling procedure, first developed in the mouse, a small volume of acidic solution is expelled from a microneedle on a small area of the zona until it is ruptured (3). Routine insemination follows the procedure. In the mouse the artificial gap in the zona is associated with increased rates of monospermic fertilization at both normal and reduced sperm numbers, and normal live young can be obtained following transfer to recipient females. Zona drilling in the mouse allows the zona to retain its other physiological functions following fertilization. Application of zona drilling to clinical IVF was not successful and demonstrated that the human oocyte differed from the mouse in its response to acidic Tyrode's solution. It was noted that microvilli in the human oocyte flatten upon contact with acidic solution (20). This may have interfered with the ability of spermatozoa to interact with the oocyte surface. Moreover, even when fertilization occurred, embryonic development was poor, likely due to the initial detrimental effects of acidic Tyrode's solution to unfertilized oocytes (21, 22).

Alternative methods for opening the human zona have been attempted. Among the procedures proposed were zona cutting, zona cracking, and the use of enzymatic zona digestion (reviewed in 2). Cohen and coworkers, however, developed a mechanical procedure for introducing a breach in the mammalian zona that resulted in the first pregnancy from microsurgical fertilization in the human (14). This method, termed *partial zona dissection* (PZD), involves the use of a sucrose solution to shrink the oocyte so that a glass microneedle can be introduced into the perivitelline space without damaging the oocyte (2). The microneedle is threaded peripherally through two sides of the zona. The portion of the zona that is incorporated between the points is massaged, resulting in a slit in the zona. Following this procedure, the oocyte is removed from the sucrose solution and reexposed to medium of normal osmolarity before it is inseminated. It is estimated that to date over 100 live births have resulted from this method.

Though the clinical use of PZD appears promising, true success, measured as an increase in monospermic fertilization as well as normal embryonic implantation, has only been reported by a limited number of IVF programs (reviewed in 2). Basically, there are two areas of controversy. First, while there is little doubt that PZD improves sperm-oocyte fusion, excessive rates of polyspermy may reduce clinical efficiency (2, 21, 22). Monospermic fertilization is decided by the size of the hole (and hence the mode of micromanipulation), the number of spermatozoa used for insemination, and patient selection criteria. These factors must be considered carefully in order to successfully apply PZD as a clinically valid tool. Second, the implanting capacity of PZD and micromanipulated embryos in general appears to be reduced in most IVF programs. Since this has not been our experience, we can only speculate about these discrepancies (9).

There are four possible reasons for embryonic demise following micromanipulation. (i) Mechanical difficulties during the micromanipulation procedure may result in cellular damage. That this is likely to occur is illustrated by reported damage rates among programs, which vary between 2% to 30%. The damage rate in our program following micromanipulation of over 5000 oocytes and embryos is now well below 0.5% (2). Additionally, it is important to realize that damage from micromanipulation may not be immediately manifested in the embryo. (ii) Improper culture conditions such as high and low pH or temperature may incur damage during micromanipulation. (iii) Leakage of cytoplasm and blastomeres through the artificial gaps may occur during rinsing and pipetting of micromanipulated embryos. (iv) Routine clinical procedures may play a role as well. For instance, all men from couples undergoing IVF in our center receive antibiotics prophylactically during the first few days of the egg retrieval cycle. In addition, the use of antibiotics and low-dose corticosteroids in the female partners, as well as the augmentation of follicular stimulation methods, should be considered (23). Finally, it is important to recognize that micromanipulated embryos are susceptible to physical damage and should be replaced with care and precision.

Subzonal Sperm Insertion

During SZI the zona is not merely physically, but also functionally bypassed. Sperm are aspirated into a sharp, beveled microneedle and transferred into the perivitelline space. Like zona drilling methods, SZI has been investigated in both animal and human models (2). It is necessary to account for both the quality and quantity of spermatozoa used in SZI. Only sperm that have undergone capacitation and acrosome reaction are probably capable of fusing with the oolemma. It has been suggested that since the zona is no longer involved in the selection process, it may be necessary to use artificial induction of the acrosome reaction in order to ensure that the "selected" sperm will be capable of fertilization. However, despite the apparent advantages of such methods in animal models, acrosome reaction-induction has not yet proven to be advantageous in the clinical situation.

Recently, we have applied SZI to severe cases of male factor infertility (24). The data obtained from these clinical trials have clarified several biological principles and have raised interesting questions pertaining to mammalian fertilization. For example, although both PZD and SZI may be effective treatments for cases of extreme teratozoospermia, it has become clear that those embryos derived from SZI implant at a significantly higher rate than those resulting from the PZD procedure (16). A possible explanation for this phenomenon, as well as further results, is offered elsewhere (2). Most of the controversies discussed above involving the clinical use of PZD also apply to SZI. This is illustrated by two recent publications involving relatively large series of patients (17, 26). The incidence of birth varied from less than 1% to 5%. However, at least three other programs have now reported high success rates with SZI (reviewed in 2).

Controlled Studies

From a scientific point of view, one should perform all insemination alternatives (IVF, PZD, and SZI) simultaneously on sibling oocytes in each patient in order to prove that micromanipulation is effective. Though logical, this is not practical for clinical, ethical, and technical reasons. This is where scientific and clinical objectives may interfere with each other. Basically, three factors determine the choice for the three insemination procedures. These are (i) the number of mature oocytes, (ii) the number of live spermatozoa (and the volume of the final sperm suspension) following removal of seminal plasma and sperm workup, and (iii) the previous history of the patient. It would be clinically irresponsible to perform regular IVF (or even PZD) when only a few motile spermatozoa can be isolated. On the other hand, one can only perform IVF and PZD simultaneously when the volume of the final sperm suspension is sufficient. A minimum volume of approximately 30 μL is required for insemination of oocytes in microdroplets under oil (2). In addition, patients with fertilization failure following standard IVF do not generally consent to the use of zona-intact sibling oocytes. We have shown that a couple who fails to fertilize following a single IVF attempt has less than a 25% chance of fertilization when conventional IVF is reapplied. This figure obviously differs among programs and should be evaluated prior to the introduction of assisted fertilization by micromanipulation.

Despite these issues we have conducted two controlled studies under the supervision of institutional review boards, comparing IVF and PZD in 84 patients prior to routine introduction of PZD (Table 15.1). Since these studies were performed during the early phases of our microsurgical investigations, the patients had better sperm profiles than those referred to our program in the last 24 months. Moreover, only a minority of them had a failure with conventional IVF prior to participation in the studies. This is reflected in the relatively high rates of fertilization (34%) when zona-intact oocytes were inseminated. Nevertheless, the incidence of monospermic fertilization improved significantly to 50% when the spermatozoa were incubated with micromanipulated oocytes.

When SZI was added to the program, we conducted a small controlled study (n = 22), comparing PZD with SZI in patients who were not acceptable for regular IVF due to poor semen profiles (see below for definitions). The incidence of monospermic fertilization in these patients was significantly enhanced when SZI was performed (Table 15.2). However, it must be noted that the sperm suspensions were very poor and dilute, diminishing the likelihood that the spermatozoa would find

TABLE 15.1. Controlled studies involving PZD using sibling oocytes of patients who had sufficient numbers of spermatozoa for two types of insemination.

Source (ref.)	No. of patients	Incidence of fertilization (%)		Cycles with monospermic fertilization (%)	
		IVF	PZD	IVF	PZD
Cohen et al., 1990 (49)	47	42/129 (33)	75/138 (54)	26/47 (55)	33/47 (72)
Cohen et al., 1991 (2, 16)	37	68/195 (35)	82/179 (46)	17/37 (46)	33/37 (89)
Combined	84	110/324 (34)	157/317 (50)	43/84 (51)	66/84 (79)
P-value		<0.05		<.005	

Note: These patients belonged to groups A and C (discussed in "Patient Selection").

TABLE 15.2. Summary of microsurgical fertilization results obtained during 1990 in couples with few spermatozoa in whom PZD was compared with SZI using sibling oocytes.

	Fertilization (%)	Cycles with fertilization (%)
SZI	37/125 (30)[a]	15/22 (68)
PZD	11/86 (13)[a]	9/22 (41)

Note: Most of these patients belonged to group B (discussed in "Patient Selection").
[a] $P < 0.01$.

TABLE 15.3. Frequency of sperm fusion after SZI for insemination of oocytes in male factor patients and reinsemination of oocytes that failed to fertilize after standard IVF in male and nonmale factor patients.

Type of infertility	SZI performed at	No. oocytes	No. sperm (avg.)	No. male pronuclei	F_f (%)
Male	Insemination	184	921 (5.0)	130	14
Male	Reinsemination	50	228 (4.6)	17	7
Nonmale	Reinsemination (all oocytes)	81	132 (1.6)	49	37
Nonmale	Reinsemination (normal oocytes)	19	69 (3.6)	41	59

Source: Data from Alikani, Adler, Reing, Malter, and Cohen (25).

the artificial gaps in the PZD oocytes. Actual fertilization percentages following the application of both techniques are very similar when other patients are included in the evaluation (2).

Frequency of Sperm-Egg Fusion Following SZI

Considering the lack of a fast block to polyspermy on the level of the human oocyte membrane, it is very likely that the success rates of these procedures, measured as the frequency of sperm fusion, could be enhanced if one increased the number of spermatozoa (25). However, there is a general consensus that one should limit the number of spermatozoa in order to avoid polyspermy. A surprisingly large proportion (>10%) of spermatozoa from subfertile men is able to form pronuclei after deposition into the perivitelline area. This figure increases when sperm cells from fertile men are placed into the perivitelline area (Table 15.3).

Patient Selection

Assisted fertilization offers treatment for patients with impaired sperm function due to oligozoospermia, asthenozoospermia, and/or terato-

TABLE 15.4. Selection criteria for microsurgical fertilization and regular IVF in instances of male factor infertility and idiopathic failure of fertilization.

Group	Selection criterion	<50,000 motile sperm recovered	Semen cutoff values			Treatments		
			Count (×10⁶/m)	Motility (%)	Normal forms (%)	PZD	SZI	IVF
A	Previous failure of fertilization	Occasionally	0.1–10	<1–20	0–10	+	+	Op[a]
B	Semen analysis unacceptable for regular IVF	Always	<0.1–5	<1–10	0–2	Op[a]	+	–
C	Semen analysis acceptable for IVF, but reduced prognosis	Not applicable low progression and survival	0.5–20	1–30	3–10	+	+	+
D	Semen analysis acceptable for IVF	Not applicable	2.0–20	5–30	4–10	–	–	+
E	Previous IVF failure and normal semen	Not applicable	≥20	≥30	>10	+	+	Op[a]

[a] Optional.

zoospermia (2, 16). However, certain oocyte abnormalities can also be potentially treated through micromanipulation. We have performed microsurgical fertilization in four groups of patients according to criteria outlined elsewhere and in Table 15.4 (16, 24). The first two groups—groups A (n = 170 cycles) and E (n = 58 cycles)—comprise 47% of the total number of patients (n = 487 cycles) (Fig. 15.2). Patients from group A had severely abnormal semen analyses (0%–10% normal sperm forms according to Kruger's strict criteria, $\leq 20\%$ motile and/or $\leq 10 \times 10^6$ spermatozoa/mL) and failed to fertilize all oocytes in a previous IVF cycle (27). Some of these patients had repeated cycles of fertilization failure. Patients from group E had idiopathic failure of fertilization in a previous IVF cycle with apparently normal semen analyses.

The third group of couples—group B (n = 111)—had not been accepted for regular IVF by any other programs, including our own. These patients' semen analyses were considered highly abnormal,

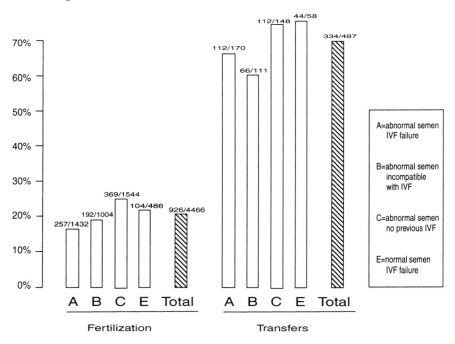

FIGURE 15.2. Incidences of monospermic fertilization and replacement in 487 patient cycles in which microsurgical fertilization (PZD, or SZI, or both) was attempted. Alphabetical numbers represent patient groups; for further definitions, see "Patient Selection" (The New York Hospital–Cornell University Medical College).

consisting of less than 2% normal sperm forms (Kruger's strict criteria) in combination with either extreme oligozoospermia ($\leq 5 \times 10^6$/mL) or extreme asthenozoospermia ($\leq 10\%$ motile). Few motile spermatozoa could be retrieved from their semen even if the last sperm pellet was resuspended in a volume of less than 50 µL. Actual sperm counts were not available for a number of these patients since the sperm had first to be centrifuged in order to retrieve a few in the counting chamber. In others motile spermatozoa were not seen, and an eosin stain was used to detect live spermatozoa. In such couples SZI was performed exclusively.

The fourth group—group C (n = 148)—had not attempted IVF previously (some were not acceptable for other IVF programs), and all had male factor infertility of intermediate severity. All these men had teratozoospermia ($\leq 10\%$ normal sperm forms). In addition, their semen was considered oligozoospermic ($<20 \times 10^6$/mL) and/or asthenozoospermic ($\leq 30\%$ motile with reduced linear progression). Based on these semen parameters, fertilization failure was anticipated in this group (group C, Table 15.4). However, standard IVF was not completely ruled out. Micromanipulation was therefore performed in some of these instances, while most oocytes were left zona intact. The results of group D (regular IVF/male factor) patients are not presented here.

During the first 28 months of our assisted fertilization program, 487 micromanipulation cycles were performed, resulting in 109 clinical pregnancies (22%). The implantation results of the patients with abnormal semen analyses (groups A, B, and C) were significantly higher ($P < 0.01$) than that of the patients with normal semen (group E) (Fig. 15.3). However, monospermic fertilization in all groups varied between 18% and 24% with a mean of only 21% (Fig. 15.2). Seventy of the 109 clinical pregnancies resulted from replacement of microsurgical embryos exclusively. The largest group of 42 pregnant patients had only SZI embryos replaced. Seven pregnancies were established in patients in whom only zona-intact embryos were transferred. The remaining pregnancies were from mixed (microsurgical and zona-intact) embryo replacements. The majority of these pregnancies were established in patients from group C, indicating that microsurgical fertilization was more effective in the other groups. However, 8 truly microsurgical pregnancies were established in group C patients, while regular IVF failed. This suggests that the use of microsurgery will be moderately useful in male factor patients who are also acceptable for regular IVF.

Semen Factors and Limitations of Assisted Fertilization

It has been shown that SZI and PZD are complimentary techniques (2, 16). Fertilization following SZI was higher than with PZD in patients who had few spermatozoa available for insemination. However, spermatozoa from group E patients seemed only to be successful following the

Microsurgical Fertilization Cornell: 1989 - 1992

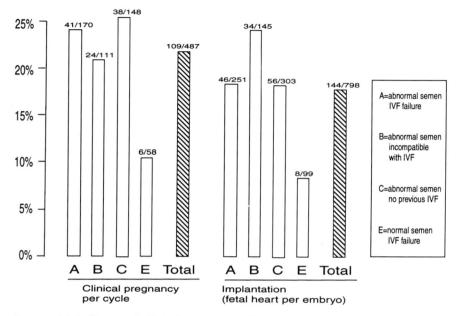

FIGURE 15.3. Rates of clinical pregnancy (fetal heart activity) and embryonic implantation in 487 patient cycles in which microsurgical fertilization (PZD, or SZI, or both) was attempted. Alphabetical numbers represent patient groups; for further definitions, see "Patient Selection" (The New York Hospital–Cornell University Medical College).

application of PZD. Furthermore, it was shown that the results of these two methods were additive in terms of implantation. PZD embryos preferentially implanted in patients whose partners had moderate teratozoospermia, whereas SZI embryos implanted more frequently in instances of severe teratozoospermia. Moreover, a comparison of our results before and after the introduction of SZI has demonstrated that the overall pregnancy rate would be considerably reduced if one of the methods were discontinued. There are, however, major differences between the two micromanipulation methods. One of the advantages of SZI is that it allows for assisted fertilization to be performed when only a few spermatozoa can be retrieved. Consequently, potential patients could include those with complete absence of motility, extreme oligozoospermia ($<1 \times 10^6$/mL), and teratozoospermia. This policy of not employing restrictive criteria has been in effect in this program for over a year (24).

Recently, we analyzed the results of the first 250 microsurgical fertilization cycles (groups A, B, and C) in order to investigate the prognosis for patients with severely abnormal semen analyses and to determine whether low cutoff limits should be implemented for application of assisted fertilization. The results of the micromanipulation procedures in the first 250 cycles were correlated with fertilization and implantation using a variety of analyses. For the first breakdown absolute figures for motility, sperm concentration, and number of normally shaped spermatozoa were correlated with the outcome of assisted fertilization. In a second analysis patients with a single abnormality of semen (either oligozoospermia, asthenozoospermia, or teratozoospermia) were compared to those with two or all three abnormalities.

No significant correlation was found between percentage normal sperm forms and fertilization. Similarly, pregnancy and embryonic implantation results were not affected by sperm morphology. Consequently, the severity of teratozoospermia cannot be used as a prognostic factor for predicting the outcome of microsurgically assisted fertilization. It is likely that this is due to the efficiency of SZI in extreme teratozoospermic cases. Only 7% of PZD eggs were fertilized in patients with 0% normal sperm forms, whereas 20% of eggs were fertilized in such patients when SZI was performed. Moreover, it was shown previously that the ability of SZI embryos to implant from patients with extreme teratozoospermia (0%–2% normal forms) was not impaired, whereas PZD embryos from similar patients rarely implanted (16). Indeed, 64 of such cycles attempted in this series led to 14 (22%) clinical pregnancies. In further analyses we were not able to correlate fertilization and implantation with sperm concentration or motility.

The results of microsurgical fertilization expressed as a function of the number of semen abnormalities (oligozoospermia, asthenozoospermia, and teratozoospermia) are presented in Figure 15.4. It is obvious from these results that the outcome of microsurgical fertilization cannot be predicted on the basis of the WHO criteria. The presence of one, two, or three abnormalities does not affect the rates of fertilization and pregnancy.

A substantial number of viable pregnancies were established well below the normal cutoff values for regular assisted conception procedures. Fertilization and pregnancy occurred following the use of spermatozoa without progressive motility or normal morphology. In some patients sperm counts were reduced, and spermatozoa could only be visualized after centrifugation. These findings provide evidence that spermatozoa from extremely oligo-astheno-teratozoospermic men can produce normal offspring after the application of micromanipulation techniques even when fertilization previously failed following standard IVF. Subgroups of patients who would not benefit from microsurgical techniques could not be identified by analyzing semen analyses. Though spermatozoa from approximately 1/3 of the patients failed to fertilize altogether, this could also not be correlated with semen analysis.

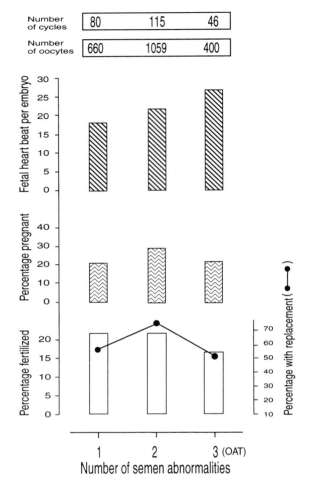

FIGURE 15.4. Outcome of the microsurgical procedures (PZD and SZI are combined) according to the number of semen abnormalities (WHO). OAT is combined oligozoospermia, asthenozoospermia, and teratozoospermia.

Augmenting Fertilizing Ability

There appear to be two major drawbacks of assisted fertilization: (i) the lack of fertilization in approximately 30%–70% of the patient population, and (ii) the lack of implantation. Though the latter phenomenon affects most of the programs, it has not played a role in our microsurgical program. The possible causes for this discrepancy in results have been discussed above. The issue of lack of fertilization seems to be of a more global nature. It is likely that both SZI and PZD should be practiced simultaneously since fertilization is not affected by the severity of semen profiles in our program (24). However, the monospermic fertilization rate is rarely in excess of 20%. One factor that can affect the chance of having a replacement is the quality and quantity of oocytes. However, even if microsurgical fertilization patients have a sufficient number of oocytes, fertilization rates are usually lower than 30%.

The lack of well-defined methods for improving the IVF ability of sperm is a direct consequence of the paucity of our knowledge concerning the fundamental nature of defective spermatozoa. Progress in this particularly elusive area of human reproduction is dependent on a systematic analysis of the molecular basis of sperm function. Currently, there are three routes one can follow regarding the enhancement of the fertilizing capacity of deficient human sperm. These are (i) the appropriateness of each method of semen preparation, (ii) the induction of capacitation and the acrosome reaction, and (iii) the inhibition of reactive oxygen species production (24). The first two factors have been investigated in some detail in our center with emphasis on the application of microsurgical fertilization. Preparative techniques involve the use of Percoll gradients, proteolytic enzymes, swim-up procedures, and sperm sedimentation methods (2). Methods are combined depending on the semen profile and the previous history of the patient. This aspect of the technology is consequently more individualized. The application of technology involving the artificial enhancement of capacitation and the acrosome reaction has not yielded standardized methods for increasing the success rates of assisted fertilization despite their successful application in hamster egg assay or mouse egg studies. We have performed pilots with electroporation and cryoshock of spermatozoa using sibling oocytes during PZD and SZI. This did not result in a consistent improvement of the fertilizing ability of impaired spermatozoa.

Several research groups (28, 29) have indicated that phosphodiesterase inhibitors, such as *pentoxyfylline* (px) and *deoxyadenosine* (dAdn) are effective in promoting fertilization. We have investigated the possible use of a combination of these two compounds (29) in a study involving 116 microsurgical fertilization patients (30). Both compounds were used in equivalent amounts at a concentration of 3 mM or 1 mM. Forty-seven of the patients yielded insufficient numbers of spermatozoa to perform the sibling study. In these patients px-dAdn was used exclusively, and comparisons with conventional semen preparation methods were not made. The incidences of fertilization and implantation did not appear to be higher than those reported in other studies (2, 9, 24).

In a second group of 69 patients from whom sufficient spermatozoa were retrieved, the investigations were performed using sibling oocytes. Spermatozoa were prepared in the presence or absence of px-dAdn. The sperm suspensions were then used for SZI, PZD, or regular IVF depending on the criteria outlined in the section on patient selection

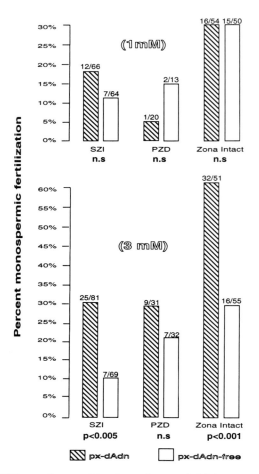

FIGURE 15.5. Effects of a combination of px and dAdn on assisted fertilization. This study involved 116 microsurgical fertilization patients in whom sibling oocytes were treated with sperm that was preincubated with or without both compounds. Both phosphodiesterase inhibitors were used in equivalent amounts at a concentration of 3 mM or 1 mM as described in reference 30.

above. The fertilization results are presented in Figure 15.5. The use of px-dAdn at a high concentration of 3 mM appeared successful following SZI and regular IVF. For unknown reasons it was not successful when PZD was applied or when the doses were reduced to 1 mM. It is recommended that these compounds be used with care despite these promising results. One hundred and seventy-one embryos were replaced in these patients, and only 27 (16%) implanted, which was less than we have previously obtained. A comparison between replacements involving embryos derived from spermatozoa exposed to px-dAdn alone and mixed

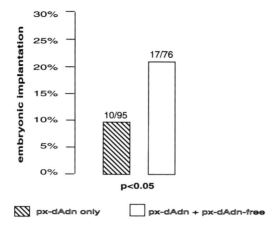

FIGURE 15.6. Rates of embryonic implantation following replacements involving embryos derived from sperm that were exposed to a combination of px and dAdn prior to assisted fertilization compared to replacements with embryos of mixed origin.

batches of embryos (treated and nontreated sperm) shows that px-dAdn embryos may be impaired in their capacity to implant normally (Fig. 15.6).

Offspring from Microsurgical Fertilization

One of the most important aspects associated with the evaluation of a novel assisted reproductive technique is that of pregnancy outcome. Thus far, 66 healthy babies were born following 50 deliveries of couples in whom either SZI, PZD, or a combination thereof was attempted at our center (Table 15.5). Of the first 109 clinical pregnancies obtained following the use of these microsurgical techniques, 12 have now miscarried (11%), but this figure is expected to increase since many of the pregnancies are still within the first 20 weeks of gestation. The miscarriages do not appear to be associated with specific semen profiles. The true success rate of microsurgical fertilization can only be assessed by determining the percentage of deliveries per attempt (= egg retrieval). Accordingly, we can now evaluate the first 250 patient cycles in our assisted fertilization program (Fig. 15.7). The take-home baby rate was only 10% during the first 100 attempts, but this rate doubled during the next 100 attempts. From our current results we estimate a true success rate of 18%, a figure that appears somewhat higher in comparison to the first 250 attempts (15.6%). The latter percentage is relatively low due to the natural learning curve associated with the incorporation of each new technique.

TABLE 15.5. Current (April, 1992) results of microsurgical fertilization procedures including ongoing pregnancies and live deliveries.

Patient cycles	487 (*a*)
Replacements	334 (69%) (*b*)
Positive βhCG	134 (28% from *a*) (40% from *b*)
Biochemical miscarriages	25
Clinical pregnancy	109 (22% from *a*) (33% from *b*) (*c*)
Clinical miscarriages	12 (11% from *c*)
Ongoing/delivered	97 (20% from *a*) (29% from *b*)
Delivered patients (babies)	50 (66%)*
Number of babies from PZD embryos	11
Number of babies from SZI embryos	18
Number of babies from a combination of PZD/SZI embryos	6
Number of babies from a combination of zona-intact/PZD/SZI embryos	27
Number of babies from zona-intact embryos	4

*In this group of patients, 47 pregnancies were still ongoing when this chapter was submitted.

Correction of Polyspermy

Zona drilling techniques provide the experimental embryologist with information on the efficiency of the block to polyspermy on the level of the plasmalemma and zona reaction. Breaching the zona of mouse eggs, for instance, leads to low levels of polyspermic fertilization, whereas similar procedures in hamster oocytes will induce polyspermy uniformly. It appears, though this is disputed by some, that the block to polyspermy in the human is similar to that of the hamster model (14, 15, 31). Rates of polyspermy following microsurgical procedures are elevated to approximately 25% of fertilized eggs (2). This level is especially high if one considers that the populations of spermatozoa used represent those with decreased fertilizing ability.

One group of researchers reported relatively low rates of polyspermy following subzonal sperm insertion (15). Rates of polyspermy following partial zona dissection in instances of normal sperm function may be as high as 50% (31). However, a weak block to polyspermy is likely to be in place on the level of the plasmalemma. One would expect equal levels of 3-, 4-, and 5-pronucleate embryos, and so on, after zona drilling if the membrane and cytoplasmic blocks were completely inactive. However, most multinucleate zygotes are dispermic, and the degree of multinucleation diminishes with each additional sperm being incorporated (2). The level of dispermy is lower than trispermy when 3 or more spermatozoa are deposited into the perivitelline area (2). It can therefore

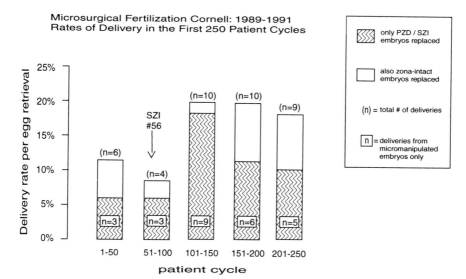

FIGURE 15.7. Baby take-home rates following completion of the first 250 assisted fertilization cycles (The New York Hospital–Cornell University Medical College).

be concluded that the block to supernumerary sperm penetration in the human primarily resides at the level of the zona pellucida (30). Also, in rare cases IVF cycles have been observed where a high degree of polyspermy occurs in the absence of zona micromanipulation. The ability to control polyspermy or to correct the genetically abnormal embryos that result from it would be of obvious value to clinical IVF.

There is a large body of research regarding genetic manipulation in rodents and large domestic species based on techniques for the vital, intact removal of pronuclei (32, 33). It would seem that these techniques could be modified and used to simply remove the extra sperm pronucleus, thereby returning the human zygote to a normal genetic complement. Preliminary experiments toward this goal have been reported by several groups (34–36).

The survival rate in our enucleation work is now well over 50% (Table 15.6). Although we have noted that a large proportion of the surviving embryos continue development, we have usually not followed their development and have opted to perform ultrastructural and genetic analyses. The central issue in using enucleation to correct polyspermy is the correct identification of the supernumerary male pronucleus. This identification is simple in rodent zygotes where nuclear size and the presence of sperm tail remnants, as well as position in relation to the *second polar body* (PBZ), are all valid criteria. In human zygotes size

TABLE 15.6. Results of microsurgical enucleation for correction of polyspermy in the human.

Source (ref.)	No. micromanipulated	No. survived (%)	No. cleaved (%)
Malter and Cohen, 1989 (36)	25	9 (36)	7 (28)
Malter and Cohen, 1991, unpublished	70	60 (87)	44 (63)

appears to be variable, and sperm tail remnants can almost never be identified by light microscopic observation (36, 37).

Pronuclei that are farthest from PBZ are currently being selected for removal; genetic analysis will hopefully prove this to be a valid criterion for identification. Currently, we perform multiplex PCR and in situ hybridization of blastomeres removed from enucleated 8-cell human embryos (Munne, Stein, Tang, Reing, Cohen, Grifo, unpublished observations). Using the postulation that the pronucleus farthest from PBZ is likely to be a male pronucleus, our laboratory has thus far completed studies on 11 enucleated embryos, of which, 4 provided signals for PCR. Two embryos were XX, whereas the 2 others were XY. If the distribution of the 3 pronuclei were random in relation to PBZ, 1/3 of enucleated zygotes would become androgenetic. Of these, 25% should have Y-signal exclusively. This approach would enable us to investigate whether the parental origin of the pronuclei would be topographically fixed, provided that a relatively large number of multiplex experiments are performed. Alternatively, one could perform X- or Y-labeling of the removed pronucleus, though this does not indicate whether that specific zygote is androgenetic. Thus far, we have not been able to perform PCR on removed pronuclei, presumably since the enucleation procedure does not remove karyoplasts, including completely intact pronuclei, but rather fragments containing nucleoplasm as well as nucleoli.

A third approach for confirmation of parental origin of human pronuclei is the use of DNA-specific vital stains preincubated with either sperm or eggs prior to insemination. Preliminary mouse experiments in our laboratory have been unsuccessful since the stain leaked from the pronuclei. Further experiments in the human were somewhat more revealing since the prestained and washed human oocyte does not lose the DNA marker as fast as the mouse egg (Dale, Cohen, unpublished observations). Twenty mature human oocytes have thus been preincubated with a vital fluorescent DNA stain and washed after 15 minutes prior to insemination with unstained sperm. Four of the monospermic zygotes had only a single fluorescing pronucleus. The stained pronuclei were always adjacent to PBZ. These preliminary experiments provide the impetus for further investigation and underline the hypothesis that the female pronucleus is in a relatively fixed position in relation to PBZ. However, until it is certain that polyspermy repair produces normal, diploid embryos with both a maternal and paternal component, clinical application would be ill-advised.

Assisted Hatching

There have been several proposals for improving IVF techniques ranging from the use of complex media supplements, such as immunosuppressants, to the application of helper cell systems (10, 11, 38, 39). However, true improvements in conventional cell-tissue culture technology among IVF laboratories have not been realized. Rarely are more than 15% of human embryos viable even in the most carefully controlled programs. This figure is frequently unclear since general IVF success rates are often expressed in terms of transfers involving sets of sibling embryos. Though this practice appears acceptable, it does not address the principle flaw of IVF, namely, that of embryonic wastage. A review of the recent literature shows that there have been only two promising proposals for enhancing embryonic viability. The first of these involves the simultaneous culture (coculture) of embryos with helper cells and is based on the vast experience of embryologists who handle early stage embryos from larger domestic species (10, 11, 39). The second method of assisted hatching is based on the hypothesis that some zonae inhibit the normal escape of otherwise normal embryos during blastocyst expansion (6, 12, 13).

Two particular findings have been crucial for the decision to implement assisted hatching clinically. First, cleaved embryos with a high prognosis produce an active component that reduces the thickness of the zona, presumably in preparation for subsequent hatching (10, 40, 41). Second, microsurgically fertilized embryos with artificial gaps in their zonae appear to have high rates of implantation (see above). Assisted hatching was first experimentally tested by introducing small incisions in the zonae of human 4-cell embryos using a mechanical method (6, 42). Though the resulting preliminary work was encouraging, routine application was not implemented at our program for several reasons. First, some of the spare embryos that were observed for prolonged periods became trapped in the narrow openings during hatching (42). This phenomenon was confirmed in a mouse model (43). Second, other embryos were possibly damaged during embryo replacement prior to the formation of structural junctions between the blastomeres (2, 44). Alternatively, it was proposed that larger openings be created in the zonae of human embryos undergoing initial compaction (2, 12, 13).

Recently, we completed three studies on the use of zona drilling (3) with acidic Tyrode's solution in 3-day-old human embryos (13, 45). The trials were performed in a completely randomized fashion in 330 IVF couples. The first and second trials included patients with normal basal *follicle stimulating hormone* (FSH) levels. During the first trial the use of assisted hatching with zona drilling in one group of patients was compared to a control group of patients whose embryos were not micromanipulated. Patients whose embryos had thick zonae derived the greatest benefit. This finding was tested prospectively in the second trial by performing assisted hatching selectively only on embryos with thick zonae in patients of the experimental group and comparing the outcome with patients from a control group without micromanipulation. The third trial was similar to the first trial, but performed in patients with elevated basal FSH levels.

Assisted Hatching: Animal Studies

One of the first requirements for the safe application of micromanipulation to human embryos is the study of these relatively invasive techniques in animal models. There is currently no other species in which the hatching mechanism is known to be dysfunctional. The hatching process of embryos from domestic species and rodents does not appear to be inhibited, though studies of preimplantation embryonic development following the application of IVF in, for instance, the bovine have not yet been performed. Hence, the use of laboratory animals for the study of assisted hatching procedures appears to be limited to the creation of specific models in which the hatching process is inhibited artificially and then reversed by micromanipulation. Mouse embryos flushed from common laboratory strains will usually develop to the blastocyst stage in vitro and hatch fully. It was shown by Malter and Cohen that micromanipulated mouse embryos hatch through the artificial gap and that hatching is initiated at an earlier stage (42). Depending on the size of the hole, embryos may become trapped in a characteristic figure-eight shape (42, 43, 46, 47). Hatching does not appear to be facilitated in embryos whose zonae have been pierced once with a sharp needle (43). One or several large holes introduced by zona drilling (3) appear to be more favorable for the integrity of the embryo than the introduction of a small slit in the zone following partial zona dissection (12, 14).

Two models have been recently suggested for the study of assisted hatching in the mouse. The first was applied in embryos with poor in vitro development. Khalifa and Tucker (48) micromanipulated morulae using acidic Tyrode's solution to either drill through the zona or to partially thin a larger area. Initiation of hatching occurred in significantly more micromanipulated than control embryos. However, the rate of completion of hatching was doubled in partially thinned embryos compared to the other groups. A second mouse model was applied in strains with normal in vitro development (47). Embryos were kept in protein-free culture medium, which inhibits hatching. Zona drilling performed at the cleavage stage followed by further culture in a protein-free environment significantly improved hatching. Both models could be used for refining assisted hatching techniques.

Assisted Hatching: Human Studies

Since the initiation of clinical assisted hatching in our laboratory three years ago, the methodology has been modified on several occasions. The first clinical experiments involved partial zona dissection of the 4-cell embryo. This method was applied prospectively in 99 consenting patients in a randomized fashion (6, 49). Though the pregnancy rate increased by 20% and the embryonic implantation nearly doubled, partial zona dissection is no longer the method utilized for assisted hatching. The holes created with this technique may be too small to allow normal completion of hatching (43). Furthermore, the results of some clinical follow-up studies have been disappointing.

High rates of implantation following biopsy of 8-cell human embryos using acidic Tyrode's have been reported (1). Biopsied embryos were replaced at the time of initial compaction (44). We therefore suggested that it may be advantageous to transfer embryos with substantially large holes in their zonae after interblastomere adherence has increased (2, 12). The use of zona drilling on day 3 of embryonic development was investigated in three fully randomized prospective trials in 330 consenting patients at The Center for Reproductive Medicine and Infertility at the New York Hospital-Cornell Medical Center. The specific methodology, embryo selection criteria, and results within each trial are being published elsewhere (13, 45).

The combined results of these trials are presented in Fig. 15.8. The incidence of clinical pregnancy (fetal heartbeat per patient) increased significantly ($P < 0.01$) from 37% (62/166) in the control group to 52% (85/164) following zona drilling. Twenty-seven percent of the zona-drilled embryos implanted (147/555) and showed fetal heart activity on ultrasound. This compared favorably ($P < 0.01$) with the control group in which 19% (104/555) of the embryos implanted. However, it must be noted that these results were obtained during three trials. Two trials were performed in consenting patients allocated to control or micromanipulation groups. The embryos from patients allocated to the micromanipulation group were always zona drilled. The first trial (n = 137) was performed in patients with an average age of 36 and normal basal FSH levels. The third trial (n = 30) included only patients with elevated basal FSH levels (>15 mIU/mL). Though embryonic implantation increased in both trials following zona drilling, the difference with the control group was significant only in the elevated basal FSH group (13).

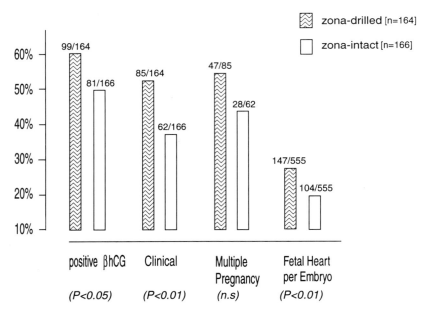

FIGURE 15.8. Combined results of 3 fully randomized assisted hatching trials involving 330 patients using zona drilling at day 3 of development as described in reference 13.

Selective Assisted Hatching

Retrospective analysis of embryos replaced during the first trial demonstrated that the zona pellucida thickness largely determined the outcome of the procedure (13). Control embryos with zonae thicker than ⩾15 μm rarely implanted, whereas zona-drilled embryos with similar zona characteristics frequently implanted (Fig. 15.9). The findings also suggested that zona drilling was detrimental in embryos with thin zonae (<12 μm). However, the latter evaluation was not statistically significant. These retrospective findings were prospectively tested during a second trial involving 163 patients. Zonae from embryos of patients allocated to the zona drilling group were measured prior to micromanipulation. Embryos with thick zonae were micromanipulated (⩾15 μm), and those with thin zonae (⩽12 μm) were left intact. The sum result of this group of patients (2/3 of their embryos were micromanipulated) was compared with a control group in which embryos were never micromanipulated. This process has been called *selective assisted hatching* (13). The retrospective conclusion from the first trial was confirmed during this selective assisted hatching trial. Both the incidences of clinical pregnancy as well as embryonic implantation increased significantly following the application of assisted hatching.

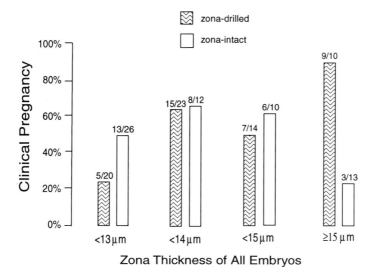

FIGURE 15.9. Effect of zona thickness on the clinical pregnancy rate of patients. The data were obtained from retrospective measurements of embryos from trial 1 patients as described in reference 13.

Figure 15.8 therefore presents the combined results of selective assisted hatching and regular assisted hatching performed in patients of varying etiology and basal FSH levels. Correlation of patients based on maternal age indicates that selective assisted hatching was most effective in the group over 38 years (Fig. 15.10). In order to evaluate the possibility of quantitative zona changes as a function of maternal age, we assessed the average zona thickness and percentage zona variation in 1023 embryos on day 3 following egg retrieval. Zona biometrics did not change in patients over 31 despite the age-related differences in response to assisted hatching (45). The mean zona thickness of patients younger than 32 was at least 1 µm greater than that of other patient groups. It can be postulated that zona deposition is an age-related process. This finding should probably be considered when evaluating the possibility of assisted hatching in embryos with thick zonae from young patients.

We have now incorporated selective assisted hatching into the IVF procedure for those embryos with thick zonae in consenting patients younger than 39 who have normal basal FSH levels. Zona drilling of all embryos, regardless of zona thickness, is being performed in all other consenting patients. Results thus far indicate that approximately 1/4 of human embryos have the ability to implant and that a clinical pregnancy rate of over 50% per transfer may be a possibility for IVF patients in the future.

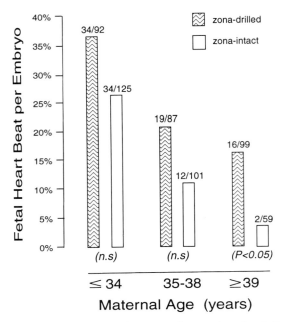

FIGURE 15.10. Effect of maternal age on the results of a fully randomized trial using selective assisted hatching of poor embryos as described in reference 13.

Partial Zona Thinning of Embryos with Thin Zonae and Good Prognosis

One of the observations based on the zona drilling trials presented above is that acidic Tyrode's solution may be detrimental to embryos with thin zonae (13). Selective zona drilling of embryos with thick zonae now results in implantation rates that are similar in both thick and thin zonae groups. However, in order to investigate whether implantation of embryos with thin zonae (<13 μm) can be improved further, a fourth trial was executed in 40 consenting patients (45). Patients were randomized in two groups. Zonae were measured prior to replacement in both groups. Embryos with zonae ⩾15 μm (poor-prognosis embryos) were zona drilled in both arms of the trial. Embryos with thin zonae were left intact in one group (control), while the outside of zonae of similar embryos in the other group were thinned with acidic Tyrode's solution using the technique of Khalifa and Tucker (48). The results thus far indicate that embryos with thin zonae do not benefit from this technique. However, their implanting potential was not jeopardized as was previously shown to occur after the zonae were pierced completely by zona drilling (13).

Embryonic implantation was high (26%–27%) in both arms of the trial, probably as a result of selective zona drilling of low-prognosis embryos with thick zonae.

Zona Hardening and Assisted Hatching

The zona reaction encompasses an increased resistance to dissolution by various chemical agents, a process that is called *zona hardening* (50). Various agents, including proteases, disulfide-bond reducing agents, sodium periodate, low pH, and high temperature, have been applied to study the phenomenon of zona hardening (51). It has been shown that the zona is more readily removed from unfertilized oocytes than from embryos. The sensitivity of the human zona to decreased pH or mechanical piercing appears to follow an inverse pattern since it is easier to drill a hole in a zygote than in an unfertilized oocyte. The human zona becomes more friable and loses its elasticity after fertilization. In addition to fertilization-induced hardening, spontaneous hardening also occurs after in vitro culture (52, 53) and in vivo aging (54).

Zona hardening may be important for three postfertilization events: the block to polyspermy (55), protection of the developing embryo (56), and oviductal transport (57). Though the process is believed to be a mechanical change, assays usually involve solubilization techniques. Recently, Drobnis et al. (51) used deformation of the zona aspirated into a micropipette (zona deformability) as an assay for zona hardening. This study revealed that mechanical deformability occurred in hamsters and mice. Hamster oocytes do not show zona hardening following regular solubilization methods. This technique quantitates zona changes without jeopardizing zona integrity and could therefore be of use clinically. Prospective quantitative studies involving zona hardening measurements have not yet been performed clinically. However, zona piercing for assisted hatching allows for indirect assessment of zona hardening. The effort involved in opening the zona pellucida with acidic Tyrode's solution can be scored by the observer as "easy to pierce" or "difficult to pierce," but this is a subjective determination.

Alternatively, we developed a method to assess zona hardening from videotaped zona drilling procedures. The videotapes of the zona drilling in embryos of 25 patients have been analyzed retrospectively. Exposure to acidic Tyrode's solution for each embryo has been expressed as a function of the time needed to pierce the outside and inside layers of the zona pellucida and the diameter of the needle used, to compensate for needle-size variation. The preliminary results presented in Table 15.7 suggest that patients whose embryos have zonae that were digested by small amounts of acidic Tyrode's solution ($<200\,\mu m\,sec$) have a higher chance of a multiple pregnancy than those patients in which zona dissolution required greater amounts of acidic Tyrode's solution. These

TABLE 15.7. Effect of acidic Tyrode's solution on zonae of embryos as a measurement of zona hardening.

Exposure to acidic Tyrode's (μm sec)	Proportion of	
	Pregnant patients (%)	Multiple pregnancies (%)
<200	10/17 (59)	9/10 (90)
≥200	4/9 (44)	1/4 (25)

Source: Data from Cohen, Alikani, Reing, Ferrara, Trowbridge, and Tucker (45).

preliminary findings suggest that embryonic viability is possibly correlated with zona hardening.

Removal of Extracellular Fragments

The importance of extracellular fragmentation has been traditionally overemphasized in clinical IVF. We have been unable to establish a correlation between moderate extracellular fragmentation (≤15% fragments) and implantation (10, 40, 41). Nevertheless, the presence of excessive amounts of degenerate material within the perivitelline area may affect embryonic viability. In order to test this hypothesis, we have removed small numbers of fragments from embryos during the assisted hatching trials. However, this pilot study was not done in a randomized fashion. In some more excessively fragmented embryos, fragments were removed, whereas other less fragmented embryos were left intact. It is therefore hard to quantify, based on our data, what the impact of fragment removal could be on further embryonic growth.

During zona drilling embryos were clamped onto the holding pipette in such a way that the acidic Tyrode's-filled microneedle is positioned at the three o'clock area either adjacent to empty perivitelline space or anucleate fragments (2). Small holes were widened mechanically by moving the microneedle through the opening in a tearing motion, and gentle suction was used in order to remove fragments. Most of the embryos from 52 patients were partially fragmented. Fragment removal was performed in those patients (n = 36) who had embryos with excessive amounts of fragments (>15%). In embryos of 16 other patients, fragments were also present, but they were not removed (Table 15.8). Though the clinical pregnancy rates were similar in these two groups (41% and 44%, respectively), it may be noted that the success rates were lower than those of the rest of the patients allocated to the zona drilling trials (Fig. 15.8). Nevertheless, it may be concluded that (i) fragment removal can be performed with relative ease, (ii) the risks of the procedure are relatively minimal, and (iii) the pregnancy rate was relatively high considering the poor morphology of the embryos involved. Additional controlled studies will be helpful in order to further extend this hypothesis.

TABLE 15.8. Effect of fragment removal following zona drilling on embryonic viability.

Procedure	Proportion of patients with	
	Positive βhCG (%)	Fetal heartbeat (%)
Fragments removed	17/36 (47)	15/36 (41)
No fragments removed	9/16 (56)	7/16 (44)

Source: Data from Cohen, Alikani, Reing, Ferrara, Trowbridge, and Tucker (45).

Future Applications of Micromanipulation

It is unlikely that breakthroughs will depend solely on the development of new, sophisticated micromanipulation procedures. As experience is gained both in large domestic animal as well as human IVF, present techniques will undoubtedly be refined, and new applications will be developed. Advances in instrumentation, particularly involving the use of the laser, may provide solutions to some existing problems. Progress in the field of assisted reproduction and that of applied reproductive physiology in general will be largely determined by the coevolution of medical ethics and molecular genetics. There are a number of areas in which micromanipulation technology will be crucial. These are (i) the application of laser technology to gamete and embryo micromanipulation (59), (ii) the use of gametes for gene therapy, (iii) genomic expansion by nuclear transplantation and artificial twinning, and (iv) placenta donation (2).

The optical trap-laser microsurgery system may play an important role in research and clinical applications involving gamete interaction (59). Human sperm have been successfully manipulated in an optical trap. Though individual sperm can be trapped and held for nearly a minute without any apparent effect on motility, it has not yet been proven that they are able to fertilize. Using a motorized stage, sperm in the trap could be moved around the slide. Preliminary reports have also been published concerning the opening of the zona pellucida using laser microbeams (2). Recently, laser micromanipulation systems have been used for opening the zona for clinical microsurgical fertilization and assisted hatching (8). Laser microbeams can drill precise holes in the zona pellucida to allow for the entry of an optical trap-directed single sperm cell that could be stabilized and fused with the oocyte using a second form of laser energy. Such a system might allow for routine fertilization using immotile and possibly even acrosome-defective sperm.

Moreover, enhancement of viability following assisted hatching may depend on the precise shape and size of the artificial gap in the zona pellucida. Laser directed energy may be quite useful in this instance.

Recently, we have applied a 308 ultraviolet laserbeam guided through the optical system of an inverted microscope (Neeve, Tadir, unpublished results). This system has enabled us to obtain hatching blastocysts following assisted hatching using laser drilling of 4- to 8-cell mouse embryos. Several 4-cell mouse embryos and a single blastocyst were also obtained following pronucleus ablation of digynic mouse zygotes. Finally, blastomere biopsy of 8-cell mouse embryos was performed successfully following laser ablation of a large portion of the zona (Gonzalez, Licciardi, Alikani, Neeve, Tadir, Cohen, unpublished results).

Once a DNA construct is devised and created, pronuclear injection should be a relatively straightforward procedure in the human zygote. Though an exponential expansion in gene characterization can be expected in the future, the requirements for successful gene transfer, including the correct expression and specificity of the gene, involve a precise placement of the transgene in relation to regulatory DNA sequences. It is unlikely that any attempts will be made to actively incorporate germ-line therapy into assisted human reproduction in the next decade. Too many important steps needed for reliable gene correction are still lacking. Obviously, the ethical, social, and political dilemmas compound already complicated issues, such as the legal status of the 4-cell embryo and the definition of motherhood. In a pluralistic society unanimity on such a major issue as germ-line therapy may be beyond the ability of any secular authority. One great advantage of germ-line therapy would be that if successful, it would terminate the heredity of the genetic lesion. A transgenically corrected genome would be theoretically passed on to future generations.

Presently, the relevance of techniques involving blastomere separation and nuclear transplantation in the human seems remote. However, it is not only the insufficiency of our knowledge of human embryonic physiology, but, again, the complex ethics involved that have prevented researchers in this field from applying it to the human. Though ethically, the delivery of more than the occasional twin following assisted human reproduction would be questionable, there may be some acceptable application both to the treatment of infertile couples and to those whose offspring may be at risk of hereditary disease. Couples with limited numbers of gametes may at any IVF attempt have only one or a few embryos for transfer. If this is due to a severe male factor infertility, it could be feasible to grow the embryo to a later stage and separate its cells. The female partner or another oocyte donor could then provide oocytes for enucleation that could then be used as conception carriers for the diploid nuclei of the original embryo. This would then create a limited number of clones of the required genetics. Cryopreservation of anucleate oocytes would facilitate this approach. Similarly, women with very few oocytes may have their embryos bisected or used in the enucleation procedure.

Surgical replacement of a diseased placenta (placenta donation) seems currently beyond the scope of obstetric medicine. However, several experimental embryology techniques can potentially be applied to women who suffer from recurrent idiopathic miscarriage of genetically normal fetuses (2). Both placental and fetal tissues originate in the preimplantation embryo. Some inner cell mass cells will develop into the fetus, whereas the trophectoderm provides most of the placental material. The two types of tisue can effectively be separated and recombined using two different embryos of the same or closely related species. Alternatively, transfer of a blastomere from an 8-cell to a 4-cell embryo leads to a chimera in which the 4-cell recipient becomes trophectoderm, and the 8-cell blastomere becomes the inner cell mass.

Acknowledgments. We are grateful to Zev Rosenwaks, Jamie Grifo, Alan Berkeley, Owen Davis, Margaret Graf, William Ledger, Myriam Feliciano, Miriam Jackson, Adrienne Reing, Alexis Adler, Janet Trowbridge, Beth Talansky, Xa-xu Tang, Hung-Chung Liu, Santiago Munne, Jonathan Stein, Frederick Licciardi, Allyson Gonzalez, and Henry Malter for their support of these studies.

References

1. Handyside AH, Kontogianni EH, Hardy K, Winston RML. Pregnancies from biopsied human preimplantation embryos sex by Y-specific DNA amplification. Nature 1990;344:378–80.
2. Cohen J, Malter HE, Talansky BE, Grifo J. Micromanipulation of human gametes and embryos. New York: Raven Press, 1992.
3. Gordon JW, Talansky BE. Assisted fertilization by zona drilling: a mouse model for correction of oligozoospermia. J Exp Zool 1986;239:347–54.
4. Laws-King A, Trounson A, Sathananthan H, Kola I. Fertilization of human oocytes by microinjection of a single spermatozoon under the zona pellucida. Fertil Steril 1987;48:637–42.
5. Surani MAH, Barton SC, Norris ML. Development of reconstituted mouse eggs suggests imprinting of the genome during gametogenesis. Nature 1984;307:548–50.
6. Cohen J, Wright G, Malter H, et al. Impairment of the hatching process following in vitro fertilization in the human and improvement of implantation by assisting hatching using micromanipulation. Hum Reprod 1990;5:7–13.
7. Verlinsky Y, Ginsberg N, Lifchez A, Valle J, Moise J, Strom CM. Analysis of the first polar body: preconception genetic diagnosis. Hum Reprod 1990;5:826–9.
8. Feichtinger W, Strohmer H, Fuhrberg P, et al. Photoablation of oocyte zona pellucida by erbium-yag laser for in-vitro fertilisation in severe male infertility. Lancet 1992;1:811.
9. Cohen J, Talansky BE, Adler A, Alikani M, Rosenwaks Z. Controversies and opinions in clinical microsurgical fertilization. J Assist Reprod Gen 1992;9:94–6.

10. Wiemer KE, Cohen J, Wiker SR, Malter HE, Wright G, Godke RA. Coculture of human zygotes on fetal bovine uterine fibroblasts: embryonic morphology and implantation. Fertil Steril 1989;52:503–6.
11. Menezo JR, Guerin JF, Czyba JC. Improvement of human early embryo development in vitro by co-culture on monolayers of Vero cells. Biol Reprod 1990;42:301–6.
12. Cohen J. Assisted hatching of human embryos. J In Vitro Fertil Embryo Transfer 1991;8:179–90.
13. Cohen J, Alikani M, Trowbridge J, Rosenwaks Z. Implantation enhancement by selective assisted hatching using zona drilling of embryos with poor prognosis. Hum Reprod (in press).
14. Cohen J, Malter H, Wright G, Kort H, Massey J, Mitchell D. Partial zona dissection of human oocytes when failure of zona pellucida penetration is anticipated. Hum Reprod 1989;4:435–42.
15. Fishel S, Jackson P, Antinori S, Johnson J, Grossi S, Versaci C. Subzonal insemination for the alleviation of infertility. Fertil Steril 1990;54:828–33.
16. Cohen J, Alikani M, Malter HE, Adler A, Talansky BE, Rosenwaks Z. Partial zona dissection or subzonal insertion: microsurgical fertilization alternatives based on evaluation of sperm and embryo morphology. Fertil Steril 1991;56:696–706.
17. Ng S-C, Bongso A, Ratnam SS. Microinjection of human oocytes: a technique for severe oligoasthenoteratozoospermia. Fertil Steril 1991;56:1117–23.
18. Hosoi Y, Miyake M, Utsumi K, Iritani A. Development of rabbit oocytes after microinjection of spermatozoa [Abstract 331]. Proc 11th Cong on Animal Reproduction, 1988.
19. Goto K, Kinoshita A, Takuma Y, Ogawa A. Birth of calves after the transfers of oocytes fertilized by sperm injection. Theriogenology 1991;35:205–21.
20. Santella L, Alikani M, Cohen J, Dale B, Talansky BE. Is the human oocyte plasma membrane polarised? Hum Reprod (in press).
21. Malter HE, Cohen J. Partial zona dissection of the human oocyte: a nontraumatic method using micromanipulation to assist zona pellucida penetration. Fertil Steril 1989;51:139–45.
22. Garrisi GJ, Talansky BE, Grunfeld L, Sapira V, Navot D, Gordon J. Clinical evaluation of three approachs to micromanipulation-assisted fertilization. Fertil Steril 1990;54:671–7.
23. Cohen J, Elsner C, Kort H, Malter H, Massey J, Mayer MP. Immunosuppression supports implantation of zona pellucida dissected human embryos. Fertil Steril 1990;53:662–5.
24. Cohen J, Adler A, Alikani M, et al. Microsurgical fertilization procedures: absence of stringent criteria for patient selection. J Assist Reprod Gen (in press).
25. Alikani M, Adler A, Reing AM, Malter HE, Cohen J. Subzonal sperm insertion and the frequency of gamete fusion. J Assist Reprod Gen 1992;9:97–101.
26. Fishel S, Timson J, Lisi F, Rinaldi L. Evaluation of 225 patients undergoing subzonal insemination for the procurement of fertilization in vitro. Fertil Steril 1992;57:840–9.

27. Kruger TF, Acosta AA, Simmons KF, Swanson RJ, Matta JF, Oehninger S. Predictive value of abnormal sperm morphology in in vitro fertilization. Fertil Steril 1988;49:112–21.
28. Yovich JM, Edirisinghe WR, Cummins JM, Yovich JL. Preliminary results using pentoxifylline in a pronuclear stage tubal transfer (PROST) program for severe male factor infertility. Fertil Steril 1988;50:179–81.
29. Fuscaldo G, Sobieszczuk D, Trounson AO. Improved fertilization rates following microinjection of human spermatozoa pretreated with 2′deoxyadenosine and pentoxifylline [Abstract 110]. 7th World Cong on IVF and Assisted Procreation, 1991:145.
30. Alikani M, Adler A, Kissin E, Reing AM, Ferrara TA, Cohen J. The use of pentoxyfylline and 2′deoxyadenosine in microsurgically assisted fertilization improves fertilization but reduces implantation (submitted).
31. Gordon JW, Grunfeld L, Garrisi GJ, Talansky BE, Richards C, Laufer N. Fertilization of human oocytes by sperm from infertile males after zona drilling. Fertil Steril 1988;50:68–73.
32. McGrath J, Solter D. Nuclear transplantation in the mouse embryo by microsurgery and cell fusion. Science 1983;220:1300–2.
33. Willadsen S. Nuclear transplantation in sheep embryos. Nature 1986;320:63–5.
34. Rawlins RG, Binor Z, Radwanska E, Dmowski WP. Microsurgical enucleation of tripronuclear human zygotes. Fertil Steril 1988;50:266–72.
35. Gordon JW, Grunfeld L, Garrisi GJ, Navot D, Laufer N. Successful microsurgical removal of a pronucleus from tripronuclear human zygotes. Fertil Steril 1989;52:367–72.
36. Malter HE, Cohen J. Embryonic development after microsurgical repair of polyspermic human zygotes. Fertil Steril 1989;52:373–80.
37. Wiker S, Malter H, Wright G, Cohen J. Recognition of paternal pronuclei in human zygotes. J In Vitro Fertil Embryo Transfer 1990;7:33–7.
38. Collier M, O'Neill C, Ammit AJ, Saunders DM. Measurement of human embryo-derived platelet-activating factor (PAF) using a quantitative bioassay of platelet aggregation. Hum Reprod 1990;5:323–8.
39. Bongso A, Soon-Chye N, Sathananthan H, Lian NP, Rauff M, Ratnam S. Improved quality of human embryos when co-cultured with human ampullary cells. Hum Reprod 1989;4:706–13.
40. Cohen J, Inge KL, Suzman M, Wiker S, Wright G. Videocinematography of fresh and cryopreserved embryos: a retrospective analysis of embryonic morphology and implantation. Fertil Steril 1989;51:820–7.
41. Wright G, Wiker S, Elsner C, et al. Observations on the morphology of human zygotes, pronuclei and nucleoli and implications for cryopreservation. Hum Reprod 1990;5:109–15.
42. Malter HE, Cohen J. Blastocyst formation and hatching in vitro following zona drilling of mouse and human embryos. Gamete Res 1989;24:67–80.
43. Cohen J, Feldberg D. Effects of the size and number of zona pellucida openings on hatching and trophoblast outgrowth in the mouse embryo. Mol Reprod Dev 1991;30:70–8.
44. Dale B, Gualtieri R, Talevi R, Tosti E, Santella L, Elder K. Intercellular communication in the early human embryo. Mol Reprod Dev 1991;29:22–8.

45. Cohen J, Alikani M, Reing AM, Ferrara TA, Trowbridge J, Tucker M. Selective assisted hatching of human embryos. Arch Med Sing (in press).
46. Talansky BE, Gordon JW. Cleavage characteristics of mouse embryos inseminated and cultured after zona drilling. Gamete Res 1988;21:277–88.
47. Alikani M, Cohen J. Micromanipulation of cleaved embryos cultured in protein-free medium: a mouse-model for assisted hatching. J Exp Zool (in press).
48. Khalifa EAM, Tucker MJ. Partial thinning of the zona pellucida for more successful enhancement of blastocyst hatching in the mouse. Hum Reprod (in press).
49. Cohen J, Malter H, Talansky B, Tucker M, Wright G. Gamete and embryo micromanipulation for infertility treatment. In: Speroff L, Marrs RP, eds. In vitro fertilization—seminars in reproductive endocrinology; vol 8. New York: Thieme Medical: 290–5.
50. Bleil JD, Wassarman PM. Structure and function of the zona pellucida: identification and characterization of mouse oocyte's zona pellucida. Dev Biol 1980;76:185–203.
51. Drobnis EZ, Andrew JB, Katz DF. Biophysical properties of the zona pellucida measured by capillary suction: is zona hardening a mechanical phenomenon? J Exp Zool 1988;245:206–19.
52. DeFelici M, Siracusa G. "Spontaneous" hardening of the zona pellucida of mouse oocytes during in vitro culture. Gamete Res 1982;6:107–13.
53. Downs SM, Schroeder AC, Eppig JJ. Serum maintains the fertilizability of mouse oocytes matured in vitro by preventing hardening of the zona pellucida. Gamete Res 1986;15:115–22.
54. Longo FJ. Changes in the zonae pellucidae and plasmalemmae of aging mouse eggs. Biol Reprod 1981;25:399–411.
55. Austin CR. The mammalian egg. Oxford: Blackwell Scientific, 1961:89–97.
56. Gwatkin RBL. Fertilization mechanisms in man and mammals. New York: Plenum Press, 1977:91–108.
57. Betteridge KJ, Flood PF, Mitchell P. Possible role of the embryo in the control of oviductal transport in mares. In: Pauerstein CJ, Adams CE, Coutinho EM, Croxatto HB, Paton DM, eds. Ovum transport and fertility regulation. Copenhagen: Harper MJK, Scriptor, 1976;381–9.
58. Tadir Y, Wright WH, Vafa O, Liaw LH, Asch R, Berns MW. Micromanipulation of gametes using laser microbeams. Hum Reprod 1991;6:1011.

16
Embryo Coculture and the Regulation of Blastocyst Formation In Vitro

Yves Menezo, Laurent Jany, and Chaque Khatchadourian

In most mammalian species embryos grown in vitro undergo developmental arrest at the approximate time of genomic activation. Moreover, despite improvements in embryo culture conditions, embryo metabolism is obviously depressed in simple culture media (1), even if enriched with serum. In vitro culture leads to loss of viability (2) except in certain cases (3). In order to overcome this block or to maintain viability, several coculture systems have been designed. In farm animals, overcoming blocks to embryo development is especially important because embryo transfer cannot be performed before the morula/blastocyst stage, when the embryo would normally be in the uterus. Otherwise, if embryos are transferred earlier, they are rapidly expelled. The motility effect, leading to expulsion, may in some cases be associated with uterine hostility. This contrasts with the situation in humans where embryos can be successfully transferred early in development. The first system that resulted in living calves starting from the 1-cell stage resulted from coculture of embryos with trophoblastic cells, presumably working through a paracrine effect (4, 5).

Oviductal cells, based on a similarity with the in vivo paracrine effect in the female genital tract, were then investigated for coculturing sheep and cattle embryos (6, 7). Coculture of embryos with fibroblasts led to less convincing results. In primates success was achieved using uterine epithelial cells for coculture (8, 9).

In the human, the problems seem less complicated because embryos can be transferred into the uterus on the second day postfertilization at a time when they would normally be in the fallopian tube. Also, blastocysts can be obtained, at low rates, in conventional culture media, and there is no apparent development block. This may explain why fibroblasts can be useful in culturing human embryos.

We have developed a system for the culture of early human embryos up to the blastocyst stage on established cell lines (10). This should permit selection of the best embryos and provide better synchrony between embryonic development and the surrounding environment. The number of embryos transferred could, therefore, be reduced. This method will also encourage study of later stage embryos by biopsy for genetic analysis and cryopreservation. In this chapter we summarize the state of the art on the use of cocultures, analyze the future possibilities of this technique, and present our recent data on early human embryology.

Cocultures with Trophoblastic Vesicles: Autocrine Effect

In the bovine, coculture of embryos with trophoblastic cells was the first system supporting embryo culture from the 1-cell stage. Living calves were obtained from frozen morulae/early blastocysts after more than 5 days of coculture (4, 5). This observation was confirmed for cattle embryos after IVF (11, 12). This type of coculture was also shown to be efficient for ovine embryos. Conditioned medium was also shown to be active. Initially, we found that the *embryotrophic effects* were related to small molecular weight (<5000 d) molecules (13). Preliminary experiments using human trophoblastic tissue for coculture of human embryos also gave encouraging results (14).

Paracrine Effect: Hormonal Independence of the Oviduct

Following the first in vivo work of Papaioannaou and Ebert (15), we cultured 1-cell random-bred mouse embryos in prepuberal oviducts and in PMS/hCG-primed oviducts (16). Blastocyst formation was similar for both treatments, indicating no hormonal dependence on the oviduct's ability to sustain early embryonic development. We even observed blastocyst implantation in prepuberal oviducts. This model was later used to confirm the nonspecies specificity of the oviduct. Porcine embryos were able to develop in an explanted mouse oviduct in vitro (17).

Similar experiments in which 1-cell embryos were transferred in vivo in prepuberal heifer oviducts gave the same type of results. Early stage embryos were able to develop to the blastocyst stage in nonhormonally primed oviducts (18).

Cocultures with Oviduct and Uterine Cells

Gandolfi and Moor (6) initiated coculture work with oviductal epithelial cells. They demonstrated that primary cultures were able to overcome the

TABLE 16.1. Effect of oviduct monolayer aging on egg development in coculture.

		No.	M + B	(B)
Mouse/mouse	First	99	67 (67)	40 (40)
(w)(L)	Second	74	19 (26)	4 (5)
Mouse/bovine	First	118		45 (38)*
(w)(L)	Second	57	7 (13)	2 (3)

Note: w = embryo; L = layer; M = morula; B = blastocyst.
* Transfer of these blastocysts led to viable litters.

early developmental block, in the sheep, with a concomitant preservation of embryo viability. In the bovine where low successes have been obtained without cells (19, 20), the percentage of viable blastocysts after IVF was much higher with oviductal cells (20). Moreover, after coculture the chronology of development and the cell number per embryo were very similar to the in vivo situation.

Several authors have focused their work on oviductal cell cultures in different species including the human (21, 22). The first coculture of human embryos on oviductal epithelial cells was reported by Bongso et al. in 1989 (23); however, effects on embryo quality and pregnancy rates have yet to be determined.

Three important points were raised at this time: the presentation of cells (suspension, monolayers, and primary and/or subpassages), aging of the cells, and the coculture media. It seems that all presentations of cells can fit with embryo development even if for some laboratory teams cell suspension under oil is better (20) while for others monolayers are superior even for early subpassages (24). It also seems that it is better to subpassage the monolayers rather than to have aging-related deleterious effects. We observed that aging of monolayers severely impaired results in the mouse (Table 16.1). This can also explain some discrepancies in the results and the need for the association between oviductal and uterine cells described by Marquant le Guienne et al. (25) (which is generally not observed by the majority of authors). Uterine cells seem to work as well (24) as oviductal cells. The endocrine state of the cell donors has no influence upon the final results of coculture (26, 27). In the rhesus monkey, coculture with homologous uterine cells gave better embryonic development with increased mean cell numbers when compared to conventional culture media alone (8, 9). This conclusion seems to represent a general rule for good-quality coculture systems (28).

Importance of the Coculture Medium

The need for a suitable coculture medium was clearly shown by Rexroad and Powell (29) who, initially, were unable to find any effect of coculture on early sheep embryonic development using either trophoblastic vesicles or oviduct cells. This was related to the use of Ham's F10 medium. At the present time B2 medium (30) and Medium 199 are best for coculture (20, 22).

Ubiquity of Embryotrophic Factors

The tissue specificity of the embryotrophic effect is also an important point since it has been claimed that fibroblasts could sustain embryonic development over the block. This is not true for early embryos of farm animals. Gandolfi and Moor (6) clearly showed that fibroblasts were not able to sustain sheep embryo viability even if the cell block was no longer present and morphology was respected. However, coculture with fibroblasts may help at later stages. In the human Wiemer et al. (31) described a positive effect of this type of coculture; however, this could not be repeated, as Verdagher (unpublished data) did not find any improvement using MRC5 cells (fibroblasts of human origin). Moreover, fibroblast growth factor did not seem to have any effect, either on early embryonic development or on cell number (32). There is also a technical aspect related to coculture with fibroblasts, namely, that it is very difficult to maintain correct pH: An acidification of the medium occurs very rapidly, which may impair embryonic development. This is not the case with epithelial cells that are susceptible to the process of *contact inhibition*.

The possibility remained of using epithelial cells originating from tissues other than the female genital tract. We checked several epithelial cell types from kidney origin (10, 33). In the mouse a high percentage of blastocysts was obtained using MDBK cells (bovine kidney). Vero cells led to less clear results since in our experiments they did not work as well while several other labs found them to be effective (34). The results obtained with trophoblastic vesicles, kidney cells, embryonic chicken cells (35), and epididymal epithelial cells (Pollard, personal communicated, cited in [20]) argue against tissue specificity. Embryotrophic factors may be more closely related to transport epithelia.

Human Embryo Coculture

Vero cells are a good support for early human embryos; conditioned media are also active (36). When embryos are cocultured, more than 55% of them reach the blastocyst stage independent of the stimulation protocol. The mean cell number of these cocultured blastocysts is higher than that obtained in classical culture media (37). We found 3 interesting points concerning blastocyst formation: (i) In coculture, blastocyst formation was not dependent upon the presence of serum (56% with

16. Embryo Coculture and Regulation of Blastocyst Formation In Vitro 283

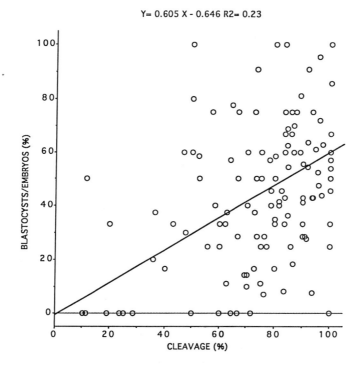

FIGURE 16.1. Relation between fertilization rate and blastocyst formation after IVF in human.

serum, 50% without, not significant); (ii) there was a direct relationship between the fertilization rate (no. of zygotes/no. of oocytes × 100 = X) and blastocyst formation (no. of blastocysts/no. of zygotes × 100 = Y) where Y = 0.605X − 0.645; t = 6.5; and P < 0.0001 for fresh sperm) (Fig. 16.1); and (iii) this model was no longer valid when frozen sperm were used. This is probably due to some alteration of chromatin stability after freezing and thawing (38) that could include asynchrony in male pronucleus formation and subsequent delays in cell cycling and genomic activation with obvious implications to blastocyst formation.

There was an increase (P < 0.05) in the implantation rate per embryo in pregnant patients when transfers were performed at the blastocyst stage (Table 16.2). The pregnancy rates per transfer at the blastocyst stage were increased only with patients who had already had several failures of embryo transfer (39). When at least one blastocyst was transferred, the pregnancy rate reached 30.5% (40/131); this is higher than can be expected with blastocysts obtained in simple culture media (40). Several other indications are now under study. One that is promising involves selection of embryos in aged patients. These

TABLE 16.2. Implantation rates per embryo in pregnant women according to the embryonic stage at transfer.

	D2		D5
No. of ongoing pregnancies (patients)	47		38
No. of embryos transferred	151		108
(range: 1–4)	(m = 3.2)		(m = 2.8)
No. of implantation sites	61		57
Yield per embryo transferred	61/151	$P < 0.05$	57/108
	40.4%		52.8%

Note: Significant between D2 and D5 ($P < 0.05$; $E > 1.96$; arc sine transformation for comparison of proportion).

TABLE 16.3. Metabolites derived from [^{35}S] methionine radiolabeling of different cell lines during the first 48 h of culture period in MEM medium.

Amino acids	MOEC (%)	BOEC (%)	MDBK (%)	Vero (%)
[^{35}S] methionine conversion rate	9	8	26	4
Cysteic acid, taurine	6.8	4.3	11.7	1.3
Glutathione, Red + Ox	0.4	0.3	0.4	0.2
Homocysteine, homecystine	0.1	1.3	1.2	—
Cystathionine	0.3	1.1	3.8	1.2

Note: MOEC = mouse oviduct epithelial cells; BOEC = bovine oviduct epithelial cells; MDBK = bovine kidney epithelial cells; Vero = monkey kidney epithelial cells.

blastocysts obtained in coculture can be frozen. Out of the first 141 transfers of frozen-thawed blastocysts, 30 pregnancies (25 ongoing) were obtained.

How Coculturing Works

While it has been claimed that coculture systems remove toxic compounds from the culture medium, it is difficult to prove, and we would rather think that two other mechanisms are involved. Initially, before and around the time of genomic activation, the cell monolayers provide a continuous supply of small molecular weight metabolites that allow the cell machinery to work normally and avoid metabolic locks. A good example is when monolayers are cultured with radiolabeled methionine, and the release of methionine-derived metabolites (Table 16.3), such as glutathione and taurine, is observed.

When metabolic locks exist (41, 42), if molecules downstream from the locks are supplied, the cell machinery still works without delays (43). Genomic activation occurs when the zygote metabolic reserves have not reached a low, irreversible point. Moreover, this avoids the storage of compounds that are not normally produced, such as glycogen (44), processes that divert embryo energy. As a second mechanism, growth factors are produced by the cell monolayers, and those embryos that have the appropriate receptors at that time (45) grow faster and increase their cell number when compared to culture in simple media.

Conclusion

We have demonstrated that coculture can be effective in supporting early embryonic development up to the blastocyst stage. Various cell types are suitable either from embryonic origin (trophoblastic tissue) or from the genital tract. However, a positive effect can be obtained using other transport epithelia cells. The utility of fibroblasts remains questionable. Coculture must not be considered as a goal, but as a transitory tool for the study of embryo metabolism in vitro and in efforts leading to a better design of the culture medium. Analysis of conditioned media, once they have been proven to be efficient, will give access to a better understanding of the embryo's needs. This study can be carried out even in the absence of embryos. Dynamic changes can be monitored by radiolabeled tracers on the monolayers themselves (33).

It is generally assumed that the feeder layer removes toxins; this could be the case if we consider the metabolic locks at the level of hypoxanthine, glucose, or mevalonate, for example. The cells are able to metabolize these compounds and/or provide metabolites downstream from the locks. There is no evidence supporting a benefit from toxin removal, such as heavy metal ions.

The feeder cells can provide unsaturated lipids (not peroxidized) that are not easily solubilized in classical culture media even if a part of them can be supplied by serum or albumin that carries various forms of lipid-soluble compounds. Compounds preventing superoxidation, such as gluthathione, are synthesized by epithelial cell feeder layers (33). For vitamin E the question remains open.

There is not a simple answer to the question of whether or not a role exists for growth factors. Several authors (32, 46) were unable to demonstrate any effect of these compounds on mouse embryo growth. On the contrary, Paria and Dey (47) described a positive effect of transforming growth factors and EGF. The degradation in vitro of added growth factors, dose effects, and specificity may be involved in this discrepancy. It is interesting to note that the paracrine effect we described with trophoblastic vesicles (4) was rediscovered and confirmed by the work of Paria and Dey (47). Serum contains various growth factors. Normally, increasing the percentage of serum should improve embryo culture; however, this is commonly not the case (48). Moreover, in coculture serum addition is generally superfluous in the presence of serum albumin. The cell numbers of the blastocysts obtained in coculture are higher. It is possible also that cocultures provide a "permissive" environment (with several small, useful metabolites) that allows embryos to reach a critical cell number more rapidly and permits the synthesis and use (through a loop system) of embryo-derived growth factors.

In the human a 55%–60% blastocyst formation rate can be obtained from IVF-produced embryos. An improvement of the pregnancy rates after transfer of blastocysts can be obtained in patients with repeated failures of embryo transfers at earlier stages. This could be related to a peculiar uterine motility behavior in this population. It seems that ectopic pregnancies can be avoided with such delayed transfers at the blastocyst stage. These blastocysts can be frozen with good yields. However, the overall beneficial effect of the coculture technique has to be demonstrated, whatever the cell support, with a double-blind, randomized study.

References

1. Jung TH. Protein synthesis and degradation in non-cultured and in vitro cultured rabbit blastocysts. J Reprod Fertil 1989;86:507–12.
2. Massip A, Van der Zwalmen P, Puissant P, Camus F, Leroy F. Effect of in vitro fertilization, culture, freezing and transfer on the ability of mouse embryos to implant and survive. J Reprod Fertil 1984;71:199–204.
3. Techakumphu M, Wintenberger-Torres S, Sevelec C, Menezo Y. Survival of rabbit embryos after culture or culture/freezing. Anim Reprod Sci 1987;13:221–8.
4. Camous S, Heyman Y, Meziou W, Menezo Y. Cleavage beyond the block stage and survival after transfer of early embryos cultured with trophoblastic vesicles. J Reprod Fertil 1984;72:479–85.
5. Heyman Y, Menezo Y, Chesne P, Camous S, Garnier V. In vitro cleavage of bovine and ovine early embryo: improved development using coculture with trophoblastic vesicles. Theriogenology 1987;27:59–68.
6. Gandolfi F, Moor RM. Stimulation of early embryonic development in the sheep by coculture with oviduct epithelial cells. J Reprod Fertil 1987;81:23–8.
7. Eyestone WH, First NL. Coculture of cattle embryos to the blastocyst stage with oviductal tissue or in conditioned medium. J Reprod Fertil 1989;85:715–20.
8. Goodeaux LL, Voelkel SA, Anzalone CA, Menezo Y, Graves KH. The effect of rhesus epithelial cell monolayers on in vitro growth of rhesus embryos. Theriogenology 1989;39:197.
9. Goodeaux LL, Thibodeaux JK, Voelkel S, et al. Collection, coculture and transfer of rhesus preimplantation embryos. ARTA 1990;1:370–9.
10. Menezo Y, Guerin JF, Czyba JC. Improvement of human early embryo development in vitro by coculture on monolayers of Vero cells. Biol Reprod 1990;42:301–6.

11. Aoyagi Y, Fukui Y, Iwazumi Y, Urakawa M, Minegishi Y, Ono H. Effect of culture system on development of in vitro fertilized ova into blastocyst. Theriogenology 1989;31:181.
12. Scodras JM, Pollard JW, Betteridge KJ. Effect of somatic cell type on bovine embryonic development in coculture. Theriogenology 1991;35:269.
13. Heyman Y, Menezo Y. Interaction of trophoblastic vesicles with bovine embryos developing "in vitro." In: Bavister BC, ed. The mammalian preimplantation embryo. New York: Plenum Press, 1987:175–91.
14. Menezo Y, Plachot M, Heyman Y, et al. Culture of human trophoblastic tissue: a potential tool for improvement of early embryo culture and transfer. In: Feichtinger W, Kemeter P, eds. Future aspects in human in vitro fertilization. Springer-Verlag, 1987:77–81.
15. Papaioannou VE, Ebert KM. Development of fertilized embryos transferred to oviducts of immature mice. J Reprod Fertil 1986;76:603–8.
16. Menezo Y, Hamidi J, Khatchadourian CH, Nardon CH. The murine prepuberal oviduct supports early embryo development in vitro. Dev Growth Differ 1989;31:551–5.
17. Krisher RL, Petters RM, Johnson BH, Bavister BD, Archibong AE. Development of porcine embryos from the one-cell stage to blastocyst in mouse oviducts maintained in organ culture. J Exp Zool 1989;249:235–9.
18. Chupin D, Hamidi J, Vallet JC, Menezo Y. Capacity of prepuberal oviducts to support early embryonic development in mouse and cow. IVth scientific meet European Embryo Transfer Soc, Lyon, 1988:35.
19. Bavister BD, Rose TA, Pinyopummintr T. Development of in vitro matured/in vitro fertilized bovine embryos in morulae and blastocysts in defined culture media. Theriogenology 1992;37:127–44.
20. Xu KP, Yadav BR, Rorie RW, Plante L, Betteridge KJ, King WA. Development and viability of bovine embryos derived from oocytes matured and fertilized in vitro and cocultured with bovine epithelial cells. J Reprod Fertil 1992;94:33–43.
21. Ouhibi N, Menezo Y, Benet G, Nicollet B. Culture of epithelial cells derived from the oviduct of different species. Hum Reprod 1989;4:229–35.
22. Carney EW, Tobback C, Foote RH. Coculture of rabbit one-cell embryos with rabbit oviduct epithelial cells. In Vitro Cell Dev Biol 1990;26:629–35.
23. Bongso A, Ng SC, Sathanathan H, Ng PL, Rauff M, Ratnam SS. Improved quality of human embryos when cocultured with human ampullary cells. Hum Reprod 1989;4:706–13.
24. Pritchard JF, Pool SH, Blekewood EG, Menezo Y, Godke RA. Culture of early stage caprine embryos using goat oviductal cell monolayers [Abstract]. Theriogenology 1991;31:259.
25. Marquant-Le Guienne B, Gerard M, Solari A, Thibault CG. In vitro culture of bovine eggs fertilized either in vivo or in vitro. Reprod Nutr Dev 1989;29:559–68.
26. Thibodeaux JK, Goodeaux LL, Roussel JD, et al. Effects of stage of the bovine estrous cycle on in vitro characteristics of uterine and oviduct epithelial cells. Hum Reprod 1991;6:751–60.
27. Thibodeaux JK, Menezo Y, Roussel JD, et al. Coculture of in vitro fertilized bovine embryos with oviductal epithelial cells originating from different stages of the estrous cycle. J Dairy Sci 1992;75.

28. Sakkas D, Trounson A. Coculture of mouse embryo with oviduct and uterine cells prepared from mice at different days of pseudopregnancy. J Reprod Fertil 1990;90:109–13.
29. Rexroad CE, Powell AM. Coculture of ovine eggs with oviductal cells and trophoblastic vesicles. Theriogenology 1988;29:387–97.
30. Menezo Y, Testart J, Perrone D. Serum is not necessary in human in vitro fertilization, early embryo culture, and transfer. Fertil Steril 1984;42:750–5.
31. Wiemer KE, Cohen J, Amborski G, et al. In vitro development and implantation of human embryos following culture on fetal bovine uterine fibroblast cells. Hum Reprod 1989;4:595–600.
32. Colver RM, Howe AM, McDonough P, Boldt J. Influence of growth factors in defined culture medium on in vitro development of mouse embryos. Fertil Steril 1991;55:194–9.
33. Ouhibi N, Hamidi J, Guillaud J, Menezo Y. Coculture of 1-cell mouse embryos on different cell supports. Hum Reprod 1990;5:537–43.
34. Lai YM, Stein DE, Soong YK, et al. Evaluation of Vero cell coculture system for mouse embryos in various media. Hum Reprod 1992;7:276–80.
35. Blakewood EG, Jaynes JM, Godke RA. Culture of pronuclear mammalian embryos using domestic chicken eggs. Theriogenology 1988;29:226.
36. Follet KL, Hammit DG, Bennett MR, Syrop CH. Enhanced development of human embryos cocultured with Vero cells. Fertil Steril program suppl of the 47th AFS meet, 1991:184.
37. Hardy K, Handyside AH, Winston RML. The human blastocyst: cell number, death and allocation during late preimplantation development in vitro. Development 1989;107:597–604.
38. Royere D, Hamanah S, Nicolle JC, Barthelemy C, Lansac J. Freezing and thawing alter chromatin stability of ejaculated human spermatoza: fluorescence acridine orange staining and Feulgen-DNA cytophotometric studies. Gamete Res 1988;21:51–7.
39. Menezo Y, Dumont J, Hazout A, Herbaut N, Nicollet B. Regulation and limitation of in vitro blastocyst formation in human. Ares Serono int symp on Implantation in Mammals, Geneve, April 23–24, 1992.
40. Bolton VN, Wren ME, Parsons JH. Pregnancies after in vitro fertilization and transfer of human blastocysts. Fertil Steril 1991;55:830–2.
41. Menezo Y, Khatchadourian CH. Involvement of glucose 6 phosphate isomerase (EC 5319) in the mouse embryo 2-cell block in vitro. C R Acad Sci 1990;310:297–301.
42. Schini SA, Bavister BD. Two cell block to development of cultured hamster embryos is caused by phosphate and glucose. Biol Reprod 1988;39:1183–92.
43. Poueymirou WT, Conover JC, Schultz RM. Regulation of mouse preimplantation development: different effect of CZB medium and Whitten's medium on rates and pattern of protein synthesis in 2-cell embryos. Biol Reprod 1989;41:317–22.
44. Khurana NK, Wales RG. Effects of coculture with uterine epithelial cells on the metabolism of glucose by mouse morulae and early blastocysts. Aust J Biol Sci 1987;40:389–95.
45. Werb Z. Expression of EGF and TGF-alpha genes in early mammalian development. Mol Reprod Dev 1991;27:10–5.

46. Caro CA, Trounson A, Kirby C. Effect of growth factors in culture medium on the rate of mouse embryo development and viability in vitro. J In Vitro Fertil Embryo Transfer 1987;4:265–9.
47. Paria BC, Dey SK. Preimplantation embryo development in vitro: cooperative interactions among embryos and role of different growth factors. Proc Natl Acad Sci USA 1990;87:4756–60.
48. Russer-Long JA, Dickey JF, Richardson ME, Ivey KW. Culture of ovine embryos in the absence of bovine serum albumin. Theriogenology 1991;35:383–91.

17

Proteoglycans as Modulators of Embryo-Uterine Interactions

DANIEL D. CARSON, ANDREW L. JACOBS, JOANNE JULIAN, AND LARRY H. ROHDE

The process by which mammalian embryos attach to and invade the uterine endometrium is both fascinating and complex. From early conception until parturition, embryonal and maternal tissues must exist symbiotically without imposing detrimental effects on the other. It remains unclear why the immunologically competent mother fails to reject the embryo in spite of histocompatibility differences. It also is unclear why the highly invasive trophoblast tissue of the embryo normally halts its progress within the endometrium although it clearly has the capacity to invade a wide range of tissues (1, 2). In this regard, the interesting feature of the uterus may not be that it supports embryo implantation, but that it has the unique ability to prevent and limit embryo invasion.

Whether or not the uterus will permit embryo attachment and invasion is determined by ovarian steroid hormone influences (3). Under the appropriate hormonal stimuli, the uterus is transiently converted to a state receptive for embryo attachment. Following this receptive phase the uterus returns to a nonreceptive, refractory state until the steroid hormone levels decline and eventually reestablish themselves during the next ovarian cycle. This decline in steroid levels must occur before a receptive state can be reestablished. If embryo implantation occurs, the steroid hormone levels are maintained, and the uterus remains nonreceptive. The state of receptivity is characterized not only by marked differences in the ability of the uterus to support embryo attachment, but also by the ability of the uterus to regulate progression of various intraluminally presented tumors (4–8). This regulation appears to be manifest at the apical surface of the uterine lumenal epithelium and is retained by polarized uterine epithelial cells in vitro (9).

Histochemical studies have shown that the types of glycoconjugates expressed at the external surface of the blastocyst change as the embryo becomes attachment competent (10). Coordinated with the development of the embryo is the differentiation of the apical cell surface of the uterine epithelium. A number of histological studies have shown that glycoconjugate expression at this uterine cell surface also changes markedly as the uterus develops to a receptive state (10–12). Consequently, it has been suggested that cell surface glycoconjugates participate in or influence the interactions that occur between the blastocyst and the apical cell surface of the uterine epithelium (13).

As a consequence of embryo attachment as well as the hormonal status of the animal, the stromal tissue underlying the implantation site is triggered to differentiate in a process called the *decidual cell response* (14). This response occurs in a defined temporal and spatial pattern permitting the identification of primary and secondary decidual zones by morphological and biochemical criteria. Decidual cells express distinct cytoskeletal proteins as well as different extracellular matrix components than stromal precursors (15–18). Many of these extracellular matrix components have been shown to support embryo outgrowth in vitro (15, 9–22). Finally, the embryo penetrates the decidua and then stops its invasion. Eventually, both trophoblastic and uterine cells will intermingle in hemochorial animals to form a placenta (23).

A recurring aspect of implantation-related events is the alteration in expression of cell-surface glycoconjugates. A series of studies from our laboratory and others have indicated that proteoglycans and proteins that bind proteoglycans participate in multiple aspects of the implantation process. Most of this chapter focuses on information drawn from rodent and human model systems. This information will be put in the context of how such changes might impact on the cell-cell interactions that occur during implantation.

Proteoglycans and Proteoglycan Binding Proteins

Proteoglycans are defined as proteins containing one or more covalently attached glycosaminoglycan chains. *Glycosaminoglycans* are generally characterized as relatively large molecular weight (i.e., >10,000 d), highly negatively charged, linear polysaccharides composed of alternating residues of a hexosamine (N-acetylglucosamine or N-acetylgalactosamine) and a uronic acid (glucuronate or iduronate). In most cases the polysaccharides also are sulfated. Glycosaminoglycan subclasses are separated on the basis of the types of hexosamine and uronic acid present in the chain. These subclasses include chondroitin sulfates, dermatan sulfate, hyaluronate, heparan sulfate, and keratan sulfate. Substitution of proteins with these polysaccharides markedly changes both the biophysical and biochemical properties of the molecule, and many functions are directly associated with the glycosaminoglycan chains themselves (reviewed in 24).

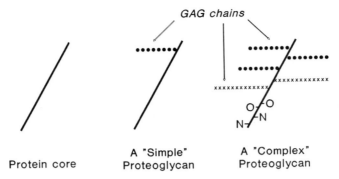

FIGURE 17.1. Generalized proteoglycan structures. The figure depicts simplified models of proteoglycan structures. The protein core is depicted as a diagonal straight line. A simple proteoglycan is depicted as a protein core containing only 1 covalently attached glycosaminoglycan (GAG) chain. A complex proteoglycan is depicted as a protein core containing more than one type of glycosaminoglycan chain. The filled circles and x's represent disaccharide units of GAG chains. The letter N refers to the presence of an N-linked oligosaccharide chain and O refers to the presence of an O-linked, mucin-type oligosaccharide.

As shown in Figure 17.1, proteoglycans may occur as relatively simple structures containing only one glycosaminoglycan chain or as complex molecules containing multiple glycosaminoglycan chains of different varieties (hybrid proteoglycans, e.g., syndecan [24]) as well as additional types of oligosaccharide chains in covalent or even non-covalent linkage. Moreover, there are examples of proteins that either do not contain glycosaminoglycans under all conditions or have different types of glycosaminoglycans depending upon the cell type in which the protein is expressed, the state of cell differentiation, or environmental influences. Therefore, control of proteoglycan expression and function may be exerted at the level of expression of the protein core or the assembly of the glycosaminoglycan chains. A few compounds are available that selectively inhibit glycosaminoglycan assembly in vivo and provide useful tools to study proteoglycan function (26, 27). Several recent review articles discuss proteoglycan structure and function in more detail (24, 28, 29).

An array of proteins with diverse functions have been shown to bind certain proteoglycans or glycosaminoglycan chains. The types of proteins displaying such interactions can be subdivided into several categories depending upon their localization or function. This classification is summarized in Table 17.1. A number of growth factors and cytokines have been shown to interact with proteoglycans, particularly those of the heparan sulfate variety (30). In some cases the growth factor itself may be

TABLE 17.1. Proteoglycan-protein interactions.

Type	Examples[a]	Reference
Soluble	bFGF	75
	Antithrombin 3	76
Extracellular matrix	Laminin (HSPG)	32
	Fibronectin (HSPG)	33
	Collagen I (CSPG)	35
	Tenascin (CSPG)	37
Cell surface	N-CAM (HSPG)	77
	Uterine epithelium (HSPG)	54, 56
	CD44 (HA)	78
Proliferative or metabolic	Binding proteins unknown, but actions on cell cycle progression and protein kinase C demonstrated (HSPG)	24

[a] Proteins or binding sites for proteoglycans or glycosaminoglycans are presented. The type of proteoglycan or glycosaminoglycan involved in the interaction is indicated in parentheses. (bFGF = basic fibroblast growth factor; CSPG = chondroitin sulfate proteoglycan; HA = hyaluronic acid; HSPG = heparan sulfate proteoglycan; N-CAM = neural cell adhesion molecule.)

a proteoglycan; that is, secreted forms of colony stimulating factor 1 (31). In addition, many extracellular matrix proteins can bind proteoglycans in vitro and codistribute with them in situ (24, 28, 29).

In general, the interactions between extracellular matrix components (e.g., laminin and fibronectin) and proteoglycans of the heparan sulfate variety appear to promote cell adhesion events (32, 33). In contrast, interactions between extracellular matrix components (e.g., collagen type I and tenascin) and proteoglycans of the chondroitin sulfate variety are associated with interruption of cell adhesion (34–37). Specific cell-surface binding sites for proteoglycans and glycosaminoglycans have been described in many systems, although few of the binding proteins have been identified. Proteoglycan interactions with cell-surface binding proteins may account not only for certain aspects of cell adhesion, but also proteoglycan modulation of cell proliferation (38, 39) and signal transduction events (40, 41). Finally, glycosaminoglycans have been shown to bind to or modify the activities of a number of intracellular proteins (reviewed in 24); however, it is not clear if or how such interactions take place in intact cells.

In the discussion below, the emphasis is placed upon how interactions of proteoglycans with corresponding binding proteins influence cell-cell and cell-extracellular matrix interactions during the implantation process. Proteoglycans also may impact implantation via their influences on the other cell biological processes described above.

TABLE 17.2. Evidence for HSPG involvement in embryo/trophoblast interactions with uterine epithelium.

1. HSPG synthesis increases markedly (4 to 5-fold) during periimplantation period.
2. Expression of HSPG core protein (perlecan) and polysaccharides at external surface of trophectoderm increases in parallel with acquisition of attachment competence in vitro and in utero. HSPGs are major proteoglycans of JAR cell surfaces.
3. Inhibition of glycosaminoglycan synthesis inhibits blastocyst attachment to uterine epithelium. Inhibition of glycosaminoglycan synthesis inhibits JAR-RL95 cell binding.
4. Certain forms of HS (and DS), but not other glycosaminoglycans, inhibit blastocyst attachment to uterine epithelia as well as JAR-RL95 cell binding.
5. Enzymatic removal of cell-surface HS (DS), but not other glycosaminoglycans, inhibits blastocyst attachment to uterine epithelia and extracellular matrix as well as JAR-RL95 cell binding.
6. Proteins with HS binding domains (platelet factor 4) support embryo attachment.
7. Mouse uterine epithelia and human uterine epithelial cell lines (RL95 and HEC) display specific cell-surface HS binding sites.

Heparan Sulfate Proteoglycan Involvement in Blastocyst and Trophoblast Attachment

A number of lines of evidence are consistent with the idea that *heparan sulfate proteoglycans* (HSPGs) participate in adhesive interactions occurring between trophectoderm and uterine epithelial cells. These findings are summarized in Table 17.2. Sulfated proteoglycan synthesis by mouse embryos increases 4- to 5-fold during the period in which blastocysts acquire the ability to attach to various substrates in vitro (42). Biochemical analyses have revealed that the majority of the sulfated glycosaminoglycans synthesized during this interval are of the *heparan sulfate* (HS) variety. Antibodies to the core proteins of particular HSPGs have been used to study their expression in mouse blastocysts (43, 44). The results indicate that syndecan expression decreases at the external surface of trophectoderm prior to acquisition of attachment competence. Surprisingly, expression of perlecan, an HSPG usually associated with basal lamina, increases at this surface over this same period both in vitro and in utero. Laminin also appears at the external surface of the blastocyst at this time; however, neither collagen type IV nor a morphologically discernible, organized, basal laminar structure is seen at this surface at this time. Therefore, it is unclear if perlecan and laminin are complexed with each other or how they are retained at the external trophectodermal surface. HS-specific staining techniques also have demonstrated the existence of HS at the external surface of attachment-competent blastocysts (21, 43). Thus, both the protein and polysaccharide constituents of HSPGs are expressed in a temporal and spatial pattern consistent with a role in adhesive interactions occurring at the trophectodermal cell surface.

Examination of HSPG expression at the surface of human blastocysts has not been performed; however, some information on proteoglycan expression by human trophoblastic cell lines (JAR [45]) is available. Biochemical analyses reveal that approximately 50% of the glycosaminoglycans at the surface of JAR cells are HS (46 and Rohde, Carson, unpublished observations). The remainder are predominantly dermatan sulfate, a glycosaminoglycan class that displays heparin-like activities in other bioassays (47). These observations do not necessarily mean that HSPGs are displayed at the surfaces of human blastocysts since trophoblasts are derivatives of trophectoderm. Furthermore, trophoblastic tumor cell lines may not precisely mimic their normal counterparts. Nonetheless, the available information is consistent with the existence of HSPGs at the cell surface of human trophoblast. It would be necessary to demonstrate the occurrence of HSPGs at human implantation sites to support this hypothesis. While this may not be feasible for ethical reasons, the development of sensitive histochemical procedures for the detection of HS should permit a more thorough examination of HSPG expression in human trophoblast and placenta.

Inhibition of proteoglycan assembly has been shown to inhibit blastocyst adhesive functions. Arylxylosides inhibit HSPG synthesis by mouse blastocysts as well as blastocyst adhesion to various substrata, including uterine epithelial cells (42) and outgrowth on uterine stromal cell extracellular matrix (Julian, Carson, unpublished observations). Treatment with chlorate ion, an inhibitor of glycosaminoglycan sulfation reactions (27), also inhibits binding of JAR cells to a human uterine epithelial cell line (RL95 [48]). In these studies it was shown that treatment of both cell lines with the inhibitor was required to inhibit cell-cell binding (Rohde, Carson, unpublished observations). HSPGs were found to be the predominant (>90%) proteoglycans displayed at RL95 cell surfaces, consistent with the observation that HSPGs are the predominant proteoglycans expressed by mouse uterine epithelial cells in vivo and in vitro (49, 50). Thus, proteoglycans of both cell surfaces may participate in the binding events.

The inclusion of soluble forms of HS and related polysaccharides to media of adhesion assays inhibits binding of mouse blastocysts to a variety of substrates, including uterine epithelial cells (21). Other glycosaminoglycans (e.g., chondroitin sulfate) do not display this activity. Soluble forms of HS also inhibit blastocyst outgrowth on uterine stromal cell extracellular matrix (Julian, Carson, unpublished observations). In this case, dermatan sulfate proteoglycans and hyaluronic acid may also participate in blastocyst interactions with this complex matrix (15–18, 51).

More detailed studies have been performed using a heterologous cell-cell adhesion assay in which single-cell suspensions of JAR cells are allowed to bind to RL95 cell monolayers (Rohde, Carson, unpublished observations). Specificity for HS and glycosaminoglycans with HS-like properties is observed in this assay as well. These studies also indicate that certain structural modifications of HS that would drastically reduce charge density (i.e., N-desulfation followed by N-acetylation) do not alter the activity in this regard. Other modifications (e.g., removal of O-sulfates) destroy biological activity. Collectively, these studies demonstrate that HS and related polysaccharides can inhibit binding activities of blastocysts and trophoblast cell lines. Presumably, this effect is due to the competition between the soluble competitor and cell-surface HS for relevant binding sites on the complementary surface.

Another test of HSPG function has been to exploit the use of specific HS-degrading enzymes to remove HS from embryo and cell surfaces. Subsequently, effects on cell adhesion events can be observed. This procedure inhibits blastocyst binding to various substrates, including uterine epithelial cells (21) and blastocyst outgrowth on uterine stromal cell extracellular matrix (Julian, Carson, unpublished observations). Similar studies have been performed using the JAR-RL95 cell-cell adhesion assay described above. The results again indicate a requirement for cell-surface HSPGs. In addition, dermatan sulfate proteoglycans appear to participate in the binding events. These studies indicate that the ongoing expression of cell-surface proteoglycans is important in the maintenance of the normal adhesive properties of embryos and trophoblast-derived cell lines.

A number of proteins with HS binding domains (e.g., fibronectin, laminin, and vitronectin) support blastocyst attachment (52). In this regard, even platelet factor 4, a small protein for which the only known functional domain is for HS binding (53), supports embryo attachment (21). Nonetheless, there are a number of examples of proteins that support embryo attachment and outgrowth in an HS-independent fashion (e.g., collagens) (20). It is well established that integrin-dependent systems are involved in important aspects of embryo adhesion (20–22, 52). Furthermore, embryos will attach and outgrow on hyaluronic acid, a pure polysaccharide, in an HS-independent manner (15). Thus, other classes of adhesion-promoting molecules exist at embryo and trophoblast cell surfaces; however, HSPGs appear to be important components of the cell adhesion systems involved in embryo attachment to uterine epithelial cells and outgrowth on uterine stromal cell extracellular matrix; that is, substrates that embryos would encounter in vivo.

HS Binding Sites at Uterine Epithelial Cell Surfaces

The above discussion has summarized evidence for the existence of HSPGs at the relevant surfaces of blastocysts and trophoblastic cell lines as well as support for HSPG function in adhesive interactions. A corollary of this model is that the complementary cell surfaces and substrates should display HS binding sites with comparable specificity. This initially was examined in primary cultures of mouse uterine epithelial cells (54). These studies demonstrated that approximately 1×10^6 binding sites with an apparent Kd of 50 nM for ^{3}H-heparin existed on these cell surfaces. These sites are very specific for heparin and certain forms of HS; however, only 9%–14% of these sites are expressed at the apical surface of polarized uterine epithelial cells. The remainder may be situated either between cells or, as seems more likely, at the basal aspect of the cells where binding to HSPGs in basal lamina would occur. Subsequent studies have demonstrated that polarized uterine epithelial cells are non-receptive with regard to embryo attachment, perhaps due to insufficient access to these sites (9). Proteins of M_r 70,000 have been isolated from primary cultures of mouse uterine epithelia by heparin-agarose affinity chromatography that support embryo attachment in vitro (55). Nonetheless, it is not clear if these proteins represent the cell-surface components that bind HS or interact with embryos. Many practical difficulties were encountered in isolating sufficient amounts of these proteins from primary cultures of mouse uterine epithelia to generate specific probes. Therefore, an alternative approach using human uterine epithelial cell lines was pursued.

The ^{3}H-heparin binding studies demonstrated that human uterine epithelial cell lines (i.e., RL95 and HEC-1) also display specific HS binding sites at their cell surfaces (56). In this case, the number of binding sites per cell is about 20-fold higher than detected on mouse uterine epithelial cell surfaces. Approximately half of these sites are not available in attached cells; however, predigestion of the cells with heparinases seems to expose these sites. Consequently, many of the sites appear to be occupied with endogenous ligand. In addition, the dissociation constant is substantially higher (Kd = 800 nM); that is, the sites are of lower affinity than those of mouse epithelia. These sites also are highly specific for heparin and certain forms of HS. Structural modifications of HS that alter its activity in the JAR cell-RL95 cell binding assay described above also comparably alter HS activity as a competitor in the cell-surface ^{3}H-heparin ligand binding assay. Consequently, the specificities of the HS binding sites and JAR cell binding activity are similar.

Photoaffinity labeling was used to identify components of RL95 cell surfaces that directly participated in HS binding. Under all conditions major components with M_r of 14,000–18,000 and 31,000 were detected; however, additional components were identified when photoaffinity labeling was performed on nonenzymatically detached intact cells in suspension. Proteolytic fragments can be released from the cell surface of RL95 cells that retain HS binding activity by several criteria. The amino termini of three of these peptides have been sequenced and shown to bear homology to other saccharide binding proteins, including histone-like HS binding lectins isolated from human placenta (57, 58). Similar M_r peptides are released from the surfaces of primary cultures of mouse uterine epithelial cells by this procedure; however, it is not clear if the mouse and human peptides are related. Development of specific antibody and cDNA probes to the human proteins should permit thorough examination of the structure of these proteins and their relationship to mouse HS binding proteins.

If HS binding sites correspond to embryo binding sites used during the periimplantation period, then expression of the HS binding sites should be regulated in a fashion that can account for the pattern of uterine receptivity seen in vivo. Thus, it is predicted that these binding sites should only be accessible at the apical surface of uterine epithelial cells under the appropriate influence of steroid hormones. Under these conditions increased expression of genes encoding these binding proteins may occur. Alternatively, retrafficking of proteins from other sites (e.g., basal surfaces) may increase relative levels of expression at the apical surface. In this regard, steroid hormones have been observed to modulate membrane trafficking in cell lines (59) as well as uterine epithelia (60). Recently, it has been observed that polarized uterine epithelial cells retain at least some of their steroid hormone responsiveness in vitro (61–63). Therefore, it should be possible to examine aspects of hormonal regulation of HS binding site expression.

A third possibility is that the number of HS binding sites expressed at the apical cell surface is maintained constitutively; however, under most conditions these sites are not accessible for embryo binding. Inaccessibility may result from occupation of HS binding sites by endogenous HSPGs expressed at the apical cell surface (49, 50, 64). Furthermore, increased HSPG degradation by uterine epithelia appears to occur under conditions leading to uterine receptivity (49). As discussed above, a large fraction of the cell-surface HS binding sites of RL95 cells are normally occupied with endogenous ligand. Inaccessibility also may result from steric hindrance. Recent work from our lab indicates that heavily glycosylated mucin glycoprotein are major components of the apical cell surface of uterine epithelia (9). These glycoproteins are extremely resistant to enzymatic attack and may provide a protective coating at this cell-surface, a function suggested for mucins in other systems (65). Further support for this idea is provided by the observation that inhibition of mucin biosynthesis by uterine epithelial cells in vitro increases access to ^{3}H-heparin and trypsin-releasable peptides at the apical cell surface (9); however, the partial reduction in mucin expression (50%) achieved in these experiments was insufficient to increase access of either embryos or melanoma cells to the apical surface.

Indirect immunofluorescence of cryostat sections of uterine tissue has been performed using antibodies directed at the cytoplasmic tail of one mucin glycoprotein, muc-1 (66). These studies indicate that mucin glycoproteins are abundantly expressed at the apical cell surface and microvilli of mouse uterine epithelia during the estrous cycle, as well as the first few days of pregnancy; however, staining is lost prior to the onset of uterine receptivity (Surveyor, Pemberton, Gendler, Carson, unpublished observations). Collectively, these observations are consistent with a role for mucin glycoproteins as antiadhesive molecules at the apical cell surface of uterine epithelia. Experiments in which mucin expression can be efficiently and selectively ablated in vivo as well as in vitro need to be performed to test this hypothesis more rigorously.

Proteoglycan and Proteoglycan Binding Protein Function in Stromal Extracellular Matrix

In addition to interactions with epithelial cell surfaces, preliminary studies from our lab indicate that proteoglycans and glycosaminoglycans participate in embryo binding and outgrowth on stromal extracellular matrix in vitro. The major proteoglycans/glycosaminoglycans produced by stroma are *chondroitin sulfate proteoglycans* (CSPGs) and *hylauronic acid* (HA) (51). A large increase in HA synthesis is observed in uteri on the day during which embryo implantation would normally take place (15). These same studies also demonstrated that HA supports embryo attachment and outgrowth in vitro. HA is associated with cell migratory activities in other systems and increases hydration of extracellular matrix due to its large hydrodynamic volume (67). Thus, HA may cooperate with other stromal extracellular matrix components to facilitate embryo invasion. CSPGs produced by uterine stroma bind tightly to collagen type I fibrils in vitro (35) and are found in stromal extracellular matrix in vivo (68). Therefore, CSPGs appear to be maintained in the stromal environment through which the embryo must migrate. In contrast to a facilitory role, CSPGs are associated with inhibition of cell adhesion and migration in other systems (34). Purified CSPGs inhibit embryo outgrowth on both collagen type I and fibronectin matrices in vitro (35). Furthermore, enzymatic removal of CSPGs from stromal extracellular matrix generated in vitro enhances both the rate and extent of embryo outgrowth (35). Consequently, CSPGs may function to restrict embryo invasion of the uterine wall in vivo.

In response to the implanting embryo, the uterine stroma differentiates in a programmed fashion referred to as the *decidual cell reaction*. This reaction results in the formation of several morphologically and functionally distinct zones (69). Among the biochemical responses associated with this differentiation are induction of expression of cytoskeletal (16), cell-surface (70), and extracellular matrix (15–18) components that are not expressed in stromal fibroblasts under other conditions. Some of the extracellular matrix molecules induced in decidual cells are likely to facilitate embryo adhesive interactions (e.g., laminin and hyaluronate). HSPGs also appear to be induced in decidual cells (17, 71), although functions for decidual HSPGs have yet to be defined. Many of the extracellular matrix components observed in decidual tissue appear to accumulate following successful implantation.

An interesting contrast to this pattern is observed for *tenascin* (TN). TN is a large extracellular matrix glycoprotein that is transiently expressed during organogenesis in many systems (72). TN expression also has been associated with reduced cell-matrix adhesion (36, 37). Examination of implantation sites reveals that TN is rapidly, locally, and transiently expressed in the uterine stroma immediately subjacent to the basal lamina at the site of embryo attachment (Julian, Chiquet-Ehrismann, Carson, unpublished observations). This induction is not observed prior to hatching of the embryo from the zona pellucida and also requires an appropriate hormonal milieu. When an artificial decidual stimulus (oil) is administered to a hormonally primed mouse, TN is induced in stromal extracellular matrix subjacent to the basal lamina along the entire length of the lumen at both the mesometrial and antimesometrial aspects; however, induction is not observed at the interface of glandular epithelia and stroma. These results demonstrate that all areas of the lumenal epithelial-stromal interface are capable of this response.

Furthermore, embryo influences are not essential. Rather, the direct signals that trigger TN production may be produced by lumenal epithelia. Support for this idea comes from studies of TN expression by primary cultures of uterine stroma in vitro. When cultured alone in serum-free media, uterine stroma produce low levels of TN that are expressed at the cell surface in patchy arrays. In the presence of certain growth factors or serum, the cell-surface TN becomes organized into fibrillar arrays reminiscent of the arrays detected at implantation sites in utero. This response is most striking when uterine stroma are cocultured with polarized uterine epithelial cells using an established system (9, 61–64). In this case not only is the fibrillar organization of TN enhanced, but TN levels are also increased several-fold. These observations demonstrate that uterine epithelial cells are capable of secreting factors from their basal aspect that can stimulate TN expression by stroma. These epithelial factors have not been identified; neither is it clear why the epithelial cells secrete these factors constitutively in vitro when this process seems to be so restricted in vivo.

The function of TN at implantation sites, as well as other biological contexts, also is uncertain. Fibrillar TN is always detected at sites containing hatched blastocysts, but is not detected at sites containing unhatched blastocysts. Since hatching is believed to occur over a relatively brief period in utero, TN induction must occur very rapidly. Nonetheless, the local accumulation of TN essentially disappears within

24–48 h; that is, prior to the time at which trophoblast would be invading the area. These observations suggest that TN functions are more likely to be manifest at the level of uterine cells rather than to act directly on the embryo.

TN has been proposed to interrupt cell-matrix adhesion (36, 37). In this regard, the association of lumenal epithelial cells with their basal lamina appears to be disturbed during the periimplantation period (73). Local induction of TN may augment this effect and facilitate early stages of embryo penetration. The distribution of TN is similar to the area destined to become the primary decidual zone. Cells of the primary decidual zone are the first stromal cells to differentiate during early pregnancy. Stromal cell-cell contacts in the primary decidual zone are increased to the extent that this tissue forms an effective permeability barrier (69). Interruption of cell-matrix adhesion by TN may initiate differentiation of these cells and perhaps drive cell-cell binding. The primary decidual zone itself is a transient structure. Consequently, TN may not be required once this cascade has been initiated.

TN also has proteoglycan (glycosaminoglycan) binding domains, most notably for CSPGs (37). This may account for TN retention by the CSPG-enriched environment of stromal extracellular matrix (51, 68). In addition, proteoglycan binding properties may enable TN to retain growth factors. Growth factors bound to hybrid-type proteoglycans may be able to complex with TN and be concentrated. In this example, HS-binding growth factors (e.g., bFGF) would associate with the HS chains of the proteoglycan. In turn, the CS chains of the proteoglycan would permit complexing with TN. Alternatively, certain growth factors may bind directly to TN. Colony stimulating factor 1 appears to be secreted primarily as a CSPG (31). In addition, expression of this growth factor by uterine epithelial cells increases several-hundred-fold during pregnancy (74). It should be noted that the time course of colony stimulating factor 1 induction does not match that observed for TN; however, TN may serve to concentrate this cytokine at a time of low production.

Summary

A series of studies has indicated that proteoglycans and proteins that bind to proteoglycans serve important roles during early stages of embryo implantation. Much of this information is summarized in Figure 17.2. Embryo attachment to the apical surface of the uterine epithelium requires acquisition of attachment competence by the embryos, a process that seems to occur shortly following hatching from the zona pellucida, as well as conversion of the normally nonreceptive epithelium to a receptive state. In the latter case, this conversion is under the direct or indirect

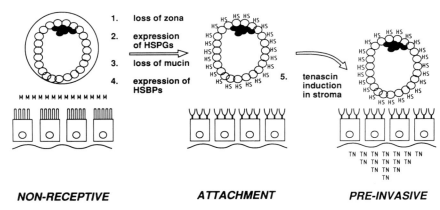

FIGURE 17.2. Model of embryo implantation. The embryo arrives in the uterus while still encased in the zona pellucida. Following hatching from the zona pellucida (step 1), the embryo becomes attachment competent, a state associated with increased expression of HSPGs at the external surface of the trophectoderm (step 2). The apical surface of the uterine epithelium is nonreceptive under most conditions. The nonreceptive state is characterized by a lack of expression or access to apically disposed embryo receptors. Access may be prevented by the expression of apically disposed mucin glycoproteins (M) that coat this surface and protect it against lytic attack and cellular invasion. Loss of the mucin coat (step 3) and reduction of apical microvilli expose appropriate embryo attachment molecules, e.g., HS binding proteins (HSBPs) (step 4). Upon binding of the embryo to the uterine epithelium, signals are transmitted to the underlying stroma that trigger the initial events of uterine differentiation, i.e., the decidual cell reaction. A very early response is an increase in TN expression (step 5). In turn, local production of TN may participate in the formation of the primary decidual zone. For further discussion, see text.

control of steroid hormones. Glycoproteins that interfere with adhesive interactions (e.g., mucins) characterize the apical cell surface of the epithelium under most conditions and are down-regulated prior to the time of embryo attachment. HSPGs and their binding proteins appear to participate in adhesive interactions occurring between trophectoderm/ trophoblast and uterine epithelial cells at initial phases of embryo attachment. HSPGs also may facilitate binding to stromal extracellular matrix during penetration of the endometrium. It will be important to identify the uterine HSPG binding proteins to understand how their expression is regulated relative to the state of uterine receptivity.

In contrast, CSPGs of the uterine stromal matrix restrict embryo outgrowth and may help limit trophoblast invasion of the uterine wall. TN, a CSPG binding protein, is elaborated locally and transiently at implantation sites. Rather than interacting directly with trophoblast, TN may modulate interactions of uterine cells with extracellular matrix. As a result, uterine cellular functions may change and initiate a cascade of events leading to the development of specialized tissues of the uteroplacental unit.

Acknowledgments. The authors appreciate the interest of Serono Symposia, USA, who supported the presentations of this work. In addition, we are indebted to David Farrar, Gulnar Surveyor and Drs. Carole Wegner and Mary C. Farach-Carson for their many helpful discussions during the course of these studies. We thank Ms. Ellen Madson for her excellent secretarial assistance. These studies were supported by NIH Postdoctoral Training Grant HD-07207 (A.L.J.) and NIH 25235 (D.D.C.).

References

1. Cowell TP. Implantation and development of mouse eggs transferred to the uteri of non-progestational mice. J Reprod Fertil 1969;19:239–45.
2. Gurchot C. The trophoblast theory of cancer (John Beard, 1857–1924) revisited. Oncology 1969;31:310–33.
3. Psychoyos A. Uterine receptivity for nidation. Ann NY Acad Sci 1986; 476:36–42.
4. Maharajan P, Maharajan V. Behavioral pattern of tumour cells in mouse uterus. Gynecol Obstet Invest 1986;21:32–9.
5. Maharajan P, Rosato F, Maharajan V. Response of mouse embryos to tumour extract treatment. Gynecol Obstet Invest 1987;23:208–16.
6. Maharajan P, Rosato F, Mirabella N, Pelayalli G, Maharajan V. Influence of the mouse uterus on the metastatic pattern of tumour cells. Cancer Lett 1988;43:33–6.
7. Schleich AB, Frick M, Mayer A. Patterns of invasive growth in vitro. Human decidua gravitadis confronted with established human cell lines and primary human explants. J Natl Cancer Inst 1976;56:221–37.
8. Schlessinger M. Uterus of rodents as sites for manifestation of transplantation immunity against transplantable tumors. J Natl Cancer Inst 1962;56:221–34.
9. Valdizan M, Julian J, Carson DD. WGA-binding, mucin glycoproteins protect the apical cell surface of mouse uterine epithelial cells. J Cell Physiol 1992.
10. Enders AC, Schlafke S. Surface coats of the mouse blastocyst and uterus during the preimplantation period. Anat Rec 1974;180:31–46.
11. Anderson TL, Hoffman LH. Alterations in epithelial glycocalyx of rabbit uteri during early pseudopregnancy and pregnancy, and following ovariectomy. Am J Anat 1984;171:321–34.
12. Anderson TL, Olson GE, Hoffman LH. Stage-specific alterations in the apical membrane glycoproteins of endometrial epithelial cells related to implantation in rabbits. Biol Reprod 1986;34:701–20.
13. Schlafke S, Enders AC. Cellular basis of interaction between trophoblast and uterus at implantation. Biol Reprod 1975;12:41–65.

14. Finn CA. Implantation, menstruation and inflammation. Biol Rev 1986;61:313–28.
15. Carson DD, Dutt A, Tang JP. Glycoconjugate synthesis during early pregnancy: hyaluronate synthesis and function. Dev Biol 1987;120:228–35.
16. Glasser SR, Lampelo S, Munir MI, Julian J. Expression of desmin, laminin and fibronectin during in situ differentiation (decidualization) of rat uterine stromal cells. Differentiation 1987;35:132–42.
17. Wewer UM, Damjanov A, Weiss J, Liotta LA, Damjanov I. Mouse endometrial stromal cells produce basement membrane components. Differentiation 1986;32:49–58.
18. Wu TC, Wan YJ, Chung AE, Damjanov I. Immunohistochemical localization of entactin and laminin in mouse embryos and fetuses. Dev Biol 1983;100:496–505.
19. Armant DR, Kaplan HA, Lennarz WJ. Fibronectin and laminin promote in vitro attachment and outgrowth of mouse blastocysts. Dev Biol 1986;116:519–23.
20. Carson DD, Tang JP, Gay S. Collagens support embryo attachment and outgrowth in vitro: effects of the Arg-Gly-Asp sequence. Dev Biol 1988;127:368–75.
21. Farach MC, Tang JP, Decker GL, Carson DD. Heparin/heparan sulfate is involved in attachment and spreading of mouse embryos in vitro. Dev Biol 1987;123:401–10.
22. Sutherland AE, Calarco PG, Damsky CH. Expression and function of cell surface extracellular matrix receptors in mouse blastocyst attachment and outgrowth. J Cell Biol 1988;106:1331–48.
23. Hernandez-Verdun D. Morphogenesis of the syncytium in the mouse placenta. Cell Tissue Res 1974;148:381–96.
24. Carson DD. Proteoglycans in development. In: Fukuda M, ed. Cell surface carbohydrates and cell development. Boca Raton, FL: CRC Press, 1992:258–83.
25. Rapraeger A, Jalkanen M, Bernfield M. Integral membrane proteoglycans as cell surface receptors: role in cytoskeleton and matrix assembly at the epithelial cell surface. In: Wight TN, Mecham RP, eds. New York: Academic Press, 1987:105–26.
26. Okayama M, Kimta K, Suzuki S. The influence of p-nitrophenyl β-D-xyloside on the synthesis of proteochondroitin sulfate by slices of embryonic chick cartilage. J Biochem 1973;74:1069–73.
27. Baeuerle PA, Huttner WB. Chlorate—a potent inhibitor of protein sulfation in intact cells. Biochem Biophys Res Commun 1986;141:870–7.
28. Hassell JR, Kimura JH, Hascall VC. Proteoglycan core protein families. Annu Rev Biochem 1986;55:39–67.
29. Ruoslahti E. Structure and biology of proteoglycans. Annu Rev Cell Biol 1988;4:229–55.
30. Vlodavsky I, Korner G, Ishai-Michaeli R, Bashkin P, Bar-Shavit R, Fuks S. Extracellular matrix-resident growth factors and enzymes: possible involvement in tumor metastasis and angiogenesis. Cancer Metastasis Rev 1990;9:203–26.
31. Price LKH, Choi HU, Rosenberg L, Stanley ER. The predominant form of secreted colony stimulating factor-1 is a proteoglycan. J Biol Chem 1992;267:2190–9.

32. Skubitz APN, McCarthy JB, Charonis AS, Furcht LT. Localization of three distinct heparin-binding domains of laminin by monoclonal antibodies. J Biol Chem 1988;263:4861–8.
33. McCarthy JB, Skubitz APN, Zhao Q, et al. RGD-independent cell adhesion to the carboxy-terminal heparin-binding fragment of fibronectin involves heparin-dependent and -independent activities. J Cell Biol 1990;110:777–87.
34. Perris R, Johansson S. Amphibian neural crest cell migration on purified extracellular matrix components: a chondroitin sulfate proteoglycan inhibits locomotion on fibronectin substrates. J Cell Biol 1987;105:2511–21.
35. Carson DD, Julian J, Jacobs AL. Uterine stromal cell chondroitin sulfate proteoglycans bind to collagen type I and inhibit embryo outgrowth in vitro. Dev Biol 1992;149:307–16.
36. Chiquet-Ehrismann R, Kalla P, Pearson CA, Beck K, Chiquet M. Tenascin interferes with fibronectin action. Cell 1988;53:383–90.
37. Murphy-Ullrich JE, Lightner VA, Aukhil I, Yan YZ, Erickson HP, Hook M. Focal adhesion integrity is down-regulated by the alternatively spliced domain of human tenascin. J Cell Biol 1991;115:1127–36.
38. Fritze LMS, Reilly CF, Rosenberg RD. An antiproliferative heparan sulfate species produced by postconfluent smooth muscle cells. J Cell Biol 1985;100:1041–9.
39. Fedarko NS, Ishihara M, Conrad HE. Control of cell division in hepatoma cells by exogenous heparan sulfate proteoglycan. J Cell Physiol 1989;139:287–94.
40. Knaus HG, Scheffauer F, Romanin C, Schindler HG, Glossmann H. Heparin binds with high affinity to voltage-dependent L-type Ca^{2+} channels. J Biol Chem 1990;265:11156–66.
41. Castellot JJ, Pukac LA, Caleb BL, Wright TC, Karnovsky MJ. Heparin selectively inhibits a protein kinase C-dependent mechanism of cell cycle progression in calf aortic smooth muscle cells. J Cell Biol 1989;109:3147–55.
42. Farach MC, Tang JP, Decker GL, Carson DD. Differential effects of p-nitrophenyl-D-xylosides on mouse blastocysts and uterine epithelial cells. Biol Reprod 1988;39:443–55.
43. Carson DD, Tang JP, Julian J. Heparan sulfate proteoglycan (perlecan) expression by mouse embryos during acquisition of attachment competence. Dev Biol 1992.
44. Sutherland AE, Sanderson RD, Mayes M, Seibert M, Calarco PG, Bernfield M, Damsky CH. Expression of syndecan, a putative low affinity fibroblast growth factor receptor, in the early mouse embryo. Development 1991;113:339–51.
45. Patillo RA, Gey GO, Delfs E, et al. The hormone-synthesizing trophoblastic cell in vitro: a model for cancer research and placental hormone synthesis. Ann NY Acad Sci 1971;172:288–98.
46. Frenette GP, Roddon RW, Kzesicki RF, Naser JA, Peters BP. Biosynthesis and deposition of a noncovalent laminin-heparan sulfate proteoglycan complex and other basal lamina components by a human malignant cell line. J Biol Chem 1989;264:3078–88.
47. Casu B, Petitou M, Provasoli M, Sinay P. Conformational flexibility: a new concept for explaining binding and biological properties of iduronic acid-containing glycosaminoglycans. Trends Biochem Sci 1988;13:221–5.

48. Way DL, Grosso DS, Davis JR, Surwit EA, Christian CD. Characterization of a new human endometrial carcinoma (RL95-2) established in tissue culture. In Vitro Cell Dev Biol 1983;19:147–58.
49. Morris JE, Potter SW, Gaza-Bulseco G. Estradiol-stimulated turnover of heparan sulfate proteoglycan in mouse uterine epithelium. J Biol Chem 1988;263:4712–8.
50. Tang JP, Julian J, Glasser SR, Carson DD. Heparan sulfate proteoglycan synthesis and metabolism by mouse uterine epithelial cells cultured in vitro. J Biol Chem 1987;262:12832–42.
51. Jacobs AL, Carson DD. Proteoglycan synthesis and metabolism by mouse uterine stroma cultured in vitro. J Biol Chem 1991;266:15464–73.
52. Armant DR, Kaplan HA, Mover H, Lennarz WJ. The effect of hexapeptides on attachment and outgrowth of mouse embryos in vitro: evidence for the involvement of the cell recognition tripeptide Arg-Gly-Asp. Proc Natl Acad Sci USA 1986;83:6751–5.
53. Handin RI, Cohen HJ. Purification and binding properties of human platelet factor four. J Biol Chem 1976;251:4272–82.
54. Wilson O, Jacobs AL, Stewart S, Carson DD. Expression of externally-disposed heparin/heparan sulfate binding sites by uterine epithelial cells. J Cell Physiol 1990;143:60–7.
55. Carson DD, Wilson OF, Dutt A. Glycoconjugate expression and interactions at the cell surface of mouse uterine epithelial cells and periimplantation-stage embryos. Troph Res 1990;4:211–41.
56. Raboudi N, Julian J, Rohde LH, Carson DD. Identification of cell surface heparin/heparan sulfate binding proteins of a human uterine epithelial cell line (RL95). J Biol Chem 1992;267.
57. Ceri H, Hwang WS, Cheung H. Endogenous heparin-binding lectin activity in human placenta: purification and developmental expression. Biochem Cell Biol 1990;68:790–5.
58. Kohnke-Godt B, Gabius HJ. Heparin-binding lectin from human placenta: further characterization of ligand binding and structural properties and its relationship to histones and heparin-binding growth factors. Biochemistry 1991;30:55–65.
59. Firestone GL, John NJ, Haffar OK, Cook PW. Genetic evidence that the steroid-regulated trafficking of cell surface glycoproteins in rat hepatoma cells is mediated by glucocorticoid-inducible cellular components. J Cell Biochem 1987;35:271–84.
60. Tung HN, Parr EL, Parr MB. Endocytosis in the uterine luminal and glandular epithelial cells of mice during early pregnancy. Am J Anat 1988;182:120–9.
61. Julian J, Carson DD, Glasser SR. Polarized rat uterine epithelium in vitro: responses to estrogen in defined medium. Endocrinology 1992;130:68–78.
62. Mani SK, Carson DD, Glasser SR. Steroid hormones differentially modulate glycoconjugate synthesis and vectorial secretion by polarized uterine epithelial cells in vitro. Endocrinology 1992;130:240–8.
63. Jacobs AL, Sehgal PB, Julian J, Carson DD. Secretion and hormonal regulation of interleukin-6 production by mouse uterine stromal and polarized epithelial cells cultured in vitro. Endocrinology 1992.

64. Carson DD, Tang JP, Julian J, Glasser SR. Vectorial secretion of proteoglycans by polarized rat uterine epithelial cells. J Cell Biol 1988;107:2425-35.
65. Jentoft N. Why are proteins O-glycosylated? Trends Biochem Sci 1990; 15:291-4.
66. Spicer AP, Parry G, Patton S, Gendler SJ. Molecular cloning and analysis of the mouse homologue of the tumor-associated mucin, MUC1, reveals conservation of potential O-glycosylation sites, transmembrane and cytoplasmic domains and a loss of minisatellite-like polymorphism. J Biol Chem 1991;266:15099-109.
67. Toole BP. Glycosaminoglycans in morphogenesis. In: Hay ED, ed. Cell biology of extracellular matrix. New York: Plenum Press, 1985:259-94.
68. Cidadao AJ, Thorsteinsdottir S, David-Ferreira JF. Immunocytochemical study of tissue distribution and hormonal control of chondroitin, dermatan and keratan sulfates from rodent uterus. Eur J Cell Biol 1990;52:105-16.
69. Parr MB, Parr EL. The implantation reaction. In: Wynn RM, Jollie WP, eds. Biology of the uterus. New York: Plenum Medical, 1989:233-77.
70. Kadokawa Y, Fuketa I, Nose A, Takeichi M, Nakatsuji N. Expression pattern of E- and P-cadherin in mouse embryos and uteri during the periimplantation period. Dev Growth Differ 1989;31:23-30.
71. Kisalus LL, Herr JC. Immunocytochemical localization of heparan sulfate proteoglycan in human decidual cell secretory bodies and placental fibrinoid. Biol Reprod 1988;39:419-30.
72. Ekblom P, Aufderheide E. Stimulation of tenascin expression in mesenchyme by epithelial-stromal interactions. Int J Dev Biol 1989;33:71-9.
73. Schlafke S, Welsh AO, Enders AC. Penetration of the basal lamina of the uterine luminal epithelium during implantation in the rat. Anat Rec 1985;212:47-56.
74. Bartocci A, Pollard JW, Stanley ER. Regulation of colony stimulating factor 1 during pregnancy. J Exp Med 1986;164:956-61.
75. D'Amore PA. Modes of FGF release in vivo and in vitro. Cancer Metastasis Rev 1990;9:227-38.
76. Atha DH, Stephens AW, Rosenberg RD. Evaluation of critical groups required for the binding of heparin to antithrombin. Proc Natl Acad Sci USA 1984;81:1030-4.
77. Reyes AA, Akeson R, Brezina L, Cole GJ. Structural requirements for neural cell adhesion molecule-heparin interaction. Cell Reg 1990;1:567-76.
78. Culty M, Nguyen HA, Underhill CB. The hyaluronan receptor (CD44) participates in the uptake and degradation of hyaluronan. J Cell Biol 1992;116:1055-62.

18
A Primatologist's Perspective on Assisted Reproduction for Nonhuman Primates

R.G. RAWLINS

Assisted reproduction for nonhuman primates has recently been heralded as a new and important means of preserving endangered primate species (1). For the reproductive biologist this development is a logical extension of years of work on primate reproductive cycles and gamete interaction—begun by Lewis and Hartman in the 1930s (2)—that has culminated in the birth of baboons (3), macaques (4), and marmosets (5) produced by *in vitro fertilization* (IVF) in the 1980s. The hope is that these remarkable achievements in reproductive biology can make a lasting contribution toward the conservation of the order Primates and at the same time provide new insight and understanding into the underlying physiological parameters of mammalian reproduction in general.

The merit of assisted reproduction for nonhuman primates, so obvious to the reproductive biologist, is less clear to the primatologist. Although primatology cuts across many scientific disciplines, over half of the members of the American Society of Primatologists are behavioral biologists largely unfamiliar with the technical aspects of assisted reproduction, typically ecologically oriented, and skeptical that a species-specific approach to saving primates in such small numbers—as opposed to habitat preservation—provides a realistic approach to conservation of the order given the rate at which global change is taking place. Reproductive biologists must make a convincing case that precious financial and animal resources be utilized for the research needed to advance primate assisted reproduction as quickly as possible. Time is running out for many primate species.

Concerns of the Primatologist

Historically, the 1985 conference "Primates: the Road to Self-Sustaining Primate Populations," organized by Kurt Benirschke in San Diego, California, introduced field biologists to the utility of assisted reproduction for conservation of endangered species (1). Implementation of the techniques has already begun, and much has been accomplished in a wide variety of mammalian species; yet, results for nonhuman primates are slow in coming. This may be due in part to the fact that the interests of the primatologist in the field frequently conflict with the manipulative requirements of the laboratory biologist or the intervention of the veterinarian charged with delivering optimal health care to the animals.

Primatologists are also well aware that human assisted reproduction, largely market driven, has reached a plateau with respect to success rates and that investigators are now returning to nonhuman primates for embryo studies not ethically possible using human materials in order to improve development and enhance implantation. Primatologists are concerned that the sudden interest in preserving primates may be a mercenary approach to securing new funding and animals for research truly directed at solving problems of human rather than nonhuman primate reproduction. Animals already endangered should not be trapped out of the natural habitat only to be placed in programs for reproductive research. Reproductive biologists must assure the primatological community that environmental concern is not being used as a pretext and that their interests are sincere. A multidisciplinary approach combining knowledge of behavior and habitat with a basic understanding of reproductive biology will be essential, from the standpoint of both field and bench biology, if the goal of conservation is to be accomplished. Human and nonhuman primate assisted reproduction will both benefit.

It is also important for the primatologist to know that development of assisted reproduction for nonhuman primates is really not a recent phenomenona. The birth of Louise Brown, the world's first human test-tube baby in 1978, was preceded by years of work on gamete biology and embryogenesis in nonhuman primates. Both the interest and the methods for doing such work with primates have been there all along. Their application to nonhuman primates has simply been eclipsed by the notoriety associated with human reproduction in vitro. For example, Allen in 1928 was the first to collect a human tubal ovum, but by 1933 Lewis and Hartmann were the first to surgically recover a tubal embryo from a macaque (2). Successful IVF was first reported for human beings by Rock and Menkin in 1944 (6), but did not result in a live birth until 1978 at the hands of Edwards and Steptoe (7). However, by the 1970s Gould's group, as well as Dukelow's, had all completed IVF in the squirrel monkey (8, 9).

There has always been a consistent interchange of technology and ideas. Indeed, much of modern human infertility therapy employs methods derived from veterinary medicine and early studies of the physiology and cell biology of reproduction in nonhuman primates. It is ironic that following transfer into human medicine, these techniques have come full circle and are now being proposed to aid nonhuman primates.

Assisted Reproduction for Nonhuman Primates

The majority of technologies used for human infertility therapy today can readily be applied to nonhuman primates. Gametes and embryos can be cryopreserved (10), embryos can be produced in vivo or in vitro (11–13), micromanipulated for cell biopsy and genetic analysis (14), or split to induce monozygotic twinning (15). Intact or manipulated embryos can be replaced in either the original oocyte donor or transferred to host females of the same or different species (12) to propagate endangered species or biologically unique animal lines useful for the study of human disease. Cloning through embryo splitting or nuclear transfer opens the possibility for creating large numbers of isogenic animals as well (16). While a good deal of the foregoing remains a technical possibility only, much has already been accomplished.

In advocating the use of assisted reproduction for conserving the order Primates, an important fact is frequently overlooked. To date, the hormonal cycles of only 25 taxa within the order have been characterized (17). Of these, 19 are Old World species, 9 of which are macaques. Only 6 New World taxa have been described in detail despite the fact that they are among the most endangered. It makes little sense to apply high technology in the absence of the basic endocrine profile for these same species.

The task for the reproductive biologist is to complete the characterization of the hormonal cycles of the primates, to improve capabilities for long-term storage of primate gametes and embryos, and to develop nonantigenic agents for primate superovulation. Assisted reproduction should be undertaken for the primate species at greatest risk and coordinated with captive breeding programs; stock lines to be used for biological reservoirs should be selected; and commercial demand on indigenous populations should be halted. Together, these steps may help preserve the order Primates. Realistically, assisted reproduction will probably be most useful in preservation of genetic stock. At present, only 20–25 nonhuman primates have been produced by various applications of assisted reproduction. Even with the disclaimer that efforts have not been directed at production, scientists will not be able to produce the numbers of animals needed to offset the annual loss of thousands.

Major Threats to Nonhuman Primates

Today, over 200 species have been identified within the order Primates, but 1 in 7 (14%) is now endangered, and more than 50% are in jeopardy of entering the endangered list soon. The major threats to primate species are habitat destruction, hunting for food, and live capture (18, 19).

Habitat destruction is particularly critical because more than 90% of the primate species live in the tropical forests. In Asia, Africa, and South America, most of the destruction is due to agriculture and ranching. Forests are cut for housing materials and sources of fuel as well. In addition, mining, hydroelectric projects, and road construction are principal causes of the loss of habitat as underdeveloped countries seek to exploit their available resources. Overpopulation accelerates these losses and promotes the use of primates as food sources as well (18, 19). Alternatives to deforestation through equitable distribution of world resources among nations seem the only possibility for halting further destruction (20).

Zoo trade and the biomedical community share much of the responsibility for maintaining the live capture market (18, 19). The biomedical community, particularly in the United States, has been a principal consumer of nonhuman primates for more than 50 years. At least 30,000 macaques were consumed yearly during the late 1970s in the United States alone. Many were used for vaccine testing in spite of the fact that alternative cell culture systems had been used for many years in Europe to test for vaccine safety.

Primate consumption continues, but today relies primarily on captive breeding programs instead of on trapping animals out of habitat countries. This is largely the historical result of a change in policy of the Indian government with respect to shipping rhesus monkeys out of the country rather than the result of an enlightened global approach to primate conservation. Due to the enormous pressure on its indigenous population of macaques from uncontrolled harvesting and political unrest rooted in the deep religious beliefs of its own people, the Indian government embargoed shipments of rhesus monkeys in the late 1970s (21). Now, although habitat destruction continues, many of the primate populations have stabilized, and some are growing again (22).

While the Indian embargo has helped the rhesus monkey, it has done little for other macaque species. Resourceful importers quickly turned to other countries, such as Indonesia and the Philippines, as unrestricted sources for nonhuman primates. A variety of macaques have subsequently been used, and the impact of live capture continues relatively unabated today. The rewards for capturing primates can be great financially. Prices for animals continue to increase, and animals continue to be shipped.

Clearly, unregulated live capture must be stopped. It is an enormous problem that has decimated primate populations worldwide. Its long-term demographic effects on primate populations have yet to be evaluated, but the immediate results appear to be a bottlenecking of disrupted social groups due to overharvesting of the youngest age classes and an accelerated loss of genetic diversity. Uncontrolled trapping, subsidized by the zoo trade or biomedical markets, promotes exploitation, not conservation of an invaluable resource.

On the positive side, captive breeding has reduced trapping of animals in habitat countries to some extent by eroding the commercial market. For example, the regional primate centers in the United States are largely self-sufficient and do not promote trapping in habitat countries. Demand has also been reduced by the high cost of doing research with nonhuman primates. Both purchase price and per diem costs for housing are effectively cutting back the numbers of animals used. The real issue is the appropriate choice for the animal model. Nonhuman primates should be used for research, but they must be used wisely and conservatively.

Ultimate Cause of Modern Species Extinctions

In exploring the proximate mechanisms of primate reproduction, we see that the ultimate cause of the enormous problems threatening the planet is human population growth. Today, one species—our own—has for the first time in history gained the ability to alter habitats on a global scale, and habitat destruction has always preceded species extirpation. With that in mind, recall the recent and frightening detection of planetary ozone depletion complete with undefined consequences of increased ultraviolet exposure for all species, the long-standing problem of acid rain due to industrial pollution, and the greenhouse effect responsible for accelerated global warming. Any or all of these changes have the potential for enormous alteration or destruction of habitats worldwide and, therefore, the loss of species (20, 23).

While pollution is increasing, the lungs of the planet—its forests—are being cut out. Current global destruction of tropical rainforest alone exceeds 17 million hectares annually. An area equivalent in size to the country of Switzerland or the state of Georgia disappears each year. As these forests are cut down, 17,500 species of plants and animals are eliminated with them yearly. The current rate of species loss exceeds that of 65 million years ago when the dinosaurs became extinct, presumably by habitat destruction as well (23). Stopping the global reduction of biodiversity should be an immediate challenge to all scientists.

Underlying all these global changes is continued human population growth. From 4000 B.C. to 1000 A.D., growth of the global human population has been estimated to have reached a maximum at about 1 billion people. As of 1992 the world contains 6 billion people, and by 2050 the population will have reached 8 billion. That is approximately 274,000 births per day, or 11,415 per hour. The total number of human IVF deliveries to date does not exceed 30,000 in 14 years of effort. If real progress is to be made toward saving the order Primates, as well as the rest of the planet, the problem of human overpopulation must be addressed. More than a billion people will be added to the globe per decade over the next 30 years. In the underdeveloped countries growth will be 3% per annum, 4 times the rate of developed countries, such as the United States. The Third World doubled in just 35 years to over 5 billion people (20, 23). These are the same countries with the highest concentration of nonhuman primates! Existing agricultural lands simply will not be able to support the increasing population, and the fate of the primates is sealed unless something is done now.

What Can Reproductive Biologists Do?

The last two presidential administrations in the United States have chosen to collectively ignore population growth and have stopped funding for population control. Even when ignored, however, the population bomb has continued to tick. As reproductive biologists we must demand the restoration of financial support for the United Nations and World Health Organization's efforts at global birth control and family planning. Primate assisted reproduction, while technically facscinating as a means of producing animals, should also be viewed as a means of improving contraception as well. The august body of reproductive scientists represented in this book is largely responsible for the discovery of the physiological requirements for fertilization, embryogenesis, and implantation in vitro. Surely the same talent and knowledge can now be applied to develop new and practical methods of contraception as well. Now we must treat the cause, rather than the symptoms, of the ecological problems facing not only the order Primates, but the planetary ecosystem as well.

References

1. Benirschke K. Primates: the road to self-sustaining populations. New York: Springer Verlag, 1986.
2. Lewis WH, Hartman CG. Early cleavage stages of the egg of the monkey (rhesus). Contracept Embryol Carnegie Inst 1933;24:187.
3. Clayton O, Kuehl TJ. The first successful in vitro fertilization and embryo transfer in a nonhuman primate. Theriogenology 1984;21:28.
4. Bavister BD, Boatman D, Collins K, Dierschke D, Eisele S. Birth of a rhesus monkey infant after in vitro fertilization and nonsurgical embryo transfer. Proc Natl Acad Sci USA 1984;81:2218–22.
5. Lopata A, Summers P, Hearn J. Births following the transfer of cultured embryos obtained by in vitro and in vivo fertilization in the marmoset monkey (*Callithrix jacchus*). Fertil Steril 1988;50:503–9.
6. Rock J, Menkin MF. In vitro fertilization and cleavage of human ovarian eggs. Science 1944;100:105.

7. Steptoe P. Historical aspects of the ethics of in vitro fertilization. Ann NY Acad Sci 1988;442:602.
8. Gould KG, Cline EM, Williams WL. Observation on the induction of ovulation and fertilization in vitro in the squirrel monkey (*Saimiri sciureus*). Fertil Steril 1973;24:260–8.
9. Kuehl TJ, Dukelow WR. Maturation and in vitro fertilization of follicular oocytes of the squirrel monkey (*Saimiri sciureus*). Biol Reprod 1979;21:545.
10. Balmaceda JP, Heitman T, Garcia MR, Pauerstein C, Pool T. Embryo cryopreservation in cynomolgus monkeys. Fertil Steril 1986;45:403–6.
11. Pope CE, Pope VZ, Beck LR. Nonsurgical recovery of uterine embryos in the baboon. Biol Reprod 1980;23:657–62.
12. Balmaceda JP, Pool T, Arana J, Heitman T, Asch R. Successful in vitro fertilization and embryo transfer in cynomolgus monkeys. Fertil Steril 1984;42:791–5.
13. Wolf DP, Vandevoort C, Meyer-Haas G, et al. In vitro fertilization and embryo transfer in the rhesus monkey. Biol Reprod 1989;41:335–46.
14. Handyside AH, Penketh RJA, Winston RML. Biopsy of human preimplantation embryos and sexing by DNA amplification. Lancet 1989;1:347–9.
15. Willadsen SM, Lehn-Jensen H, Fehilly CB, Newcomb R. The production of monozygotic twins of preselected parentage by micromanipulation of nonsurgically collected cow embryos. Theriogenology 1981;15:23–9.
16. Willadsen SM. Cloning of sheep and cow embryos. Genome 1989;31:956–62.
17. Robinson JA, Goy RW. Steroid hormones and the ovarian cycle. In: Dukelow WR, Erwin J, eds. Reproduction and development. New York: Alan R. Liss, 1986:63–91.
18. Mittermeir RA. Strategies for the conservation of highly endangered primates. In: Benirschke K, ed. Primates: the road to self-sustaining populations. New York: Springer Verlag, 1986:1013–21.
19. Mittermeir RA, Cheney DL. Conservation of primates and their habitats. In: Smuts B, Cheney DL, Seyfarth R, Wrangham R, Struhsaker T, eds. Primate societies. University of Chicago Press, 1987:477–90.
20. Sigma Xi Forum Proceedings. Global change and the human prospect: issues in population, science, technology and equity. Research Triangle Park, NC: Sigma Xi, Scientific Research Soc, 1992.
21. Dukelow WR, Erwin J. Reproduction and development. New York: Alan R. Liss, 1986:xi.
22. Southwick CH, Siddiqui MF. Partial recovery and a new population estimate of rhesus monkey populations in India. Am J Primatol 1988;16:187–97.
23. World Resources Inst, World Conservation Union, United Nations Environment Program, Food and Agriculture Organization, United Nations Education, Scientific and Cultural Organization. Global biodiversity strategy: a policy maker's guide. Baltimore, MD: World Resources Inst Publications, 1992.

Part V

Gamete/Embryo Applications

19

Nuclear Transfer in Mammals

N.L. First and M.L. Leibfried-Rutledge

The transplantation of nuclei from one mammalian cell to another has provided a valuable tool for understanding developmental biology and for production of nearly identical animals useful in food production or research. The initial use of nuclear transfer was in amphibians 40 years ago when Briggs and King (1) reported that blastula stage nuclei transplanted to oocytes could direct development to the tadpole stage. They were testing the original hypothesis of Spemann (2), who suggested that cell totipotency and its loss with differentiation could be tested by transfer of nuclei from progressively advanced stages of development into enucleated oocytes. Continued study of cell totipotency in amphibia resulted in production of fertile frogs from blastula nuclei (3, 4), thereby demonstrating their totipotency. In rare cases even older embryonic and young larval cell nuclei from *Xenopus* directed formation of fertile frogs (5). However, nuclei from differentiated frog cells could not direct complete development (6). Completion of development failed in spite of the fact that transplanted nuclei were reprogrammed in nuclear composition by translocation of proteins from the oocyte cytoplasm (7, 8), and in a few cases individual genes turned off by differentiation were again expressed (9, 10). Nuclear transfer continues to be a useful tool in amphibia for the study of nuclear-cytoplasmic interactions, cell differentiation, and totipotency, as well as mechanisms regulating gene expression.

Early attempts to extend these studies to mammals were in mice and usually without much success. The first attempt in mice repeated the amphibian studies except that it involved direct injection of either blastocyst *inner cell mass* (ICM) or trophoblast cells into pronucleus-removed eggs. The level of success was low, but three mice were born from use of ICM cells and none from the differentiated trophoblasts (11). Until recently, using cell fusion to introduce the nucleus, the only embryo cleavage stages that resulted in subsequent development were from early 2-cell stage nuclei (12, 13) or from transfer of 4- or 8-cell nuclei into an enucleated 2-cell embryo (13, 14). Recently, however, a low frequency of blastocysts (10%), but no offspring, was obtained when blastocyst ICM cells were injected into enucleated mouse oocytes (15). The reasons for failed development after nuclear transfer of mouse embryos as compared to amphibian are unknown. Differences in the stage of embryo development at which commitment to differentiation and loss of totipotency occur have been suggested.

More recently, nuclear transfers via whole-cell fusion into enucleated oocytes have been successful in species where the onset of embryonic transcription and onset of commitment to differentiation and cell polarization are at stages much later than the mouse. The species where major effort has been placed on development of nuclear transfer are food-producing animals, largely because of the expected economic value of phenotypically selected rather than strictly genetically selected animals. Offspring have resulted from fusion of late stage blastomeres with enucleated secondary oocytes in sheep (16, 17), cattle (18–20), swine (21), and rabbits (22). This chapter focuses primarily on the development and use of nuclear transfer in mammals.

Nuclear Transfer Procedure

Nuclear transfer, as illustrated in Figure 19.1, is usually accomplished by fusion of a blastomere from a preblastocyst or ICM stage embryo with an

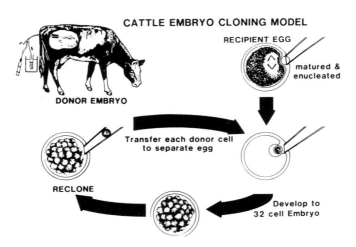

FIGURE 19.1. The steps in nuclear transfer as commonly practiced in cattle. Embryos are flushed from uteri of cows or produced in vitro. Their cells are isolated and then transferred by electrofusion or other fusion methods into the cytoplasm of enucleated abattoir-recovered oocytes. The fusion product is cultured to morulae or blastocyst, and these cells are used as nuclear donors in repeated rounds of nuclear transfer.

enucleated oocyte, followed by large-scale multiplication as a result of repeating the procedure using blastomeres from the embryos produced from nuclear transfer. This allows the production of clonal lines that, when appropriately selected for performance in a given trait, can be reproduced to capture expression in the offspring of both additive and nonadditive inheritance. The genetic gain realized by incorporating cloning into the selection procedures has been estimated to be 5 times the genetic gain per year as compared to the best procedures currently used to improve cattle genetics (23).

There have been few pure nuclear transfers. Procedures used in amphibia (5, 6) and the mouse experiments of Illmensee and Hoppe (11) involved microinjection of a nucleus accompanied by a small amount of cytoplasm directly into the cytoplasm of enucleated oocytes. This procedure damages the oocyte if the microinjection pipette is more than 7–8 µ in diameter, but the donor cells or nucleus are often too large for such a small pipette, and subsequent embryo development is reduced. In 1984 McGrath and Solter (12) showed that nuclei could be introduced into mouse oocytes with high efficiency using inactivated Sendai virus to fuse a karyoplast into the oocyte. In cattle this Sendai virus method or use of herpes virus gave little success, whereas electrofusion resulted in good fusion efficiency (24). This method was subsequently modified by use of electrofusion to fuse an entire blastomere into enucleated oocytes of sheep (16), cattle (18), rabbits (22), and swine (21, 25). This has become the common method for nuclear transfer in mammals.

Electrofusion is not without problems. It is inefficient when the difference in size of the cells to be fused is great, such as fusion of blastula, somatic cells, or sperm cells into oocytes (26). It is damaging to cell membranes since it induces membrane pores and long-term destabilization of the oocyte plasma membrane (27). Because electroporation also causes Ca^{++} elevation within the cell, the oocyte is partially or completely activated by the electrofusion. Elevation of intracellular Ca^{++} at an inappropriate time is undesirable (Stice, personal communication). Therefore, less damaging and more effective fusion systems need to be developed.

Enucleation of the Oocyte

Complete removal of all oocyte chromatin appears to be essential to prevent polyploidy and a reduced frequency of subsequent embryo development, as shown by Willadsen (16) with sheep oocytes in which nuclei were transferred into enucleated and nucleated oocytes. Enucleation has been performed by bisection of the oocyte with rejection of the half showing evidence of polar body or chromatin or by aspiration of the polar body and adjacent cytoplasm (17, 18, 21, 22). Subsequent development of the embryo may be affected by the amount of cytoplasm removed and reduced if as much as half is removed (28). The assurance of complete enucleation is greatly improved by first staining the chromatin with a fluorescent DNA-specific dye, such as DAPI or Hoechst 33258 (29, 30), or by staining the removed portion. Subsequent development of embryos derived from nuclear transfer was not affected by fluorescent enucleation (30).

Oocyte Activation

Central to nuclear transfer is the ability to activate the oocyte used as recipient cytoplasm for a donated nucleus. Activation events turn on the program for meiotic/mitotic transition and development that has been previously stored in the oocyte during growth and differentiation prior to ovulation. In the mouse where either pronuclear stage zygotes (11, 12) or 2-cell blastomeres (13, 14) are used as recipient cytoplasm, activation has been caused by sperm penetration prior to manipulations and is therefore not an issue. Where metaphase II oocytes are used in nuclear transfer as recipient cytoplasm, and this includes domestic species and the rabbit, then activation becomes a major concern since there is no evidence in the literature to indicate that a donated blastomere or karyoplast can "jump-start" the oocyte's developmental program. Fortuitously, the electroporation used for fusing the oocyte/blastomere unit does induce activation of metaphase II oocytes, as evidenced by pronucleus formation (cattle: 31; pigs: 32; and rabbits: 33). In fact, it would be most useful to have a nonactivating fusion method for use in domestic animal species similar to the Sendai virus-induced fusion used in murine nuclear transfer (12). This would allow a more definitive examination of the interaction between oocyte and blastomere during nuclear transfer and the role of activation in successful cloning.

As suggested by Kubiak (34), murine oocytes after completion of meiosis gradually acquire the ability to respond to activation stimuli, either sperm induced or parthenogenetically mediated, but oocytes are activated by sperm earlier than by artificial stimuli. We have verified this in cattle since sperm are capable of fertilizing oocytes 24 h of age (35), but these same-age oocytes do not respond to common parthenogenetic stimuli, such as electrical current, Ca^{++} ionophore, or ethanol, by forming pronuclei until a later time period (31, 36). What is most interesting is that bovine oocytes 24 h of age activated by either electrostimulation (37) or ionomycin (Leibfried-Rutledge, Susko-Parrish, First, unpublished observations) do undergo meiosis II, but rather than forming pronuclei, enter a third meiotic arrest. This was also described in mice (34). We need to reevaluate the type and/or quality of current methods of oocyte activation in mammals since our artificial methods do not induce activation and parthenogenetic development during periods when sperm can successfully interact with oocytes.

With time, oocytes become responsive to artificial stimuli and form pronuclei and frequently even parthenogenetic blastocysts in some species. Competence for activation has long been known to be an age-dependent phenomenon in rodents (reviewed in 38), and this has been verified for cattle (31, 36), pigs (32), and rabbits (33). In general, the older the secondary oocyte is following *first polar body* (PB1) expulsion, the more sensitive to parthenogenetic stimuli it becomes. The threshold for activation responsiveness becomes so lowered in some species that their oocytes spontaneously activate upon aging (hamster: 39 and rat: 40). Thus, as we wait until secondary oocytes become amenable to our activation methods for use in nuclear transfer, the danger of senescent changes setting in becomes prominent. This is discussed in more detail in the following section. Cattle oocytes are used at >40 h of age in many reported nuclear transfer schemes (19, 41–43), yet bovine oocytes fertilized in vitro at >40 h of age result in only 3% becoming blastocysts (44).

In summary, while we do not completely understand how a donated embryonic nucleus is modified by the cytoplasm of an activated oocyte, it is apparent that current parthenogenetic stimuli are probably less than optimal. It is also apparent that the age of the oocyte that responds to conventional activation stimuli may not be the age of the optimum recipient for a donated nucleus.

Age and Stage of Recipient Cytoplasm

The pronuclear ovum and metaphase II secondary oocyte have been the two popular types of recipient cytoplasm used in cloning mammalian embryos. A low frequency of success using pronuclear ova in mice (11) and cattle (24) has been achieved. The low success rate using pronuclear zygotes as recipient cytoplasm may be due to the difficulty in precisely controlling the age postfertilization of the oocyte used as recipient cytoplasm. If somatic cell nuclei are fused to oocytes late in the first cell cycle of activated oocytes, then remodeling of the donated nucleus does not occur (45–47) and the nucleus may produce fewer blastocysts after nuclear transfer (33). Sperm nuclei are also not remodeled into pronuclei if penetration occurs late after activation (48, 49). Currently, much emphasis is placed on the need for remodeling and swelling of the donated nucleus to occur in order to develop a successful nuclear transfer system in mammals. Since Willadsen (16) first reported success in cloning sheep embryos with metaphase II oocytes as recipient cytoplasm, most laboratories studying nuclear transfer with mammalian embryos use secondary oocytes. As mentioned earlier, there has been limited success in mice (with very early stages of embryo donors) using 2-cell blastomeres as recipients for the introduced nucleus (13, 14). This has not been examined for potential use in domestic animals. The idea is intriguing since it places a mitotic cell cycle nucleus into mitotic cytoplasm, rather than into the meiotic cytoplasm of the secondary oocyte.

The frequent use of the secondary oocyte as recipient cytoplasm in cloning necessitates consideration of activation as previously discussed. Since responsiveness to activation stimuli increases with the age of the oocyte, some concern for gamete senescence also becomes pertinent. Postovulatory aging of the oocyte is known to cause developmental abnormalities and a decrease in pregnancy and/or litter size (50). The interaction of aging cytoplasm from the oocyte with components of the cytoplasm brought in with the embryonic nucleus becomes of interest, yet has not been examined.

In the bovine using oocyte cytoplasm either 24 or 42 h of age, we find several major differences in the nuclear transfer embryos produced (51). Since we have developed a method of activating younger stage cattle oocytes (37), these studies are possible, yet the activation protocol is still not completely satisfactory. In the comparison by age of recipient cytoplasm, the timing of events leading to the first cleavage division is advanced in cloned embryos made with 42-h cytoplasm. This is reminiscent of the speed of early developmental events reported after fertilization of senescent oocytes in laboratory species (52, 53). A second difference between cloned embryos made from 24- or 42-h oocyte cytoplasm concerns the method of processing the donated nucleus in the cytoplasm. If oocyte chromatin was allowed to remain, the 24-h oocyte forced the embryonic nucleus to break down and undergo an anaphase/telophase situation with expulsion of a polar body. In the older cytoplasm the intact oocyte chromatin also underwent meiotic division, but the introduced nucleus was not broken down. When enucleated, the 42-h cytoplasm again did not break down the new nucleus, while the 24-h cytoplasm still caused nuclear envelope breakdown in a significant portion of the oocytes, followed by an anaphase/telophase event.

Older oocytes have been found to give a reduced frequency of blastocyst in cloning in the rabbit and also reduced fusion success (33). A decrease in the development of clones with age of oocyte has been reported in cattle (18). Younger cattle oocytes also fuse more successfully with blastomeres (unpublished observations). Older cattle oocytes degenerate more in response to electrical stimulation than do younger oocytes, although an introduced blastomere provides a protective effect (51). We have already discussed the difference in susceptibility to activation between older and younger oocytes of a number of species. Further research in this area seems warranted both to improve efficiency of cloning and to shed greater light on nuclear/cytoplasmic interactions and normal oocyte function.

Totipotency and Cell-Cycle Stage of the Donor Nucleus

Experiments in amphibia described earlier show that little development to adult occurs when the nuclei used in nuclear transfer are from differentiated cells, even though a small amount of genomic reprogramming can be caused to occur (5, 6, 9, 10). Similar evidence is accumulating for mammals. Thus far, the latest stages used successfully in mammals for nuclear transfer have been the 4-cell stage in swine and blastocyst ICM in sheep (17), rabbits (33), mice (11), and cattle. While nondifferentiated or stem cells appear to be totipotent, differentiated cells of mammals, as in amphibians, appear not to be totipotent. In rabbits (33) and in mice (11, 54), the use of trophoblast cells in nuclear transfer has failed to produce blastocysts in experiments where blastocysts and pregnancies resulted from ICM cells. Differentiation of a trophoblast depends on a commitment to differentiate when embryonic cells polarize before compaction.

In cattle the use of polarized morula cells in nuclear transfer resulted in only 7% morula and blastocysts, whereas the frequency was 30% for unselected cells and 47% for nonpolarized cells (55). From this it might be expected that a higher frequency of embryo development should occur from the use of ICM cells rather than morula. Thus far, this has not been the case in rabbits. Where cells from morula and ICM were compared, development to morula and blastocysts was highest for morula cells (26).

Several laboratories have been attempting to isolate and culture embryonic stem cells from cattle, sheep, and swine (56, 57). This is being done in order to use stem cells as a means for gene transfer in which desired genes could be introduced by electroporation and transfection or by viral vector into the cultured stem cells, as is presently being done to make transgenic mice. Moreover, since stem cells are progenitors of all other cells, and if they can be identified in a species, cultured to large numbers, and found to be totipotent, as in mice, they will be useful as nuclear donor cells in nuclear transfer. This would allow clonal multiplication of embryos in the culture dish to numbers limited only by the multiplication rate of the cultured stem cells, perhaps resulting in as many as 10^3 to 10^6 clones from one embryo. The multiplication to large numbers from one cell line has not yet been accomplished. Thus far, the largest number of cloned domestic animals born has been 11 calves at Granada Genetics in 1990 (20, 58). Pluripotent embryonic stem cell lines have been established for swine (56, 59, 60), cattle (56), sheep (59), mice (61, 62), and hamsters (63). However, only in mice (57, 62) has it been published that these pluripotent lines remained totipotent, as tested by chimerization or nuclear transfer. The principle problem in most species seems to be failure to prevent differentiation of the cultured stem cells (57).

A potentially major variable affecting embryo development after nuclear transfer is the stage of the cell cycle of the nuclear donor cell at the time of nuclear transfer. The optimal cell cycle stage may depend on the age or stage of the enucleated metaphase II oocyte that is fused with a donor cell. If a nucleus has completed DNA synthesis and is in G2, then for normal ploidy to result, the nucleus must remain in G2 and not undergo any additional DNA synthesis after nuclear transfer. Conversely, if the nucleus is in G1—that is, before DNA synthesis when transferred—then it must undergo a single round of DNA synthesis. As previously discussed, the early metaphase oocyte (24 h) attempts to remodel the introduced nucleus, including decondensation and recondensation of chromatin, while the late metaphase II oocyte (42 h) uses the introduced nucleus as is and remodels it into a pronuclear-like structure that quickly prepares for the first cleavage division (51). Thus, in theory, if the enucleated oocyte is entering metaphase II, the best embryo development should occur from the use of a G1 nuclear donor since it could complete DNA synthesis, whereas later stages may have started DNA synthesis or be in an inappropriate state to complete a full round of DNA synthesis. If the oocyte is in late metaphase and ready to initiate cleavage, the donor cells of G1 or early S will require and attempt to initiate a round of DNA synthesis, but with insufficient time for completion.

The data testing these hypotheses are limited. In rabbits where the oocyte is at early metaphase II, Collas et al. (64) have clearly shown that embryo development occurs primarily from the use of G1 donor cells, and development is progressively poorer with later cell cycle stages. If an aged oocyte were the recipient where the nucleus is driven quickly to cleavage without remodeling, the preferred nucleus might be one at G2 of the cell cycle. This interaction of oocyte and nucleus has not yet been tested. In mice when cytoplasts of approximately G1, S, and G2 stages were transferred into enucleated approximately G1, S, and G2 2-cell embryos, the greatest frequency of embryos developing to 90 h was from late 2-cell (G2) karyoplasts or cytoplasts (65).

Another variable is the influence of the developmental stage of the donor cell on cell cycle interaction of nuclear donor and recipient oocyte. In most species early stages of embryo development are characterized by the omission of G1 and G2 periods and the absence of embryonic transcription. Nuclei from stages with and without G1 and G2 have been compared in amphibians. Nuclei from more slowly dividing cells promoted development at a higher rate if in G2 when transferred (66), whereas the stage of the cell cycle was less important for nuclei from rapidly dividing cells (67, 68). Although nuclei in G1 undergo DNA synthesis after nuclear transfer to activated, enucleated metaphase II oocytes, in many instances the replication is not complete. This incomplete DNA replication in amphibians resulted in chromosomal breakage and unequal inheritance among the daughter cells (69, 70).

Later in development, the chromosomal abnormalities are manifested as developmental restriction points or stages beyond which the nuclear transfer embryo is unable to progress. These restriction points are stably inherited, as shown in amphibians with serial nuclear retransfer (71). At present, it seems that the stage of the cell cycle of the nuclear donor is important, but the optimal stage for a species may depend on the age and stage of the oocyte and the developmental stage of the donor nucleus. Mistakes in the stage matching of nuclear donor and recipient oocytes may account for the increased frequency of embryo and fetal losses from nuclear transfer.

Efficiency and Application

In animal agriculture the ability to produce large numbers of clonal offspring has value as a powerful tool in animal selection by allowing nearly phenotypic selection and by rapidly propagating the highest-producing or environmentally best-fit individuals. The frozen storage of embryos from a clonal line also provides the opportunity for extensive genetic screening of each line before its widespread use in commercial animal production. As a research tool, cloned animals would be of great value in studying genotype-environmental interactions or in studies where it is desirable to provide similar genotype across all treatments. Nuclear transfer has become a powerful tool in amphibians for studying totipotency and differentiation and for studying control of gene expression. Nuclear transfer will likely be used for similar purposes in mammals. Combined with embryonic stem cell technology, nuclear transfer could be a powerful tool in the manufacture of transgenic animals from gene-modified cultured cells. The latter uses, rather than the propagation of large numbers of clones, are likely to be the potential uses for nuclear transfer in the study of nonhuman primates.

These applications of nuclear transfer are dependent on an efficient system for nuclear transfer, the production of high-quality embryos capable of normal embryo and fetal development, and the use of cultured stem cells as nuclear donors. At present, the success in combining stem cell technology with nuclear transfer is limited. The nuclear transfer 42-day pregnancy rates in cattle (approximately 30%) (20, 72) are lower than normal, as is the survival to term. Although the calves born are anatomically and physiologically normal, an increased frequency of large calves requiring assisted parturition has been reported (73). The cause is unknown. Little is known about the interaction of cytoplasm and mitochondria from the oocyte and donor cell, but there is evidence in cattle for cytoplasmic inheritance (74, 75). While several aspects of the nuclear transfer procedure have less than perfect efficiency, great progress has been made in increasing the procedural efficiencies. For example, oocyte enucleation is now nearly 100% efficient, cell fusion

70%–80%, and development to transferable blastocysts 20%–25%. Improvements in the latter may also improve pregnancy and embryo survival because the procedural steps are less than perfect. While nuclear transfer embryos can be recloned (76) in the absence of using embryonic stem cells as nuclear donor cells, the nuclear transfer process has not produced more than 7, 8, or, maximally, 11 cloned calves at Granada Genetics (20, 58).

References

1. Briggs R, King TJ. Transplantation of living nuclei from blastula cells into enucleated frogs' eggs. Proc Natl Acad Sci USA 1952;38:455–63.
2. Spemann H. Embryonic development and induction. New York: Hafner, 1938:210–1.
3. Gurdon JB. Adult frogs derived from the nuclei of single somatic cells. Dev Biol 1962;4:256–73.
4. McKinnel RG. Intraspecific nuclear transplantation in frogs. J Hered 1962;53:199–207.
5. Gurdon JB. Nuclear transplantation in eggs and oocytes. J Cell Sci 1986;4(suppl):287–318.
6. DiBerardino MA. Genomic potential of differentiated cells analyzed by nuclear transplantation. Am Zool 1987;27:623–44.
7. DiBerardino MA, Hoffner NJ. Nucleocytoplasmic exchange of nonhistone proteins in amphibian embryos. Exp Cell Res 1975;94:235–52.
8. Leonard RA, Hoffner NJ DiBerardino MA. Induction of DNA synthesis in amphibian erythroid nuclei in *Rana* eggs following conditioning in meiotic oocytes. Dev Biol 1982;92:343–55.
9. Wakefield L, Gurdon JB. Cytoplasmic regulation of 5S RNA genes in nuclear-transplant embryos. EMBO J 1983;2:1613–9.
10. Gurdon JB, Brennan S, Fairman S, Mohun TJ. Transcription of muscle specific actin genes in early *Xenopus* development: nuclear transplantation and cell dissociation. Cell 1984;38:691–700.
11. Illmensee K, Hoppe PC. Nuclear transplantation in *Mus musculus*: developmental potential of nuclei from preimplantation embryos. Cell 1981;23:9–18.
12. McGrath J, Solter D. Inability of mouse blastomere nuclei transferred to enucleated zygotes to support development in vitro. Science 1984;226:1317–9.
13. Robl JM, Gilligan B, Critser ES, First NL. Nuclear transplantation in mouse embryos: assessment of recipient cell stage. Biol Reprod 1986;34:733–9.
14. Tsunoda Y, Yasui T, Shioda Y, Nakamura K, Uchida T, Sugie T. Full term development of mouse blastomere nuclei transplanted into enucleated two-cell embryos. J Exp Zool 1987;242:147–51.
15. Kono T, Kwon OY, Nakahara T. Development of enucleated mouse oocytes reconstituted with embryonic nuclei. J Reprod Fertil 1991;93:165–72.
16. Willadsen SM. Nuclear transplantation in sheep embryos. Nature 1986;320:63–5.

17. Smith LC, Wilmut T. Influence of nuclear and cytoplasmic activity on the development in vivo of sheep embryos after nuclear transplantation. Biol Reprod 1989;40:1027–35.
18. Prather RS, Barnes FL, Sims ML, Robl JM, Eyestone WH, First NL. Nuclear transfer in the bovine embryo: assessment of donor nuclei and recipient oocyte. Biol Reprod 1987;37:859–66.
19. Willadsen SM. Cloning of sheep and cow embryos. Genome 1989;31:956–62.
20. Bondioli KR, Westhusin ME, Looney CR. Production of identical bovine offspring by nuclear transfer. Theriogenology 1990;33:165–74.
21. Prather RS, Sims MM, First NL. Nuclear transplantation in early pig embryos. Biol Reprod 1989;41:414–8.
22. Stice SL, Robl JM. Nuclear reprogramming in nuclear transplant rabbit embryos. Biol Reprod 1988;39:657–64.
23. Smith C. Cloning and genetic improvement of beef cattle. Anim Prod 1989;49:49–62.
24. Robl JM, Prather R, Barnes F, et al. Nuclear transplantation in bovine embryos. J Anim Sci 1987;64:642–7.
25. Prather RS, Sims MM, First NL. Nuclear transplantation in the pig embryo: nuclear swelling. J Exp Zool 1990;255:355–8.
26. Collas P, Robl JM. Relationship between nuclear remodeling and development in nuclear transplant rabbit embryos. Biol Reprod 1991;45:455–65.
27. Chang D, Chaffy BM, Saunders JA, Sauers AE. Handbook of electroporation and electrofusion. San Diego, CA: Academic Press, 1991.
28. Northey DL, Leibfried-Rutledge ML, Nuttleman PR, First NL. Theriogenology 1992;37(1):266 [Abstract].
29. Tsunoda Y, Shioda Y, Onodera M, NaKamyra M, Ochioa T. Differential sensitivity of mouse pronuclei and zygote cytoplasm to Hoechst staining and ultraviolet irradiation. J Reprod Fertil 1988;82:173–8.
30. Westhusin ME, Lavanduski MJ, Scarbrough R, Looney CR, Bondioli KR. Viable embryos and normal calves after nuclear transfer into Hoechst stained enucleated demi oocytes of cows. J Reprod Fertil 1992;95:475–80.
31. Ware CB, Barnes FL, Maiki-Laurila M, First NL. Age dependence of bovine oocyte activation. Gamete Res 1989;22:265–75.
32. Hagen DR, Prather RS, First NL. Electrical activation of in vitro matured pig oocytes. Mol Reprod Dev 1991;28:70–3.
33. Collas P, Robl J. Factors affecting the efficiency of nuclear transplantation in the rabbit embryo. Biol Reprod 1991;43:877–84.
34. Kubiak JZ. Mouse oocytes gradually develop the capacity for activation during the metaphase II arrest. Dev Biol 1989;136:537–45.
35. Susko-Parrish J, Nuttleman P, Leibfried-Rutledge ML. Effect of bovine oocyte aging in vitro on development [Abstract]. Biol Reprod 1991;44(suppl 1):156.
36. Nagai T. Parthenogenetic activation of cattle follicular oocytes in vitro with ethanol. Gamete Res 1987;16:243–9.
37. First NL, Leibfried-Rutledge ML, Northey DL, Nuttleman PR. Use of in vitro matured oocytes 24 hr of age in bovine nuclear transfer [Abstract]. Theriogenology 1992;37(1):211.

38. Kaufman MH. 1. Parthenogenesis: a system facilitating understanding of factors that influence early mammalian development. In: Harrison RG, Holmnes RL, eds. Progress in anatomy; vol 1. London: Cambridge University Press, 1981:1–34.
39. Chang MC, Fernandez-Cano L. Effects of delayed fertilization on the development of pronucleus and the segmentation of hamster ova. Anal Rec 1958;132:307–19.
40. Keefer C, Schuetz AW. Spontaneous activation of ovulated rat oocytes during in vitro culture. J Exp Zool 1982;224:371–7.
41. Sims MM, Rosenkrans CF Jr, First NL. Development in vitro of bovine embryos derived from nuclear transfer [Abstract]. Theriogenology 1991;35(1):272.
42. Stice SL, Keefer C. Improved rates for bovine nucleus transfer embryos using cold shock activated oocytes [Abstract]. Biol Reprod 1992;46(1):166.
43. Wolfe BA, Kraemer DC. Methods in bovine nuclear transfer. Theriogenology 1992;37(1):5–15.
44. Aktas H, Leibfried-Rutledge ML, Wheeler MB. Bovine oocytes arrested in meiosis in vitro retain developmental capacity [Abstract]. Biol Reprod 1991;44(1):75.
45. Czolowska R, Modlinski JA, Tarkowski AK. Behaviour of thymocyte nuclei in non-activated and activated mouse oocytes. J Cell Sci 1984;69:129–34.
46. Czolowska R, Waksmundzka M, Kubiak JZ, Tarkowski AK. Chromosome condensation activity in ovulated metaphase II mouse oocytes assayed by fusion with interphase blastomeres. J Cell Sci 1986;84:129–38.
47. Szollosi D, Czolowska R, Szollosi MS, Tarkowski AK. Remodeling of mouse thymocyte nuclei depends on the time of their transfer into activated, homologous oocytes. J Cell Sci 1988;91:603–13.
48. Usui N, Yanagimachi R. Behavior of hamster sperm nuclei incorporated into eggs at various stages of maturation, fertilization and early development. J Ultrastruct Mol Struct Res 1976;57:276–88.
49. Komar A. Fertilization of parthenogenetically activated mouse eggs, I. Behavior of sperm nuclei in the cytoplasm of parthenogenetically activated mouse eggs. Exp Cell Res 1982;139:361–7.
50. Szollosi D. Mammalian eggs aging in the follopian tubes. In: Aging gametes. Basel: Karger, 1975:98–121.
51. Leibfried-Rutledge ML, Northey DL, Nuttlemann PR, First NL. Processing of donated nucleus and timing of post-activation events differ between recipient oocytes 24 or 42 hr of age [Abstract]. Theriogenology 1992;37(1):244.
52. Fraser LR. Rate of fertilization in vitro and subsequent nuclear development as a function of the post-ovulatory age of the mouse egg. Reprod Fertil 1979;55:153–60.
53. Shalgi R, Kaplan R, Kraicer PF. The influence of postovulatory age on the rate of cleavage in in-vitro fertilized rat oocytes. Gamete Res 1985;11:99–106.
54. Modlinski JA. The fate of inner cell mass and trophectoderm nuclei transferred to fertilized mouse eggs. Nature 1981;273:466–7.
55. Navara CS, Sims MM, First NL. Timing of polarization in bovine embryos and developmental potential of polarized blastomeres [Abstract 82]. Biol Reprod 1992;46:71.

56. Evans MJ, Notarianni E, Lauir S, Moor RM. Derivation and preliminary characterization of pluripotent cell lines from porcine and bovine blastocysts. Theriogenology 1990;33:125–8.
57. Anderson GB. Isolation and use of embryonic stem cells from livestock species. Anim Biotech 1992;3(1):165–72.
58. First NL, Prather RS. Genomic potential in mammals. Differentiation 1991;48:1–8.
59. Notarianni E, Laurie S, Moor RM, Evans MJ. Maintenance and differentiation in culture of pluripotential embryonic cell lines from pig blastocysts. J Reprod Fertil 1990;41:51–6.
60. Notarianni E, Galli C, Laurie S, Moor RM, Evans MJ. Derivation of pluripotent, embryonic cell lines from the pig and sheep. J Reprod Fertil 1991;43(suppl):255–60.
61. Evans MJ, Kaufman MH. Establishment in culture of pluripotential cells from mouse embryos. Nature 1981;292:154–6.
62. Stewart CL. Prospects for the establishment of embryonic stem cells and genetic manipulation of domestic animals. In: Pedersen RA, McLaren A, First NL, eds. Animal applications of research and mammalian development. New York: Cold Spring Harbor Lab Press, 1991:267–83.
63. Doetschman T, Williams P, Maeda N. Establishment of hamster blastocyst-derived embryonic stem (ES) cells. Dev Biol 1988;127:224–7.
64. Collas P, Balise JJ, Robl JM. Influence of cell cycle stage of the donor nucleus on development of nuclear transplant rabbit embryos. Biol Reprod 1992;46:492–500.
65. Smith LC, Wilmut I, Hunter RHF. Influence of cell cycle stage at nuclear transplantation on the development in vitro of mouse embryos. J Reprod Fertil 1988;84:619–24.
66. Von Beroldington CH. The developmental potential of synchronized amphibian cell nuclei. Dev Biol 1981;81:115–26.
67. McAvoy JW, Dixon KE, Marshall JA. Effects of differences in mitotic activity, stage of the cell cycle and degree of specialization of donor cells on nuclear transplantation in *Xenopus laevis*. Dev Biol 1975;45:330–9.
68. Ellinger MS. The cell cycle and transplantation of blastula nuclei in *Bombiana orientalis*. Dev Biol 1978;65:81–9.
69. Gurdon JB. The transplantation of living cell nuclei. Adv Morphol 1964;4:1.
70. DiBerardino MA, Hoffner NJ. Origin of chromosomal abnormalities in nuclear transplants—a reevaluation of nuclear differentiation and nuclear equivalence in amphibians. Dev Biol 1970;23:185–209.
71. DiBerardino MA, King TJ. Development and cellular differentiation of neural nuclear-transplant embryos of known karyotype. Dev Biol 1965;15:102–28.
72. Barnes FL, Westhusin ME, Looney CR. Embryo cloning: principles and progress. 4th World Cong on Genetics Applied to Livestock Production, Edinburgh, Scotland, 1990;16:323–33.
73. Bondioli KR. Commercial cloning of cattle by nuclear transfer. Proc Symp on Cloning Mammals by Nuclear Transplantation, Fort Collins, CO, January 15, 1992:35–8.
74. Schutz MM. Cytoplasmic and mitochondrial genetic effects on economic traits in dairy cattle [Dissertation]. Iowa State University, Ames, IA, 1991.

75. Freeman AE, Beitz DC. Cytoplasmic inheritance-molecular differences and phenotypic expression. Proc Symp on Cloning Mammals by Nuclear Transplantation, Fort Collins, CO, January 15, 1992:17–20.
76. Stice SL. Multiple generation bovine embryo cloning. Proc Symp on Cloning Mammals by Nuclear Transplantation, Fort Collins, CO, January 15, 1992: 28–31.

20

Assisted Reproduction in the Propagation Management of the Endangered Lion-Tailed Macaque (*Macaca silenus*)

M.R. CRANFIELD, B.D. BAVISTER, D.E. BOATMAN, N.G. BERGER,
N. SCHAFFER, S.E. KEMPSKE, D.M. IALEGGIO, AND J. SMART

Each year, an area of rainforest habitat estimated to be greater than the size of the state of New York is lost to human intervention. At this rate of rainforest and other ecosystem loss, some authorities believe that between 1 and 5 million species of animals and plants could become extinct in the next two decades. There are hundreds of species of known plants and animals on the Appendix 1 list of the *Convention on International Trade in Endangered Species* (CITES) at this time.

To reverse the steady loss of species and to preserve biological diversity, the cooperation of many organizations, institutions, and individuals is required. Wildlife conservation is one of the primary missions of zoological parks and aquaria, of which many conduct multifaceted research projects on wild and captive populations to supplement the overall knowledge necessary to species conservation (1).

In 1981 the American Association of Zoological Parks and Aquariums established *species survival plans* (SSP) for the most critically endangered species for which there were viable captive populations. The goal of the SSPs is to maintain the greatest genetic diversity in small populations over long periods of time. These plans consider entire captive North American or, in some species, international populations.

There are dozens of endangered primate species on Appendix 1 at this time. The *lion-tailed macaque* (*Macaca silenus* [LTM]) is one of the most severely endangered. In 1982 the wild population of this native of the Western Ghats mountains in India was estimated at only 915–1500 individuals (2). Not only was the wild LTM population small, but isolated gene pools were being formed by the creation of farm land, hydroelectric dams, and railroads, obstacles that this arboreal species would not cross (3).

The Baltimore Zoo has chosen this beautiful primate as its flagship species for conservation. In the spring of 1982, the zoo sponsored a three-day international symposium at which more than 60 delegates discussed aspects from captive rearing to the political and social ramifications of preservation of the LTM's native habitat; an LTM SSP was formed.

The North American carrying capacity for LTMs—the number of individuals that zoological institutions can house collectively while maintaining broad genetic diversity and without sacrificing the ability to house and conserve other species—is 250 individuals. A great discrepancy in genetic representation of individuals within this population was noted (4). Some individuals had reproduced well in captivity and were genetically over-represented; for these animals contraceptive measures were deemed desirable. However, 59% of wild-caught specimens had not reproduced in captivity, whether for medical, behavioral, or physiological reasons, and were genetically underrepresented; artificial reproductive techniques would be required to preserve their genetic material.

The goals of the Baltimore Zoo's LTM project were (i) to select a less severely endangered model species for the LTM; (ii) to establish normal ranges for reproductive physiologic parameters in both LTMs and the model species; (iii) to produce a successful birth using artificial

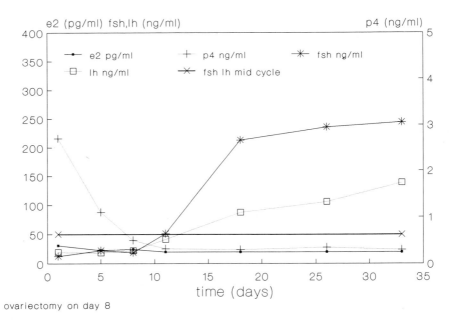

FIGURE 20.1. Serum E_2, P_4, FSH, and LH levels pre- and postovariectomy in 5 PTMs.

reproductive techniques in the model species to justify the use of these techniques in the more highly endangered LTM; (iv) to transport underrepresented, nonbreeding captive LTMs from zoological institutions throughout North America to the medical facility at the Baltimore Zoo for reproductive assessment; (v) to apply artificial reproductive techniques, such as *artificial insemination* (AI), *in vitro fertilization* (IVF), and *embryo transfer* (ET), to these nonreproducing individuals; and (vi) to explore different avenues of reversible contraception for use in overrepresented animals.

In 1984 many artificial reproductive techniques were being pioneered. The Baltimore Zoo team decided to use a model species to develop and enhance some of these techniques. The *pig-tailed macaque (Macaca nemestrina* [PTM]) was chosen because its size and behavior are similar to those of the LTM. In addition, in applying techniques previously used only in rhesus macaques (*Macaca mulatta*) and cynomolgus monkeys (*Macaca fascicularis*) (4–6) to PTMs, more physiologic data could be accumulated on the *Macaca* genus in general, and the level of confidence increased in applying these techniques to one of its most endangered species.

Normal Reproductive Values

Female

The normal menstrual cycles in LTMs and PTMs were delineated using serum *estradiol* (E_2) and *progesterone* (P_4) and assayed using a solid-phase ^{125}I rapid radioimmunoassay kit (Diagnostic Products Corp., Los Angeles). The menstrual cycles of these two species are similar, both to each other and to that of the rhesus monkey. Menstrual cycles were 28–32 days in length, with the serum E_2 peak occurring on day 12–15 (mean PTM E_2: 452 pg/mL, N = 4; mean LTM E_2: 397 pg/mL, N = 8). Serum P_4 levels reached 15 ng/mL (PTM) and 11 ng/mL (LTM) during the 15-day luteal phase.

Serum E_2, P_4, *follicle stimulating hormone* (FSH), and *luteinizing hormone* (LH) in 5 normally cycling PTM females before and after ovariectomy are presented in Figure 20.1. Using the values and the patterns of change in these hormones before and after ovariectomy as a reference point, LTM females were divided into three categories by examining as few as 3–6 serum samples during a 30-day period: *category 1*: normally cycling females, in which serum FSH and LH were consistently at baseline or peaked once within the sampling period while serum E_2 and P_4 levels fluctuated; *category 2*: amenorrheic but nonmenopausal females in which serum FSH, LH, E_2, and P_4 all remained at baseline; and *category 3*: menopausal females in which serum FSH and LH were consistently elevated around peak levels, and E_2 and

P_4 were consistently at baseline (10). LTM females in category 1 were considered to be good candidates for artificial reproductive manipulations. Further research was required to elucidate the reason(s) for amenorrhea in category 2 females. The genetic material from category 3 LTM females was lost, but these animals might be utilized as possible embryo recipients after hormonal supplementation or as behavioral models, as aunts to infant LTMs.

Male

The characteristics of semen collected from LTM and PTM males by *rectal probe electroejaculation* (RPE) are similar (Table 20.1) (11). The results of computer analyses (Hamilton-Thorne Motion Analyzer) of semen samples collected by RPE from fertile LTM males are presented in Table 20.2; analyses were performed on videotapes of samples that had been placed in a Makler chamber and examined by phase-contrast microscopy. Serum testosterone, FSH, and LH in samples from anesthetized fertile male LTMs were determined at 15-min intervals over a 2-h period (12).

As much as a 3.5-fold difference in testosterone levels within individuals over this short period of time was seen. Great variation in these values was apparent between individuals, but testosterone levels within individuals tended toward either high, low, or midrange. The overall mean serum testosterone for the LTMs was 8.16 ng/mL (± 0.86), the range of means was from 3.5 to 12.3 ng/mL, and the overall range was from 1.6 to 19.5 ng/mL. The overall mean FSH was 73.14 ng/mL (± 4.39),

TABLE 20.1. Characteristics of semen collected by RPE from proven breeder pig-tailed (*Macaca nemestrina*) and lion-tailed macaques (*M. silenus*) and from LTMs that had not reproduced successfully in captivity.

Semen characteristics	Reproductive PTM (n = 4)	Reproductive LTM (n = 4)	Reproductively unsuccessful LTM (n = 4)
Sperm concentration × 10^6/mL (mean)	4–360	38–800 (216)	6–88 (36)
Percent sperm motility (mean)	50–60	30–95 (69)	25–67 (48)
Sperm forward progression (1–5)	3–5	2–5	2–5
Sperm pH	7.0–8.5	8.0–9.0	8.0–9.0
Semen fluid volume in mL (mean)	0.7–1.5	0.01–4.0 (1.1)	0.1–1.4 (0.55)
Percent sperm morphology (total mean)		(18)	(27)
Primary abnormalities	1–55	1–25	35–80
Secondary abnormalities	1–25	2–79	3–45

Note: At least 12 samples were collected from each individual.

TABLE 20.2. Physical parameter of raw ejaculate semen samples collected from LTMs (n = 7) by RPE and analyzed using the Hamilton-Thorne Motion Analyzer.

Parameters	Range	Mean	SD
Lateral head displacement (μ)	1.51–3.34	2.58	0.61
Frequency of sperm head crossing (Hz)	11.27–14.04	12.75	1.14
Mean linearity (%)	54.00–84.88	72.82	10.57
Average path velocity (μ/sec)	35.23–113.38	69.35	29.25
Straight-line velocity (μ/sec)	26.60–101.60	62.57	28.21
Curvilinear velocity (μ/sec)	37.56–124.36	78.75	31.21
Motile concentration $\times 10^6$/mL	34.48–271.75	113.56	89.36
Percent motility (%)	27.82–84.57	54.59	21.82
Total concentration $\times 10^6$/mL	71.57–461.95	199.10	157.25
Progressive percentage (%)	4.44–54.23	29.95	18.60
Rapid percentage (%)	6.15–67.36	36.71	22.29
Medium percentage (%)	6.50–26.78	17.88	7.48
Slow percentage (%)	4.49–44.32	17.76	16.18
Static percentage (%)	2.87–60.82	27.02	22.91

Note: Raw ejaculates (N = 7).

and the range of individual means was from 43.7 to 121.6 ng/mL. The overall range was from 40.5 to 136 ng/mL. Serum LH values had an overall range of from <19.2 to 77.2 ng/mL; the overall mean could not be determined from the raw data since many values were below the sensitivity range of the test (<19.2 ng/mL) (12).

Sperm Management

The majority of available LTM males were older, intractable animals. As a result, semen collection techniques utilizing unanesthetized individuals were considered impractical. Data from trials conducted on ketamine hydrochloride-anesthetized LTMs (10 mg/kg IM) utilizing the three following methods are found in Table 20.3: (i) physostigmine and rectal massage, (ii) a flexible latex rectal probe and rectal massage during electrostimulation; and (iii) electroejaculation using a standard rectal probe with three longitudinal electrodes. The standard rectal probe clearly out-performed both alternative methods.

Retroejaculation of sperm into the bladder is one of the drawbacks of RPE. In one study utilizing the standard rectal probe, an average of 65% of sperm released was retroejaculated into the bladder during electroejaculation (13). As a result, a method for harvesting retroejaculated sperm during RPE was developed. Under ketamine hydrochloride anesthesia (10 mg/kg IM), the urinary bladder is catheterized with an 8 Fr infant feeding tube (Argyle Premature Infant Feeding Set, Sherwood Medical, St. Louis, MO 63103) and emptied of

TABLE 20.3. Comparison of physical parameters of LTM semen collected by RPE using 3 different techniques.

Parameter	Standard rectal probe	Flexible latex probe	Physostigmine
Volume 1st ejaculate (mL)	0.10–1.00 0.38	0–0.70 0.20	0.01–0.70 0.13
Volume 2nd ejaculate (mL)	0.10–1.00 0.37	0.01–0.50 0.18	0–0.80 0.13
Volume 3rd ejaculate (mL)	0–0.70 0.34	0.01–1.10 0.40	— —
Total volume (mL)	0.30–2.25 1.10	0.03–1.50 0.79	0.01–1.50 0.26
Coagulum (mg)	0–811 261	0–1040 250	0–534 90
Concentration of sperm in the raw ejaculate ($\times 10^6$/mL)	5.7–1585.0 467.6	0–80.0 18.7	0–84.8 15.6
Total no. of sperm in the raw ejaculate ($\times 10^6$)	8.6–527.5 204.0	0–72.0 18.9	0–12.6 2.5
No. of sperm in the bladder ($\times 10^6$)	1.2–652.3 166.8	0.3–140.0 27.3	0–3.6 1.3
Total no. of sperm ($\times 10^6$)	8.6–1128.9 344.9	1.6–212.0 46.2	0–16.7 3.9
Abnormals 1° (%)	4.5–30.0 14.4	7.0–67.0 38.0	2.5–95.0 48.8
Abnormals 2° (%)	8.0–53.0 25.8	5.0–42.0 21.8	2.0–57.5 29.8
No. nonsperm cells/HPF (raw ejaculate)	0.20–26.50 5.8	0.15–43.00 8.4	5.10–6.80 6.0

urine. The bladder was flushed with 20 cc of 0.9% NaCl that was then discarded. Twenty cc of TALP-HEPES medium (5) was infused into the bladder, and the catheter removed.

Electroejaculation was performed using a 1-in diameter rectal probe with three longitudinal electrodes and a Lane Pulsator III (Lane Manufacturing, Inc., Denver, CO 80220). The minimal amount of current required to obtain erection and ejaculation was administered and applied in a series of 15 peaks during each of 3 ejaculation attempts. After each attempt the pelvic urethra was massaged per rectum; then, the penal urethra was "stripped" externally to remove residual semen. Following the final ejaculation attempt and urethral massage, the urinary bladder was recatheterized, and its contents aspirated. These contents were immediately centrifuged at 500 × g for 9 min, after which the supernatant was discarded, and the pellet resuspended with TALP-HEPES. Retroejaculated spermatozoa recovered from the urinary bladder have been successfully utilized for PTM and LTM IVF, and 2 pregnancies have resulted (13). Benzathine penicillin (1 cc sc) was administered prophylactically to each male at the time of electroejaculation; no catheter-related infections have occurred.

Semen samples collected by RPE often contain large amounts of cellular debris and dead sperm, which led to swim-up techniques being utilized successfully early in this project to produce clean samples of highly motile sperm. There was a loss of motile sperm in both ejaculate (84% loss) and bladder (65.2% loss) samples due both to sperm loss in the supernatant and entrapment of sperm in the pellet (Cranfield, unpublished data). Modification of centrifugational forces and times and resuspension of the pellet prior to swim-up improved results slightly.

A study was conducted comparing swim-up techniques with a disposable filtration column (Sperm Prep Fertility Technologies, Natick, MA 01760). In our laboratory the Sperm Prep technique produced final samples with fewer sperm, more cellular debris, and, occasionally, inert beads from the column; the use of this method was discontinued. At present, we are attaining the highest percent recovery of clean motile sperm samples utilizing a double-density Percoll gradient (12). Sperm processed using the Percoll gradient have had the same fertilization rate as those processed using the conventional swim-up techniques.

Once processed, sperm was incubated in TALP medium in 5% CO_2 incubators at 37°C. One-and-one-half hours prior to insemination, it was hyperactivated by the addition of both *dibutyryl cyclic AMP* (dbcAMP) and caffeine. Hyperactivation causes progressive sperm to swim in a figure-8 motion and to agglutinate, attaching at their heads and forming masses of sperm, or *comets*. The sperm concentration for incubation leading to desirable hyperactivation after 1.5h is 20×10^6/mL (13). As in the human application, sperm samples for this project for IVF must originate from specific individuals; often these samples contain fewer than 20×10^6/mL sperm after preparation. It was noted that hyperactivation occurred earlier and with a higher degree of agglutination in these low-concentration samples. In a study conducted at the Baltimore Zoo, it was found that the time and degree of hyperactivation varied directly with the concentration of dbcAMP and caffeine and inversely with the sperm concentration. The following formula was utilized to standardize the hyperactivation event with variable sperm concentrations: ($Y/20 \times 10\,\mu L$ of 100 mM dbcAMP and caffeine) × number of mL of medium, where Y = the concentration of live sperm $\times 10^6$/mL.

Artificial Insemination

Artificial insemination has been successful in 1 of 2 attempts in our laboratory. A nonrepresented dysfunctional LTM male that had demonstrated no mounting behavior while housed with single- or multiple-cycling LTM females of known fertility was utilized as the sperm donor. Semen was collected by RPE and washed, and a swim-up was performed to obtain a clean sample of 20 million motile sperm in 0.2 mL of TALP-HEPES medium. This processed sample was placed via intracervical catheter into the uterine body of a 10-year-old, normally cycling female LTM whose serum E_2 and P_4 levels at the time of the successful insemination were 48 pg/mL and 1.5 ng/mL, respectively. After 168 days' gestation, a healthy female LTM was born. This represented the first successful artificial manipulation in LTMs. More importantly, the genetic material from this unrepresented male was reintroduced into the captive gene pool.

Oocyte Management

Early in the project, oocytes were collected either during natural cycles or during stimulated cycles using *pregnant mare serum gonadotropin* (PMSG) or an FSH/LH preparation (Pergonal, Serono Laboratories, Inc., Randolph, MA 02368). In a study of serum E_2 levels, in 7 naturally cycling PTMs, there was wide variation between animals (312–755 pg/mL), but each individual animal's E_2 peaks were consistently within 25% of each other (Cranfield, unpublished observations). Based upon this finding, these monkeys were categorized as having low, medium, or high E_2 peak levels; oocyte retrieval was scheduled for 12 h after the estimated time of the serum E_2 peak.

For PMSG stimulation the protocol was a regimen of PMSG/hCG previously described for rhesus monkeys (5); however, both PMSG (200 IU) and hCG (4000 IU) were administered IM on cycle day 14. Surgery for oocyte aspiration was performed 30 h post-hCG injection. For Pergonal stimulation, the protocol was a step-up regimen. Beginning on cycle day 3, 37.5 IU Pergonal was administered IM. Subsequently, the dosage of Pergonal was increased by 37.5 IU on any day on which the serum E_2 level had not attained at least 150% of the previous day's value. When serum E_2 levels reached 1000 pg/mL or had started to plateau after cycle day 10, 2000-IU hCG was injected IM to induce oocyte maturation (Berger, personal communication).

The results of this study revealed that although the eggs harvested from natural cycles had the highest fertilization rate, it would be necessary to use stimulated cycles to obtain the number of embryos desired (Table 20.4) (9). All PTMs were refractory to multiple stimulations with PMSG, which is consistent with findings in the rhesus monkey (7). In the present study Pergonal could be used successfully after PMSG and was used up to 4 times in the same monkey, although individuals became increasingly refractory to the drug (9).

More recently, protocols have utilized an FSH preparation (Metrodin, Serono Laboratories, Inc., Randolph, MA 02368) administered for 6 days at a dosage of 37.5 IU IM b.i.d., followed by 3 days of twice-daily treatment with 37.5-IU Pergonal; a bolus of 2000 IU of hCG is

TABLE 20.4. Comparison of oocyte yield and embryo development from naturally cycling and artificially stimulated PTM females.

		Type of cycle	
		I	II
Measure	Natural	PMSG/hCG	Pergonal/hCG
No. cases	21.00	5.00	5.00
No. oocytes recovered/cycle	1.5	27.6	20.8
No. initially mature oocytes/ cycle[a] (% of total)	0.05 (3.1)	10.00 (37.0)	2.40 (11.5)
No. in vitro matured oocytes/ cycle[b] (% of total)	1.0 (68.7)	9.4 (34.0)	7.4 (35.5)
No. matured (sum) cycle (% of total)	1.05 (71.8)	19.40 (70.1)	9.80 (47.1)
No. penetrated oocytes/cycle[c] (% of inseminated matured)	0.57 (54.5)	11.20 (57.7)	4.80 (48.9)
No. cleaved embryos/cycle[d] (% of inseminated matured)	0.38 (34.8)	6.00 (30.9)	3.40 (34.6)
Total no. quality embryos	5.00	10.00	7.00
No. quality embryos/cycle[e] (% of inseminated matured)	0.24 (22.7)	2.00 (10.3)	1.40 (14.2)

[a] PB1 within 0–8 h incubation.
[b] PB1 within 8–36 h incubation.
[c] Two or more pronuclei and 2 polar bodies or cleavage to 2-cells.
[d] Oocytes that cleaved to at least 2 nucleated cells.
[e] Cleaved to 2-cells by 24–30 h postinsemination; even-sized cells with nuclei.

administered on day 10, and surgery is performed 30 h later (20). Both the number and quality of oocytes obtained from PTMs and LTMs improved significantly with these protocols, as has the number of quality embryos produced. Serum E_2 profiles were considered good if E_2 levels rose steadily during the course of the treatment regime. In 8 procedures with good E_2 profiles, a total of 245 eggs were collected, and 51 good-quality IVF embryos were produced for an average of 6.4 embryos per cycle. Three procedures in which serum E_2 responses were poor and poor-quality eggs and no quality embryos were produced were not included in these data.

Our current stimulation regime begins on day 3 of the menstrual cycle with IM injections of 37.5 IU of Metrodin administered b.i.d. until day 5; the dosage rates of Metrodin and Pergonal are then "customized" to achieve a steady rise in serum E_2 levels until they reach approximately 2000 pg/mL. An IM injection of 1000 IU of hCG is then administered, and surgery is performed 30 h later.

Early in the project, the ovaries were assessed and oocytes collected laparoscopically, but follicles are now aspirated via laparotomy into a Monoject Company 3-cc syringe containing 0.5 cc of TALP-HEPES medium using a 22-gauge Monoject needle. The aspirates are examined under a dissecting microscope for the presence of oocytes that are then transferred to TALP medium containing 1 mg/mL FSH and 20% calf serum. Oocytes are then examined for quality and maturity of cumulus and cytoplasm using a Nikon Diaphot inverted microscope. All observations are recorded on videotape.

Upon visualization of a *first polar body* (PB1), the oocyte is transferred into medium containing 1% calf serum and inseminated with hyperactivated sperm (2×10^6/mL). Twelve hours later, the oocyte is transferred into TALP medium containing 10% calf serum. Inseminated oocytes are examined for fertilization, defined by the presence of a *second polar body* (PB2) and 2 pronuclei at 15–18 h. In our laboratory we have utilized both small volumes of medium under oil and large volumes of uncovered medium with no appreciable difference in fertilization rates. When 2-, 4-, or 6-cell embryos are observed, they are transferred into the fallopian tubes of recipient monkeys or are frozen in a 1.5 M propanediol and 0.2 M sucrose solution using a Planer R204 cell freezer (Planer Products Ltd, Windmill Road, Sunbury on Thames, Middlesex, England TW 167HD).

Recipients

Abnormally high serum E_2 and P_4 values in poststimulated macaques suggested that the uterine environment of these oocyte donors would not be a suitable recipient environment (11). Because macaques cannot be synchronized with prostaglandins, three alternative recipient options were considered: (i) a colony of monitored, normally cycling females large enough to ensure that on any given day there would be at least 1 female at the appropriate stage of the cycle to accept embryos, (ii) castrated females that would be manipulated using exogenous hormones to synchronize endometrial receptivity with fresh embryo availability, and (iii) normally cycling females that would be implanted with frozen-thawed embryos 3 days following their serum E_2 peak.

The zoo could not maintain a sufficiently large colony of recipient PTMs. Arrangements were made with the Wisconsin Regional Primate Center in Madison, Wisconsin, to utilize a large number of monitored, normally cycling rhesus macaque females housed at that location. Early in the study, 2- to 4-cell stage PTM embryos were placed in TALP medium in airtight tubes previously equilibrated with 5% CO_2 at 37°C and flown from Baltimore to Madison strapped into the axillary region of an investigator for temperature regulation. Although this procedure caused no deleterious morphologic effects and embryos divided in culture overnight, none of the subsequent intrauterine transfers resulted in pregnancies.

Six PTMs were ovariectomized, and three methods of exogenous hormone administration were tested: (i) subcutaneous silastic E_2 and P_4 implants (17), (ii) vaginal polysiloxone rings impregnated with E_2 and P_4 (18), and (iii) oral esterase and injectable progesterone. After adjustments were made in the amounts of hormone delivered by each method, serum E_2 and P_4 values within normal physiologic levels were attained. However, the project was hampered by unexplained erratic laboratory E_2 and P_4 assay results. Even when paired samples were run, marked discrepancies often occurred between values. Sexual swellings and behavior of these monkeys were as would be expected for normally cycling females. The oral esterase and injectable progesterone protocol was abandoned due to the uncertainty of ingestion of grape-disguised esterase pills, particularly when higher dosages were administered. In addition, the twice-daily progesterone injections were felt to be more stressful than the other tested methods of hormone supplementation. The vaginal ring protocol was also abandoned because polysiloxone (18) was removed from the market. Modified human vaginal rings did not fit properly and often either fell out or were removed by the monkeys and eaten. Immobilization was required to ascertain whether the vaginal rings were in place.

The method for recipient synchronization using subcutaneous silastic implants follows. One 50-mg E_2 implant was placed sc in the craniadorsal thorax of the recipient on the donor's first observed day of menses (M_1). Additional 50-mg E_2 implants were each placed sc on M_6, M_7, and M_8 of the donor's cycle. Implants placed on M_6, M_7, and M_8 were removed on M_9; the original (M_1) E_2 implant remained in place throughout synchronization and pregnancy. A single 75-mg P_4 implant was placed subcutaneously on M_9; another was placed on M_{12}, and a third on M_{15}. The third P_4 implant was removed on M_{18}, while the implants placed on M_9 and M_{12} remained throughout the pregnancy. Embryos were transferred into the synchronized recipient on M_{12}. Serum E_2 and P_4 levels released by the implants decrease with time; their effect was minimal after 90 days. Because the primate placenta maintains pregnancy after 3 weeks, further implants were not necessary. Following parturition, hormonal supplementation of ovariectomized females was necessary for initiation of lactation.

In 1987, utilizing a human protocol, the laboratory began freezing mouse embryos as a model, using TALP-HEPES medium instead of Ham's F10. The freezing rate was controlled using a Planer R204 cell freezer. Embryos were transferred into TALP-HEPES with 20% BSA and then into propanediol at concentrations of 0.5 M, 1.0 M, and 1.5 M; embryos remained in each solution for 5 min. The solution in which embryos were frozen contained 1.5 M propandiol and 0.2 M sucrose. The initial rate of freezing was $-2°C/min$ to $-7°C$. Embryos were maintained at $-7°C$ for 5 min, and then seeded with liquid nitrogen-cooled forceps.

Freezing continued at a rate of −0.3°C/min to −30°C, at which temperature the vials were plunged into liquid nitrogen for storage. Embryos were quick thawed in a 37°C water bath and then passed through TALP-HEPES and 20% BSA solutions containing decreasing concentrations of propanediol/sucrose, beginning with 1.0 M/0.2 M, and progressing through 0.5 M/0.2 M and 0.2 M sucrose, and were finally transferred into TALP-HEPES with 20% BSA alone (16).

Embryo Transfers

Two methods for ET into recipients were used. First, 2- to 6-cell embryos were aspirated into a sterile 4.5-in 3.5 Fr tomcat catheter and introduced through the fimbria into the midampullary region of the fallopian tubes during laparotomy (16). Second, later stage embryos were introduced through the cervix into the uterus.

Early in the project, intrafallopian transfer of five 2-cell natural cycle embryos into the original donors was performed. In another procedure, two 2-cell embryos collected from 2 donors were transferred into a single donor. No pregnancies resulted. Three 4- to 8-cell embryos from stimulated cycles were flown to the Wisconsin Regional Primate Research Center and placed through the cervix into the uteri of 3 monitored rhesus monkeys. A fourth 2-cell embryo was replaced into the original donor. Again, no pregnancies resulted.

TABLE 20.5. Results of transfer of fresh or frozen-thawed embryos into naturally cycling or timed recipient PTMs and LTMs at the Baltimore Zoo between 1987 and 1989.

Procedure	Type of embryo[a]	Number of embryos	Type of recipient	Placement of embryos	Results
A	F	6 (6-cell)	Natural cycle	Uterus	—
B	F	5 (2-cell) 1 (4-cell)	Timed	Fallopian tubes	Twins[d]
C	F	5 (2-cell)	Timed	Fallopian tubes	Singleton[e]
D	F	6 (2-cell)	Timed	Fallopian tubes	—
E	F/T	2 (3-cell) GM[b] 4 (2-cell) PM[c]	Timed	Fallopian tubes	—
F	F/T	4 (4-cell) GM 5 (4-cell) PM	Natural cycle	Fallopian tubes	Singleton[f]

[a] F = fresh embryo; F/T = frozen-thawed embryo.
[b] GM = good embryo morphology postthaw.
[c] PM = poor embryo morphology postthaw.
[d] Fetal heartbeats detected ultrasonically days 31 and 70 posttransfer; pregnancy aborted day 93.
[e] Live birth; death shortly afterwards.
[f] Normal pregnancy, live birth, and normal postnatal development to present time.

It is known from human research that the percent of successful ETs increases as the number of embryos introduced increases. However, if more than 3 embryos are transferred, the chance of multiple implantations is increased.

Between 1987 and 1989, transfers of multiple fresh or frozen-thawed embryos into either naturally cycling or timed recipients were performed. The outcomes of these procedures are detailed in Table 20.5. Ultrasound was utilized 30 days posttransfer to confirm pregnancy. The fetal heartbeats of the twin pregnancy (Table 20.5, procedure B) were detected by ultrasound at days 31 and 70 posttransfer. Unfortunately, the pregnancy was aborted at day 93. The first singleton pregnancy (Table 20.5, procedure C) ended in dystocia. Although the PTM neonate was alive when delivered vaginally, it died shortly thereafter. There were no morphologic lesions on postmortem examination. The surrogate mother, an ovariectomized and artificially primed female, was not lactating at the time of delivery.

The third pregnancy (Table 20.5, procedure F) resulted in the uneventful delivery of a hybrid PTM × LTM male after 162 days gestation. The genetically unrelated, intact, surrogate PTM female has exhibited excellent maternal care. During his 2 years, the hybrid has shown normal development.

Lion-Tailed Macaque

Despite its postnatal death, the birth of the first neonate PTM represented the accomplishment of our goal with the model species. Because the ovariectomized recipient did not lactate following parturition, more studies using recipients were necessary before stimulation of LTMs could begin. PTM oocytes were fertilized with LTM sperm to obtain embryos for freezing. Sperm from LTMs performed identically in the laboratory to that of PTMs. The project was hampered from the outset by variability in the quality of sperm samples collected by RPE; much of 1989 through 1991 was invested in the development of LTM sperm-handling methodology.

Following the birth of the PTM × LTM hybrid from frozen-thawed embryos, LTM stimulations for oocyte collection were begun. Of 15 stimulations, only 5 animals had adequate E_2 responses and went to surgery. Fertilization was not achieved in the first 4 procedures. One procedure failed due to incubator malfunction; in the other 3, low numbers of poor-quality eggs were believed responsible for failure of fertilization.

The lack of an appropriate response of female LTMs to administration of exogenous FSH and LH was likely due to two factors: (i) The ages of the LTM females stimulated were much higher than those of the PTMs that had been utilized, and (ii) the LTMs were highly stressed when removed from their groups and housed singly for stimulation. In addition, they were further stressed by the manipulations of IM hormone injection and immobilization for venipuncture, demonstrated by the fact that their appetites usually decreased significantly. The use of genetically desirable males with suboptimal sperm samples—necessary in order to obtain the most genetically rare offspring possible—may also have influenced the lack of fertilization. Frequently, the concentration of sperm was much less than 20×10^6/mL, necessitating the use of larger volumes of medium containing dbcAMP and caffeine in order to introduce adequate numbers of sperm to oocytes. This may have adversely affected fertilization.

In March of 1992 a young female LTM was stimulated utilizing an individualized regime of IM injections of Metrodin and Pergonal. Serum E_2 response was good with serum values of 25, 277, 1261, and 3250 pg/mL on days M_4, M_6, M_8, and M_9, respectively. At the time of surgery, serum P_4 was 2.4 mg/dL, and serum E_2 was 946 pg/mL. A total of 36 eggs were retrieved by laparotomy. On initial assessment 9 oocytes had polar bodies. These were inseminated with 2×10^6/mL hyperactivated sperm collected from an unrepresented LTM male prior to surgery. After Percoll treatment sperm motility was assessed as good; the concentration of live sperm was 20×10^6/mL. The sperm was incubated for 6 h prior to the addition of dbcAMP and caffeine. Sperm motility and concentration of live sperm decreased by approximately 50% during incubation and hyperactivation. To counteract this, the volume of medium containing sperm with dbcAMP and caffeine transferred to the egg dishes was doubled to provide the correct sperm concentration for insemination.

Sperm attached to the eggs; 1 oocyte progressed to the PB2 stage and arrested there. The remaining eggs were divided into three groups (II, III, and IV) as they developed polar bodies and were inseminated with sperm samples collected from a younger, genetically represented LTM. The sperm samples were also divided into three groups and were hyperactivated separately and utilized to inseminate 11 eggs from group II 9 h postoperatively and 6 eggs from group III 10.5 h postoperatively; the remaining 10 eggs (group IV) were inseminated 20.5 h postoperatively whether or not polar bodies were observed. Eight embryos developed: 4 from group II, 1 from group III, and 3 from group IV. These were frozen in liquid nitrogen at the 2-, 4-, or 6-cell stages and remain frozen at this time (19).

Contraception

Many captive LTMs have bred naturally, have produced several offspring, and are overrepresented genetically within this population. It has become necessary to slow their rate of reproduction to stay within the species' carrying capacity. To achieve reversible contraception in females, many zoos have utilized sc melangesterol acetate implants, IM injections of medroxyprogesterone acetate, and oral human contraceptive pills. At the Baltimore Zoo, sc silastic implants containing progesterone and IM injections of leuprolide acetate have also been successfully utilized to produce reversible anestrus in female macaques.

To date, reversible male contraceptive techniques have not been utilized in macaques (12). Reversible male contraception is desirable as males quickly become overrepresented when housed within normal multifemale troops. While removal of an overrepresented male and introduction of a less well represented LTM might be one way of controlling genetic representation of males, it is disruptive and dangerous to change troop males, which are very aggressive and often practice infanticide when first introduced to LTM females with infants. In addition, as LTM behavior and troop dynamics allow for only 1 adult male per troop at a given time, and as males comprise half the captive LTM population, many must be housed singly, often in suboptimal conditions. Reversible male contraception would be beneficial in that (i) troop behavior and dynamics would not need to be altered frequently; (ii) the need for the use of contraceptives in females, with their associated potential long-term complications, would be eliminated; and (iii) genetic management of an entire troop could be controlled through a single animal. A potential benefit of utilizing contraceptive compounds that reduce serum testosterone levels might be the ability to create artificial multimale groups.

Controlled breeding is more easily accomplished with behaviorally stable troops. Underrepresented females could be removed from the group during estrus, impregnated by underrepresented males, and reintroduced. Since the troop male would be present during gestation and parturition, the probability of infanticide would be markedly reduced.

A study involving 6 reproductively successful LTM males was conducted at the Baltimore Zoo utilizing the GnRH agonist leuprolide acetate (Lupron Depot, TAP Pharmaceuticals, Inc., Deerfield, IL 60015) at a dosage of 2 mg injected IM once monthly. After 4 months of treatment, testicular volume had decreased by an average of 32%. Serum testosterone rose temporarily at the beginning of the GnRH agonist administration and then dropped to normal values for 2 animals, fell to <30% of normal levels in 2 animals, and to castrate levels in 2 animals; no changes in behavior (aggression) or secondary sex characteristics were noted. At this time, motile sperm concentrations were decreased below pre-leuprolide acetate concentrations by 0%–94%, with an average reduction of 62.4%. Individual motile sperm concentrations decreased in 5 out of 6 treatment animals, with 4 of 6 displaying oligospermia, but none displaying azoospermia. Based on the results of similar studies using other GnRH agonists in other macaque species (21, 22), it was felt that a higher dose of leuprolide acetate might have been more effective.

A study currently being conducted uses ceramic implants containing 100 mg each of the GnRH antagonist (Acetyl-β-[2-Naphthyl]-D-Ala-D-p-Chloro-Phe-β-[3-Pyridyl]-D-Ala-Ser-Nε-[Nicotinoyl]-Lys-Nε-[Nicotinoyl]-D-Lys-Leu-Nε-[Isopropyl]-Lys-Pro-D-Ala-NA$_2$) (Antide, Sigma Chemical Co., St. Louis, MO 63178) placed sc in 5 LTM males; a sixth male, the control, was implanted with a sham ceramic implant. Data are now being collected. Data from a preliminary study using the same implants in castrated sheep (five 100-mg implants per animal) were promising.

Future Direction

It would appear that the easiest way to reintroduce rare genetic material into the population is to utilize the poor sperm samples for AI of younger normal females. Good-quality sperm samples from young LTM males should be utilized to inseminate oocytes collected following exogenous hormonal stimulation. Recipients of nonfrozen embryos will be menopausal or over-represented LTM females chemically castrated with Lupron Depot and synchronized utilizing sc silastic E$_2$ and P$_4$ implants. Lupron Depot will be discontinued upon pregnancy determination to allow normal lactation. Frozen embryos will be placed into naturally cycling multiparous LTM females with good maternal instincts.

Acknowledgments. The work contained in this manuscript was supported in part by the Maryland Zoological Society and several grants from the Institute of Museum Services (IMS), Smithsonian Institution. The contents do not necessarily represent the policy of the IMS or endorsement by the federal government.

The authors gratefully acknowledge the Andrology Department at the Greater Baltimore Medical Center for supplying the culture medium used and for performing the serum hormone assays, Catherine Rauschenberg for typing the manuscript, the staff at the Baltimore Zoo Hospital, and Steve Eisele at the Wisconsin Regional Primate Center for help with the PTM embryo/rhesus transfers in Madison. The continuing support of Diagnostic Products and past support of Sony, Nikon, and Richard Wolf Medical Instruments, Inc., are greatly appreciated. Special appreciation is due to Hamilton-Thorne Research and their staff for hours of computer time and technical assistance in completing sperm analysis.

References

1. Hutchins M. Beyond genetic and demographic management: the future of the species survival plan and related A.A.Z.P.A. conservation efforts. Zoo Biol 1991;10:285–92.

2. Ali R. An overview of the status and distribution of the lion-tailed macaque. In: The lion-tailed macaque. New York: Alan R Liss, 1985:13–26.
3. Karanth KU. Conservation prospects for lion-tailed macaques in Karnataka, India. Zoo Biol 1992;11:33–41.
4. Kempske S. *Macaca silenus*. Survey of North American and European zoo practices. In: The lion-tailed macaque. New York: Alan R Liss, 1985:221–36.
5. Bavister BD, Boatman DE, Collins K, Dierschle DJ, Eisele SG. Birth of rhesus monkey infant after in vitro fertilization and non-surgical embryo transfer. Proc Nat Acad Sci USA 1984:2218–22.
6. Bavister BD, Boatman DE, Leibfried ML, Loose M, Vernon MW. Fertilization and cleavage of rhesus monkey oocytes in vitro. Bio Reprod 1983;28:983–99.
7. Bavister BD, Dees C, Schultz RD. Refractoriness of rhesus monkeys to repeated ovarian stimulation by exogenous gonadotrophins is caused by non-precipitating antibodies. Am J Reprod Immunol Microbiol 1986;11:11–6.
8. Balmaceda JP, Pool PB, Arana JB, Heitman TS, Asch RH. Successful in vitro fertilization and embryo transfer in cynomologous monkeys. Fertil Steril 1984;42:791–5.
9. Cranfield MR, Schaffer NE, Bavister BD, et al. Assessment of oocytes retrieved from stimulated and unstimulated ovaries of pig-tailed macaques (*Macaca nemistrina*) as a model to enhance the genetic diversity of captive lion-tailed macaques (*Macaca silenus*). Zoo Biol Suppl 1989;1:33–46.
10. Cranfield MR, Berger NG, Kempske SE, Linnehan RM, Schaffer NE. Diagnosing menopause in the macaque species using serum hormone profiles. Proc Am Assoc Zoo Veterinarians Conf, Toronto, Ontario, Nov 6–10, 1988:194–5.
11. Cranfield MR, Kempske SE, Schaffer NE. The use of in vitro fertilization and embryo transfer techniques for the enhancement of genetic diversity in the captive population of the lion-tailed macaque (*Macaca silenus*). Int Zoo Yearbook 1988;27:149–59.
12. Cranfield MR, Ialeggio DM, Berger NG, England B, Kempske SE. The search for a reversible male birth control in the lion-tailed macaque as a model for other primate species. Proc Am Assoc Zoo and Wildlife Veterinarians (in press).
13. Schaffer NE, Cranfield MR, Fazdeabas AT, Jeyendran RS. Viable spermatozoa in the bladder after electroejaculation of lion-tailed macaques (*Macaca silenus*). J Reprod Fertil 1989;86:767–70.
14. McClure RD, Nunes L, Tom R. Semen manipulation: improved sperm recovery and function with a two-layer Percoll gradient. Fertil Steril 1989;51:874–7.
15. Boatman DE, Bavister BD. Stimulation of rhesus monkey sperm capacitation by cyclic nucleotide mediators. J Reprod Fertil 1984;42:143–5.
16. Cranfield MR, Berger NG, Kempske SE, Bavister BD, Boatman DE, Ialeggio DM. Successful birth of a macaque in a surrogate mother after transfer of a frozen/thawed embryo produced by in vitro fertilization. Proc Am Assoc Zoo Veterinarians, South Padre Island, TX, Oct 21–26, 1990: 305–8.
17. Hodgen GD. Surrogate embryo transfer combined with estrogen progesterone therapy in monkeys. J Am Med Assoc 1983;250:2167–71.

18. Simon JA, Rodi IA, Stumpf PG, et al. Polysiloxone vaginal rings and cylinders for physiological endometrial priming in functionally agonadal women. Fertil Steril 1986;46:619–25.
19. Cranfield MR, Berger NG, Smart J, Ialeggio DM, Done LB. Successful production of lion-tailed macaque embryos by in vitro fertilization. Proc Am Assoc Zoo and Wildlife Veterinarians (in press).
20. Van de Voort CA, Baughman WL, Stouffer RL. Comparison of different regimens of human gonadotrophins for superovulation of rhesus monkeys: ovulatory response and subsequent luteal function. J In Vitro Fertil Embryo Transfer 1989;6:85–91.
21. Mann DR, Gould KG, Smith MM, Duffey T, Collins DC. Influence of simultaneous gonadotrophin-releasing hormone agonist and testosterone treatment on spermatogenesis and potential fertilizing capacity in male monkeys. J Clin Endocrinol Metab 1987;65:1215–24.
22. Blasin L, Heber D, Steiner BS, Handelsman DJ, Swerdloff RS. J Clin Endocrinol Metab 1985;60:998–1003.

21
Genetic Abnormalities in the Human Preimplantation Embryo

ALAN H. HANDYSIDE

In the human, gross genetic abnormalities, mainly involving chromosome number, are a major cause of pregnancy failure. The incidence of chromosomal abnormalities at birth in the normal population is about 0.5% (1). However, a high proportion of fetuses examined after induced or spontaneous abortions are karyotypically abnormal, with numerical chromosome anomalies indicating that the incidence at conception is much higher (2). The development of methods associated with *in vitro fertilization* (IVF) for the treatment of infertility over the last ten years has facilitated direct study of the incidence of genetic abnormalities in gametogenesis, at fertilization, and during preimplantation development.

Genetic Abnormalities in Early Human Development

The hamster egg penetration test developed to study male infertility has been used to examine the chromosomes of human spermatozoa from fertile individuals by culturing penetrated oocytes until the sperm chromosomes condense and can be karyotyped. The incidence of abnormalities, mainly aneuploidy, in a large series of human sperm was about 8% (3). The incidence in female gametogenesis has been studied by examining the chromosomes of oocytes arrested in metaphase II that fail to fertilize in vitro. These studies demonstrate a much higher incidence of aneuploidy in oocytes, ranging between 8% (4) and 65% (5), confirming that in the majority of aneuploid embryos, the error arises during the first meiotic division of the oocyte (6). Recently, Angell (7) has reported that hyperhaploid oocytes contain extra copies of single chromatids and not chromosomes, indicating that trisomies may arise by predivision and abnormal segregation of chromatids rather than by nondisjunction of chromosomes.

In addition to genetic defects inherited from the gametes, abnormal fertilization also contributes to genetic abnormalities in preimplantation embryos. The main abnormality arising at fertilization in vitro is fertilization by more than one sperm that may be partly caused by the abnormally high sperm concentrations used to inseminate oocytes. Conversely, some oocytes fail to fertilize and are parthenogenetically activated. Both of these abnormalities are easily detected by removing the cumulus cells that surround the oocytes when they are collected and examining the oocytes for pronuclei 12-18 h postinsemination. Fertilization rates for IVF average about 60%, with a 5%-10% incidence of polypronucleate embryos resulting mainly from polyspermic fertilization and a 1% incidence of embryos with one or no pronuclei with a *second polar body* (PB2), indicating parthenogenetic activation.

Triploid fetuses can develop to advanced stages before aborting, and the majority have been shown to result from dispermic fertilization (8). A proportion of 3-pronucleate embryos develop to the blastocyst stage in vitro (9). However, in some cases vacuoles can be mistaken for pronuclei so that the embryo may have been normally fertilized (10). In other cases, during syngamy, a tripolar spindle is formed, and the embryo divides into 3 diploid cells (11, 12). Similarly, haploid parthenogenetic embryos undergo cleavage, and a minority reach the blastocyst stage, but again, errors in detecting the number of pronuclei may mean that some of these embryos are diploid.

At cleavage stages, reports of the incidence of chromosomal abnormalities in normally fertilized embryos have varied between 30% (13) and 40% (14), possibly because many of the studies were on poor-quality embryo surplus after transfer. It is known that there is a higher incidence of abnormalities in fragmenting embryos (15). The most frequently reported defects are polyploidy and aneuploidy, but structural abnormalities have also been found in some cases. Mosaicism, mainly of ploidy but also of aneuploidy, has also been reported at cleavage stages (16). How these mosaic embryos arise is not known. However, it has been suggested that the PB2 chromosomes or extra sperm nuclei are segregated between blastomeres during the early cleavage divisions. With aneuploid embryos, both trisomic and monosomic embryos have been observed (13). Some aneuploid embryos are likely to be eliminated at early stages of development since monosomies rarely develop to advanced stages with the exception of XO.

Using fluorescent methods to detect in situ hybridization (FISH), we have examined a series of embryos developing in culture between days 2 and 7 postinsemination with a probe specific for the centromeric region of chromosome 18 (17). The efficiency of hybridization with interphase nuclei was demonstrated with normal female lymphocytes in which close to 90% of the nuclei of these cells had 2 hybridization signals. With normally fertilized preimplantation embryos, the numbers of hybridization signals were much more variable. At all stages, a significant number of interphase nuclei were apparently tetraploid with 4 hybridization signals. Tetraploid cells have been identified in embryos using DNA quantitation and ISH with a radiolabeled probe specific for the Y-chromosome (18). How these blastomeres arise is not known. One possibility is that, as with cells in tissue culture, they are an artifact of culture itself. Alternatively, they could represent precursors of polyploid trophectoderm cells. Fifty-four out of 59 embryos examined were considered to be diploid on the basis that the majority of cells had 2 hybridization signals. Among the 5 remaining embryos, one 5-cell embryo was trisomic for chromsome 18, one 4-cell embryo was monosomic, and 3 other embryos were aneuploid mosaics and/or had multinucleated blastomeres.

Preimplantation Diagnosis of Inherited Disease

Current methods for the prenatal diagnosis of inherited disease involve sampling cells of fetal origin—for example, by amniocentesis in the second trimester or *chorion villus sampling* (CVS) in the first trimester of pregnancy—and using cytogenetic, biochemical, or DNA methods to detect the genetic defect. If the pregnancy is affected, however, couples face the difficult decision of whether or not to terminate the pregnancy, and some have repeated terminations before establishing a normal pregnancy. IVF and diagnosis at preimplantation stages of embryonic development in vitro, or preimplantation diagnosis, would allow only unaffected embryos to be returned to the uterus. Any pregnancy should, therefore, be unaffected by the disease, and the possibility of a termination following diagnosis at later stages of pregnancy would be avoided (19).

Cleavage-Stage Biopsy

For detection of genetic defects, one or more cells must be removed or biopsied from each embryo. Biopsy of some of the outer trophectoderm cells at the blastocyst stage has a number of advantages. Primarily, a maximum number of cells can be recovered at this advanced preimplantation stage, maximizing the chances of an accurate diagnosis. However, only about half of normally fertilized embryos reach the blastocyst stage in vitro (20), and only about half of these could be successfully biopsied (21). This would significantly reduce the likelihood of identifying unaffected embryos for transfer. Also, pregnancy rates after blastocyst transfer have been inconsistent and, at best, no more successful than earlier transfers at cleavage stages (22, 23).

The alternative is to biopsy embryos at earlier cleavage stages when pregnancy rates after transfer are more consistent, even though this restricts the number of cells that can be biopsied. During cleavage each division subdivides the zygote into successively smaller cells. To minimize

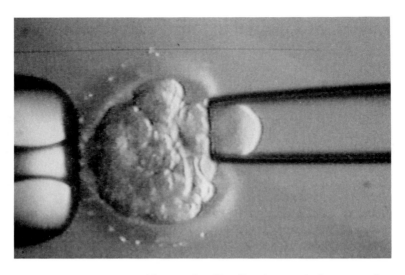

FIGURE 21.1. Cleavage-stage biopsy of a 12-cell polyspermic human embryo on day 3 postinsemination. A large micropipette is pushed through a hole made in the zona pellucida surrounding the embryo to aspirate a single nucleated cell.

the reduction in cellular mass involved in removing a single cell, therefore, embryos have been biopsied as late as possible; that is, at the 8-cell stage on the morning of the third day postinsemination (day 3), which then leaves only 8–12 h for genetic analysis before transferring selected embryos later on the same day (Fig. 21.1). Biopsy of 1 or 2 cells at the 8-cell stage does not adversely affect preimplantation development (24), and several girls have now been born following identification of sex in couples at risk for X-linked disease (25).

Detection of Genetic Defects

The time limitation following cleavage-stage biopsy and access to one or only a few cells biopsied from each embryo severely restrict the methods that can be used for the detection of genetic defects. However, several recent developments have allowed analysis of both chromosomal abnormalities and single-gene defects. For example, FISH with chromosome-specific DNA probes (26) is rapid and efficient compared with other in situ methods and can be used to detect abnormal numbers of chromosomes. For single-gene defects, the *polymerase chain reaction* (PCR) enables amplification of short fragments of DNA over a millionfold (27) within a few hours, making it possible to detect even single base changes in the DNA of single cells.

For the detection of chromosomal abnormalities, cytogenetic analysis of banded metaphase chromosomes would be ideal. However, problems with spreading of chromosomes and a tendency for the chromosomes to be too short for banding have so far prevented reliable karyotyping of human embryo nuclei by standard procedures. FISH with chromosome-specific DNA probes has several advantages (26). First, fluorescent detection of hybridization is both sensitive and efficient and allows detection in both metaphase and interphase nuclei. It also requires less time than conventional autoradiographic methods for detection of radiolabeled probes. Second, by conjugating probes to different haptens, different-colored detection systems can be used for the simultaneous detection of each probe. Finally, using a mixture of short DNA probes to a particular chromosome or chromosomal region, *chromosomal painting* is possible, which should allow the detection of various translocations and other structural abnormalities.

The potential of FISH for the simultaneous detection of probes specific for different chromosomes has led to the suggestion that the embryos of older women undergoing IVF could be screened for the common trisomies (28). The abortion rate in IVF pregnancies for women over 40 years of age is 2.5× more frequent than for women around the age of 30, probably due to a combination of chromosomal abnormalities in embryos and lower uterine receptivity (29). The most frequent autosomal trisomies found during prenatal screening are 21, 18, and 13, and their incidence rises significantly in women over the age of 35. Each of these trisomies are compatible with development to term, but may result in late abortions or perinatal death. If these could be screened in combination with trisomy 16, the most frequent autosomal trisomy in abortuses, not only would this prevent the birth of trisomic individuals, but also reduce the risk of a miscarriage. Together, these trisomies account for about 50% of trisomic abortions (30). However, the technical problems, particularly of overlapping chromosomes, are likely to be exacerbated as the number of probes is increased, and it seems unlikely that this would be feasible with the number of cells available from a cleavage-stage embryo.

Nearly 5000 conditions caused by single-gene defects have been described (31). Although many of these are rare, in couples known to be at risk, the chance of having an affected child is often as high as 1 in 4 or 1 in 2 depending on whether the condition is dominant or recessive. X-linked recessive diseases are transmitted by women carrying a defect in a gene on the female X-chromosome. One of the most common of these is *Duchenne's muscular dystrophy* (DMD) that affects 1 in 3500 male births. With these diseases boys are affected if they inherit the defect on the X-chromosome from their mother because the genes involved are not duplicated on the male Y-chromosome, whereas girls inheriting the defect are unaffected because they inherit the normal gene on the X-chromosome from their father. Therefore, the probability that a male embryo is affected is 50%, and the overall probability of having an affected child is 1 in 4.

An increasing number of X-linked diseases, notably DMD (32), have been mapped and extensively characterized at the molecular level, and prenatal diagnosis of normal or carrier females, as well as normal or affected males, is now possible by DNA analysis. Others are not so well characterized, and all that can be offered is to diagnose the sex of the fetus and to give the option of terminating males with a high probability of being affected. Similarly, for preimplantation diagnosis a number of approaches are being developed for specific diagnosis of a few X-linked diseases, but initial efforts have concentrated on identifying the sex of embryos so that normal or carrier females can be selected for transfer in any of these recessive disorders (33).

Using PCR for DNA amplification of Y-specific repeat sequences, the sex of cleavage-stage embryos can be accurately identified from single biopsied cells (34). The identification and transfer of normal or carrier female embryos was attempted in 8 couples known to be at risk for transmitting various X-linked diseases (25). These included X-linked mental retardation, Lesch-Nyhan syndrome, adrenoleukodystrophy, retinitis pigmentosa, and hereditary sensory motor neuron disease type II. Also, 2 couples at risk of DMD opted for this approach, even though a specific DNA diagnosis by conventional methods may have been possible after CVS.

After routine assessment of each couple for IVF, the women were induced to superovulate using an established protocol involving an initial period of suppression of ovarian function with an LHRH agonist followed by stimulation of folliculogenesis and administration of hCG to trigger ovulation (35). The numbers of oocytes recovered and normally fertilized in these predominantly fertile couples were similar to those obtained with infertile couples. Normally fertilized embryos developing to the 6- to 10-cell stage by the morning of day 3 were biopsied, and a single cell was removed (in one cycle 2 cells were removed). The single cells were then lysed, and the Y-specific fragment (if present) was amplified by PCR while the biopsied embryos were returned to culture. After gel electrophoresis of the amplification products and identification of the sex of the embryos on the basis of the presence of the Y-specific fragment, up to 2 of the best female embryos were selected for transfer in the evening of the same day.

Five out of the 8 women became pregnant after a total of 13 treatment cycles, 3 after one, and one each after 2 and 3 treatment cycles. The first 2 were both twin pregnancies and the other 3, singletons. The sex of each of the 7 fetuses was examined by CVS and karyotyping at about 10–11 weeks. All were female except for 1 singleton pregnancy in which the karyotype indicated that the fetus was male. Since this couple is at risk for transmitting type II hereditary sensory motor neuron disease and a specific diagnosis to determine whether a male fetus is affected is not possible in their case, the couple took the decision to terminate the pregnancy. Both of the twin pregnancies and the remaining 2 singleton pregnancies have now gone to term. Apart from the second of the second set of twins, which was stillborn, all are apparently normal, healthy girls. Detailed postmortem examination failed to reveal any gross abnormality in the stillborn twin, and the cause was probably intrapartum anoxia prior to cesarean delivery.

The misidentification of the sex in one case is now known almost certainly to have been because of amplification failure from a single cell. Amplification from each cell of disaggregated male embryos has subsequently demonstrated that amplification failure occurs with a frequency of about 15%, so that at least 2 cells amplified independently would be necessary for an acceptable level of accuracy. Although biopsy of 2 cells from 8-cell stage embryos does not appear to harm their development, the use of dual FISH with X- and Y-specific probes is currently being tested as an alternative approach.

FISH to interphase nuclei of human preimplantation embryos has been demonstrated for X- and Y-specific probes (28). With each of these probes, the majority of interphase nuclei had the appropriate number of hybridization signals, indicating that the efficiency of hybridization was high. However, hybridization failure and, conversely, artifactual signals in some nuclei, together with a relatively high incidence of tetraploid nuclei, indicate that several nuclei will need to be analyzed for a reliable result with a single probe. For example, in attempting to identify the sex of an embryo by FISH with a Y-specific probe to a single nucleus, hybridization failure would lead to the misidentification of a male as a female, and, similarly, with an X-specific probe hybridization to a tetraploid nucleus would also lead to the misidentification of a male as female. Recently, this problem has been overcome by exploiting the potential of FISH for the simultaneous detection of both X- and Y-specific probes (36). The use of two probes and the identification of either one X- and one Y-signal in males or 2 X-signals in females allows the sex of embryos to be accurately identified from a single nucleus in the majority of nuclei analyzed.

Specific Diagnosis of Single-Gene Defects

Initially, attempts at specific diagnosis are being directed at prevalent single-gene defects, especially those that are predominantly caused by one or a limited number of mutations in the genes involved, using PCR to amplify a fragment of DNA containing the defective sequence. Cystic fibrosis is a common autosomal recessive disease carried by 1 in 20 of the Caucasian population. It is caused in a majority of cases by a 3-bp deletion at position 508 of the polypeptide (ΔF508) of the *cystic fibrosis transmembrane regulator* (CFTR) gene (37). A fragment of the CFTR gene including the ΔF508 region has been successfully amplified from single cells biopsied from cleavage-stage embryos using a second round of amplification with *nested primers*; that is, primers annealing within the sequence of the first amplified fragment (38). The deletion is then detected by mixing the amplified DNA with DNA previously amplified from cells known to be either homozygous normal or homozygous deleted. The mixtures are denatured and cooled to allow *heteroduplex formation*—that is, the formation of double-stranded DNA from a normal single strand and a deleted single strand (if these are present in the mixture)—prior to gel electrophoresis. Since migration of the heteroduplex is significantly retarded, the genotype of the cell can then be deduced from the presence or absence of the heteroduplex bands in the various mixtures (Fig. 21.2).

Preimplantation diagnosis was attempted in 3 couples who all had previously given birth to a child with cystic fibrosis. In each case, both parents carried the predominant ΔF508 deletion. IVF was used to recover several oocytes, and these were fertilized with the husband's sperm. Normally fertilized cleavage-stage embryos were biopsied on the third day postinsemination; 1 cell was removed from each, and the region containing the deletion was amplified using this nested amplification protocol. Two couples in whom unaffected and carrier as well as affected

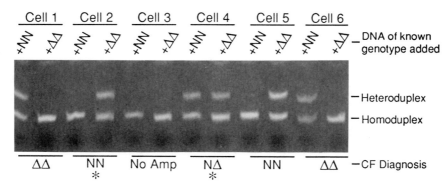

FIGURE 21.2. Detection of the predominant ΔF508 deletion in DNA amplified from single blastomeres by heteroduplex formation and gel electrophoresis for preimplantation diagnosis of cystic fibrosis (CF). By mixing the amplified DNA of unknown genotype with previously amplified DNA from cells of known genotype, as indicated at the top of the gel, it is possible to deduce the genotype (shown at the bottom of the gel) because heteroduplexes formed between normal-length fragments and those with the 3-bp deletion are significantly retarded. Two unaffected embryos marked with an asterisk were selected for embryo transfer. The embryo corresponding to cell 2 implanted, and at birth the girl was confirmed to be free of both parents' CF-causing deletions. Reprinted from Handyside, Lesko, Tarin, Winston, and Hughes (39) with permission of the New England Journal of Medicine.

embryos had been identified chose to have 1 unaffected and 1 carrier embryo transferred, as these appeared morphologically to be the best embryos. After uterine transfer on the same day as the biopsy, one of these patients became pregnant and has subsequently delivered a healthy normal girl, free of both alleles with the deletion (39).

Similarly efficient PCR protocols for 2 individual-specific mutations of the hypoxanthine phosphoribosyl transferase gene causing Lesch-Nyhan syndrome have also been developed, and preimplantation diagnosis has been attempted; so far, without pregnancy success (Lesko, Snabes, Handyside, Hughes, unpublished observations). The prospects for the detection of single-gene defects in general, therefore, are optimistic for those in which either the defect itself has been sequenced, allowing the design of the relevant primers for PCR, or there is a closely linked marker that has been similarly characterized.

References

1. Nielsen J. Chromosome examination of new-born children. Purpose and ethical aspects. Humangenetik 1975;26:215–22.
2. Burgoyne PS, Holland K, Stevens R. Incidence of numerical chromosome anomalies in human pregnancy: estimation from induced and spontaneous abortion data. Hum Reprod 1991;6:555–65.
3. Martin RH, Balkan W, Burns K, Rademaker AW, Lin CC, Rudd NL. The chromosome constitution of 1000 human spermatozoa. Hum Genet 1983;63:305–9.
4. Van Blerkom J, Henry G. Cytogenetic analysis of living human oocytes: cellular basis and developmental consequences of perturbations in chromosomal organization and complement. Hum Reprod 1988;3:777–90.
5. Wramsby H, Fredga K. Chromosome analysis of human oocytes failing to cleave after insemination in vitro. Hum Reprod 1987;2:137–42.
6. Hassold T, Jacobs P, Kline J, Stein Z, Warburton D. Effect of maternal age on autosomal trisomies. Ann Hum Genet 1980;44:29–36.
7. Angell RR. Predivision in human oocytes at meiosis I: a mechanism for trisomy formation in man. Hum Genet 1991;86:383–7.
8. Jacobs PA, Angell RR, Buchanan IM, Hassold TJ, Matsuyama A, Manuel B. The origin of triploids. Ann Hum Genet 1978;44:49–57.
9. Hardy K, Handyside AH, Winston RML. The human blastocyst: cell number, death and allocation during late preimplantation development in vitro. Development 1989;107:597–604.
10. Van Blerkom J, Bell H, Henry GH. The occurrence, recognition and developmental fate of pseudomultipronuclear eggs after in vitro fertilization of human oocytes. Hum Reprod 1987;2:217–25.
11. Kola I, Trounson A, Dawson G, Rogers P. Tripronuclear human oocytes: altered cleavage patterns and subsequent karyotype analysis of embryos. Biol Reprod 1987;37:395–401.
12. Sathananthan AH, Kola I, Osbourne J, et al. Centrioles in the beginning of human development. Proc Natl Acad Sci USA 1991;88:4806–10.

13. Angell RR. Chromosome abnormalities in human preimplantation embryos. In: Yoshinaga K, Mori T, eds. Development of preimplantation embryos and their environment. Prog Clin Biol Res 1989;294:181–7.
14. Papadopoulos G, Templeton AA, Fisk N, Randall J. The frequency of chromosome anomalies in human preimplantation embryos after in vitro fertilization. Hum Reprod 1989;4:91–8.
15. Plachot M, de Grouchy J, Junca A-M, et al. From oocyte to embryo: a model, deduced from in vitro fertilization for natural selection against chromosome abnormalities. Ann Genet 1987;30:22–32.
16. Bongso A, Fong C-Y, Ng S-C, Ratnam S, Lim J. Preimplantation genetics: chromosomes of fragmented human embryos. Fertil Steril 1991;56:66–70.
17. Schrurs B, Winston RML, Handyside AH. Preimplantation diagnosis of aneuploidy by fluorescent in situ hybridization: evaluation using a chromosome 18 specific probe. Hum Reprod 1992.
18. Angell RR, Sumner AT, West JD, Thatcher SS, Glasier AF, Baird DT. Post-fertilization polyploidy in human preimplantation embryos fertilized in vitro. Hum Reprod 1987;2:721–7.
19. Handyside AH. Preimplantation diagnosis. Curr Obstet Gynaecol 1992;2:85–90.
20. Hardy K. Development of the human blastocysts in vitro. In: Bavister B, ed. Preimplantation embryo development. New York: Springer-Verlag, 1992.
21. Dokras A, Sargent IL, Ross C, et al. Trophectoderm biopsy in human blastocysts. Hum Reprod 1990;5:821–5.
22. Dawson KJ, Rutherford AJ, Winston NJ, et al. Hum blastocyst transfer, is it a feasible proposition? Hum Reprod 1988;suppl 145:44–5.
23. Bolton VN, Wren ME, Parsons JH. Pregnancies following in vitro fertilization and transfer of human blastocysts. Fertil Steril 1991;55:830–2.
24. Hardy K, Martin KL, Leese HJ, et al. Human preimplantation development in vitro is not adversely affected by biopsy at the 8-cell stage. Hum Reprod 1990;5:708–14.
25. Handyside AH, Kontogianni EH, Hardy K, Winston RML. Pregnancies from biopsied human preimplantation embryos sexed by Y-specific DNA amplification. Nature 1990;344:768–70.
26. Trask BJ. Fluorescence in situ hybridisation: applications in cytogenetics and gene mapping. Trends Genet 1991;7:149–54.
27. White TJ, Arnheim N, Erlich HA. The polymerase chain reaction. Trends Genet 1989;5:185–9.
28. Griffin DK, Handyside AH, Penketh RJA, et al. Fluorescent in situ hybridisation to interphase nuclei of human preimplantation embryos with X and Y chromosome specific probes. Hum Reprod 1991;6:101–5.
29. Feldberg D, Farhi J, Dicker D, Ashkenazi J, Shelef M, Goldman JA. The impact of embryo quality on pregnancy outcome in older women undergoing in vitro fertilization-embryo transfer (IVF-ET). J In Vitro Fertil Embryo Transfer 1990;7:257–61.
30. Boué A, Boué J, Gropp A. Cytogenetics of pregnancy wastage. Adv Hum Genet 1985;14:1–58.
31. McKusick VA. Mendelian inheritance in man. 9th ed. Baltimore, MD: Johns Hopkins University Press, 1991.

32. Koenig M, Hoffman EP, Bertelson CJ, et al. Complete cloning of the Duchenne muscular dystrophy (DMD) cDNA and preliminary genomic organization of the DMD gene in normal and affected individuals. Cell 1987;50:509–17.
33. Handyside AH, Delhanty JDA. Cleavage stage biopsy and diagnosis of X-linked disease. In: Edwards RG, ed. Preimplantation diagnosis of human genetic disease. Cambridge, UK: Cambridge University Press, 1992.
34. Handyside AH, Pattinson JK, Penketh RJA, et al. Biopsy of human preimplantation embryos and sexing by DNA amplification. Lancet 1989;1:347–9.
35. Rutherford AJ, Subak-Sharpe RJ, Dawson KJ, et al. Improvement of in vitro fertilisation after treatment with Buserelin, an agonist of luteinising hormone releasing hormone. Br Med J 1988;296:1765–8.
36. Griffin DK, Wilton L, Handyside AH, et al. Dual fluorescent in situ hybridization for simultaneous detection of X and Y chromosome-specific probes for the sexing of human preimplantation embryonic nuclei. Hum Genet 1992;89:18–22.
37. Riordan J, Rommen JM, Kerem B-S, et al. Identification of the cystic fibrosis gene: cloning and characterisation of complementary DNA. Science 1989;245:1066–73.
38. Lesko J, Snabes M, Handyside A, Hughes M. Amplification of the cystic fibrosis ΔF508 mutation from single cells: applications toward genetic diagnosis of the preimplantation embryo. Am J Hum Gen 1991;49(4)suppl:223.
39. Handyside AH, Lesko J, Tarin J, Winston RML, Hughes M. Birth of a normal girl following preimplantation diagnosis for cystic fibrosis. New Engl J Med 1992.

22
Intra-Acrosomal Contraceptive Vaccine Immunogen SP-10 in Human, Macaque, and Baboon

John C. Herr, Richard M. Wright, Charles J. Flickinger,
Alex Freemerman, Kenneth Klotz, James Foster, and
John Shannon

SP-10 is an acrosomal protein that is first detected within the developing acrosome of round spermatids in the human testis and persists within the acrosome of mature spermatozoa (1, 2). SP-10 has been designated a *primary contraceptive vaccine candidate* by a WHO task force on contraceptive vaccines on the basis of several characteristics (3). First, current tissue specificity data suggest that SP-10 is specific to maturing germ cells within the testis (1, 4, 5). Such tissue specificity reduces the likelihood of autoimmune disease arising in females who are administered an SP-10 vaccine. Second, SP-10 has been detected in the sperm of all human males tested to date (N > 200) and thus appears to be conserved among men (2). Third, SP-10 remains associated with the sperm head after the acrosome reaction (2). Finally, a monoclonal antibody to SP-10 (MHS-10) was shown to inhibit human sperm penetration in the hamster egg penetration assay (3). Additional preliminary data have shown human IVF to be inhibited by a monoclonal antibody that reacts with a molecule considered to be SP-10 (6).

This chapter summarizes our studies of the primary amino acid structure of SP-10 in baboons and macaques. These species are possible models for testing the contraceptive potential of a recombinant human SP-10 vaccine. In addition, we present results from microsequencing native human SP-10 polypeptides and discuss possible mechanisms leading to the heterogeneity of SP-10 observed in ejaculated sperm.

22. Intra-Acrosomal Contraceptive Vaccine Immunogen SP-10 361

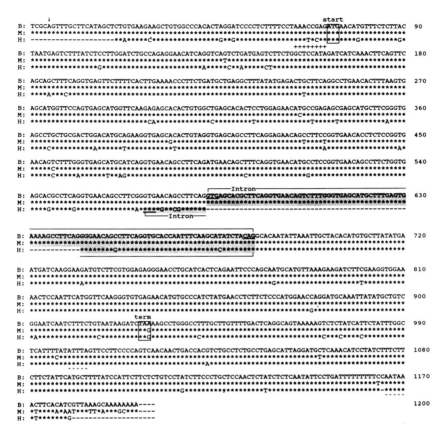

FIGURE 22.1. Complete nucleotide sequences derived from overlapping SP-10 cDNAs of baboon, macaque, and human. The cDNA sequences of macaque and human SP-10 are compared to the cDNA sequence of baboon SP-10. The numbering to the right indicates the nucleotide positions of the baboon and macaque cDNAs. Matching nucleotides are denoted by an asterisk (*). Areas lacking comparable sequence are denoted by dashes (-----). The translational start and termination codons are boxed and overscored with the words *start* and *term*, respectively. Nucleotides contained within the alternatively spliced introns are shaded. The first and last 3 nucleotides, GTG and CAG, respectively, within the shaded region are underlined, denoting the primate consensus splice nucleotides. The 5' consensus sequence flanking the ATG start codon is underscored by the symbol (+++++); the mRNA degradation consensus sequence is underscored by the symbol (^^^^^); the polyadenylation consensus sequence is underscored by the symbol (~~~~~). The major transcriptional start site is overscored by an arrow (↓) at nucleotide 4. Reprinted with permission from Freemerman, Wright, Flickinger, and Herr (28), © by Wiley-Liss, 1992.

Cloning and Sequence Analysis of Baboon and Macaque SP-10 cDNAs

Human SP-10 has been previously cloned and sequenced, and its amino acid sequence deduced from cDNAs (4). Two alternatively spliced forms of human SP-10 were isolated that encode proteins of 246 and 265 amino acids. The 2 encoded proteins are identical except for a 19-amino acid deletion in the central portion of one (4). To clone and sequence baboon and macaque SP-10, testis cDNA libraries from these species have been developed and screened with a 634-bp human SP-10 cDNA probe (4). Positive clones were isolated, and both strands of the three largest cDNAs from each library were sequenced. Sequencing revealed 2 distinct full-length SP-10 cDNAs in each species that were 1.1 kb and 1.2 kb in length, respectively. The 1.1-kb SP-10 cDNAs from baboon and macaque each contained an open reading frame of 753 bp, while the 1.2-kb SP-10 cDNAs from baboon and macaque each contained an open reading frame of 855 bp. Within the same species, the nucleotide sequence of the 1.1-kb and 1.2-kb cDNAs were identical, with the exception of a 102-bp deletion in the 1.1-kb cDNA. The cDNAs sequenced from both species also contained up to 70 nucleotides of the 5' untranslated region and 267 nucleotides of the 3' untranslated region.

The 1.2-kb SP-10 cDNA sequences from baboon and macaque were analyzed and compared to each other and then compared to the human SP-10 sequence utilizing the EuGene sequence analysis program (Baylor College of Medicine, Houston, TX) (Fig. 22.1). The 1.2-kb SP-10 cDNAs from baboon and macaque shared an overall 98% homology. No nucleotide substitutions were observed within the 5' untranslated region, 7 were found within the open reading frame, and 13 were noted within the 3' untranslated region.

Within the 5' untranslated region in the baboon and macaque, a stretch of 70 nucleotides demonstrated a 100% homology. This region contained the sequence AAACCGAG located adjacent to the translational start site. This was similar, but not identical, to the conserved motif AAATCAAA that has been found next to many eukaryotic start codons (7).

The open reading frames in the 1.2-kb SP-10 cDNAs of baboon and macaque had a 99% homology over 855 nucleotides. The initiation codons (ATG) began at nucleotide 71 in both species, and the termination codons (TAG: macaque; TAA: baboon) occurred at nucleotide 927. Within the open reading frame, both species exhibited alternative splicing that resulted in the formation of 2 distinct cDNAs: 1.1 kb and 1.2 kb. The 1.1-kb baboon and macaque SP-10 cDNAs each had identical internal deletions that were 102 nucleotides in length. The flanking sequence of these deletions encoded the 5' GTG-CAG 3' consensus splice sequence characteristic of an intron-exon border.

Following the open reading frame, the 3' untranslated region showed a 95% homology between baboons and macaques over 267 nucleotides extending from the stop codon through the beginning of the (poly)A tail. The relatively lower homology in this region resulted from sequence divergence between baboons and macaques over a span of 25 nucleotides (1172–1197) following the polyadenylation consensus sequence. In both primates a putative eukaryotic mRNA degradation sequence ATTTA was located at nucleotides 970–974, and the polyadenylation sequence AATAAA (7) was located at nucleotides 1166–1171 in the 1.2-kb cDNAs.

Sequence Comparison of Nonhuman Primate and Human SP-10 cDNAs

Like the baboon and macaque SP-10 cDNAs, human SP-10 has shown 2 alternatively spliced mRNAs (4). Sequence comparison between the largest alternatively spliced human and baboon SP-10 cDNAs revealed an overall sequence homology of 89% (Fig. 22.1). The human SP-10 cDNA contained 51 nucleotides in the 5' untranslated region and within this region showed an 88% homology with the baboon. The open reading frame of the human and baboon cDNAs shared a 79% homology. Within the open reading frame, the human SP-10 cDNA contained 60 fewer nucleotides than the 1.2-kb baboon cDNA, resulting in a lowered regional homology. However, the sequence following the deletion and extending through the end of the open reading frame exhibited a local homology of 98% between the human and baboon over 245 nucleotides. The 3' untranslated region in the human cDNA contained 250 nucleotides and shared a 96% homology to the baboon cDNA. This region of the human SP-10 cDNA contained the putative mRNA degradation sequence ATTTA and the polyadenylation sequence AATAAA at the same positions as contained in the baboon sequence. Also, the human SP-10 cDNA contained a single base deletion at nucleotide 1165 immediately 5' to the polyadenylation sequence.

Primer Extension Analysis of Baboon and Macaque mRNAs

Primer extension analysis indicated a single, major transcriptional start site and 3 possible minor start sites in the baboon and macaque (Fig. 22.2). The major start site was located at nucleotide 4 of the baboon and macaque cDNAs, 67 nucleotides 5' to the ATG codon. Minor start sites existed 69, 95, and 100 nucleotides 5' to the ATG codon. Both the major and the minor transcriptional start sites mapped to precisely the same nucleotides in baboons, macaques, and humans (12).

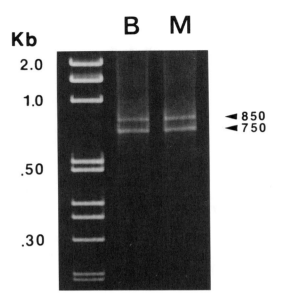

FIGURE 22.2. Primer extension analysis of baboon and macaque SP-10 mRNAs. Both baboon (B) and macaque (M) have major extension products at nucleotide 4 of the cDNA sequences and minor products 2, 28, and 33 nucleotides upstream from nucleotide 4. A synthetic oligonucleotide complementary to residues 75–94 in both species [5' d(GGGGATCCATTAGTAAGAGAAACATGTTCAT)] was used to generate the extension products and the sequence ladder. The 1.2-kb baboon SP-10 cDNA in pBluescript was used as the template for the sequence ladder. Reprinted with permission from Freemerman, Wright, Flickinger, and Herr (28), © by Wiley-Liss, 1992.

PCR Analysis of SP-10 mRNA Alternative Splicing

Alternative splicing of human SP-10 mRNA was previously demonstrated to occur within the coding region of the protein (4). PCR experiments on baboon and macaque reverse-transcribed testis (poly)A RNA confirmed that SP-10 mRNAs are alternatively spliced and that this splicing occurred within the coding region. Primate SP-10-specific oligonucleotide primers were used in the PCR such that amplified PCR products contained the entire open reading frame. The amplification products for both primates were separated on an acrylamide gel and stained with ethidium bromide (Fig. 22.3). The result was 2 bands per lane that migrated at approximately 850 and 750 nucleotides that corresponded precisely to the size of the open reading frame of the 1.2-kb and 1.1-kb cDNAs, respectively. Within the same species the bands were present in about equal ratios. This was in accord with results obtained from cDNA

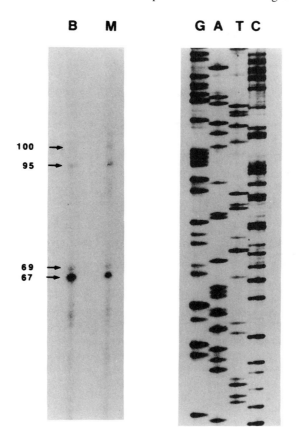

FIGURE 22.3. PCR amplification of the open reading frame of baboon (B) and macaque (M) SP-10. The PCR products have been separated in a 5% polyacrylamide gel and stained with ethidium bromide as described in "Materials and Methods". The upper band in baboon and macaque migrates at 850 bp, and the lower band migrates at 750 bp. This corresponds precisely to the open reading frames of the 1.2-kb and 1.1-kb SP-10 cDNAs, respectively. Reprinted with permission from Freemerman, Wright, Flickinger, and Herr (28), © by Wiley-Liss, 1992.

library screening, which also showed that the frequency of the individual transcripts was about 1:1.

Amino Acid Analysis of Baboon and Macaque SP-10

The 1.2-kb baboon and macaque SP-10 cDNAs contained identical open reading frames of 855 nucleotides that encoded proteins of 285 amino acids (Fig. 22.4). These proteins shared a greater than 98% (285/289)

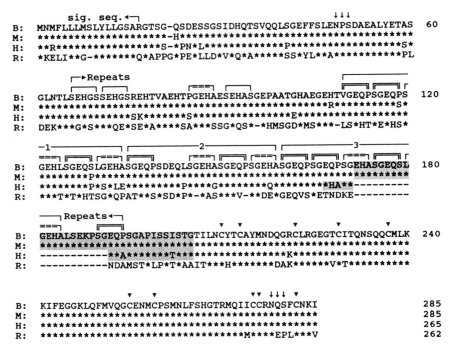

FIGURE 22.4. Deduced amino acid sequence of baboon (B), macaque (M), and human (H) SP-10 and mouse (R) MSA-63. The deduced amino acid sequence of macaque and human SP-10 and MSA-63 are compared to the deduced amino acid sequence of baboon SP-10. The numbering to the right indicates the positions of the amino acids. The hydrophobic leader sequence contains the N-terminal 18 residues and is overscored by *sig. seq.* Exact matches are denoted by an asterisk (*). Regions lacking comparable sequence are denoted by dashes (-----). Conserved cysteine residues are denoted by arrowheads (▼). The middle 50% of the sequence contains the repeat motifs. The pentapeptide repeats are overscored from above: (S, E, H, G/A, A) → ⌐⎯⎯⌐ ; (S/L, G, E, H, A/L) → ⌐===⌐ ; and (S/V, G, E, Q, P/S/A) → ⌐⎯⎯⌐ . The 3 larger 25-amino acid repeat motifs are labeled above the pentapeptide symbols ⌐⎯1⎯⌐, ⌐⎯2⎯⌐, and ⌐⎯3⎯⌐. The conserved N-linked glycosylation sites are overscored by arrows (↓ ↓ ↓). Amino acids included within the shaded region are encoded for by the alternatively spliced introns. Alternative splicing results in SP-10 proteins with internal deletions of 34 residues (baboon and macaque) and 19 residues (human). Reprinted with permission from Herr, Klotz, Shannon, Wright, and Flickinger (13) and Freemerman, Wright, Flickinger, and Herr (28).

homology—differing by only 4 conserved amino acid substitutions—and had deduced molecular weights of 30.1 kd. The alternatively spliced 1.1-kb cDNAs contained open reading frames of 753 nucleotides that encoded proteins of 251 amino acids with deduced molecular weights of

26.8 kd. The alternative splicing in both species resulted in an SP-10 variant with an internal deletion of 34 amino acids. Both baboon and macaque SP-10 contained two canonical N-linked glycosylation sites (N-X-S/T) at residues 48 and 278 (8, 9).

Hydropathy analysis of the deduced amino acid sequences of baboon and macaque SP-10 indicated that the proteins could be subdivided into a signal peptide of approximately 18 residues (10) and 2 distinct regions: a hydrophilic region that included the aminoterminal 2/3 of the protein and a hydrophobic region that included the carboxyl 1/3 of the protein.

The aminoterminal 2/3 of baboon and macaque SP-10 contained 189 amino acids that were 98% (185/189) homologous. This region contained the alternative splice sites and was characterized by 3 major repeat motifs consisting of 5 residues: (V/S, G, E, Q, P/S), (P/L/S, G, E, H, A/L), and (S, E, H, G/A, S). The pentapeptides occurred 9, 7, and 3 times, respectively. Fourteen of these repeats were arranged into 3 adjacent larger repeat motifs each containing 25 amino acids. The splice contained the last 3.6 pentapeptide repeats along with 16 other residues.

The carboxyterminal 1/3 of the protein (baboon and macaque) contained 78 residues and was 100% (78/78) homologous. This region was relatively hydrophobic and contained 10 cysteine residues.

Amino Acid Comparison of Nonhuman Primate SP-10, Human SP-10, and MSA-63

The amino acid sequence comparison between baboon and human SP-10 revealed an overall homology of 85% (242/285). This included 242 exact matches, 20 conserved substitutions, 3 nonconserved substitutions, and 20 residues for which there was no match due to a deletion in the human sequence. The percent homology was determined by dividing the number of exact amino acid matches by the number of total possible matches.

Comparisons of the mouse intra-acrosomal antigen, MSA-63 (an antigen cloned and sequenced by Greg Lee's group) (11), to primate SP-10 amino acid sequences showed a 53% (151/285) homology with baboon SP-10 and a 60% (158/265) homology with human SP-10. The two canonical N-linked glycosylation sites (N-X-S/T) at residues 48 and 278 are conserved in the nonhuman primate and human SP-10 sequences, while only the second site at residue 278 was present in MSA-63 (8, 9).

Comparison of the hydropathy plots of the deduced amino acid sequences of baboon and human SP-10 and mouse MSA-63 indicated that these proteins contained similar regions. All 3 proteins contained a hydrophobic sequence characteristic of a signal peptide (10), a distinct hydrophilic aminoterminal region, and a more hydrophobic carboxyl region.

The hydrophilic aminoterminal 2/3 of baboon and human SP-10 shared a 78% (148/189) homology, while baboon SP-10 and MSA-63 shared only a 39% (73/189) homology in this region. Human SP-10 contained the same 3 pentapeptide repeats (4) as in baboon SP-10. Eleven of these pentapeptide repeats formed 2.5-larger repeat domains of 25 residues; the third 25-residue repeat was truncated. MSA-63 did not contain the repeat motifs to the same extent, having only 2 repeats of (S, G, E, Q, P/S) and 3 repeats of (S, G/T, E, H, T/L).

The hydrophobic carboxyl region (78 residues) exhibited the greatest degree of interspecies conservation. There was a 99% (77/78) and 86% (67/78) homology between baboon SP-10 and human SP-10 and MSA-63, respectively. Of particular interest in the hydrophobic carboxyterminal region are 10 cysteine residues that were absolutely conserved in human, baboon, macaque, and mouse.

Homology Searches

Nonhuman primate SP-10 cDNAs and amino acid sequences were compared to available sequences in Genbank and the PIR protein bank using the fasta search program in Eugene. Neither the nucleotide nor the amino acid sequences showed any homology to the sequences in the banks.

Significance of Sequence Data on Baboon and Macaque SP-10

The cloning and characterization of the acrosomal protein SP-10 in the baboon and macaque were undertaken in anticipation of fertility trials using human SP-10 as a contraceptive vaccinogen in female baboons and/or macaques. The appropriateness of testing the human immunogen in nonhuman primates depends on a high level of homology between human and nonhuman primate SP-10. For example, other studies that utilized the *β-subunit of human chorionic gonadotropin* (β-hCG) as a contraceptive vaccinogen encountered difficulties because of a limited crossreactivity between antibodies raised against hCG and other nonhuman primate CGs due to differences in the β-subunit of the hormone, with low crossreactivity of antihuman CG with baboon CG, macaque CG, and marmoset CG (13). In the present work we postulate that significant homology between human SP-10 and nonhuman primate SP-10 predicts that antibody titers to human SP-10 may recognize nonhuman primate SP-10 and possibly induce infertility in these animal models. The high degree of homology between human SP-10 and both baboon and macaque SP-10 indicates that these two species are appropriate models for testing a human SP-10 vaccine.

The alternative splice sites in both human and nonhuman primate SP-10 cDNAs each began near the C-terminal end of the repeat domain and extended 11 amino acids beyond the repeat motifs to terminate at the same amino acid. The ratio of the 2 alternatively spliced transcripts in baboons and macaques was about 1:1; however, the ratio may vary between individuals. In contrast, the larger transcript is the predominant form in humans (Wright, unpublished data). The function of the alternative splicing is unclear at this time, but it probably contributes to the differences in mass for human and nonhuman SP-10 peptides on Western blots (2, 5), by forming SP-10 peptides differing by 19 (human) and 34 (baboon and macaque) amino acids.

Human, baboon, and macaque SP-10 and MSA-63 all demonstrated an aminoterminus characteristic of a leader peptide of 18 amino acids (10). This signal sequence could direct the protein into the endoplasmic reticulum, with subsequent transport through the Golgi apparatus in order to enter the developing acrosomal vesicle in the early spermatid, where immunoreactive SP-10 has been first observed (1).

An intriguing feature of human, baboon, and macaque SP-10 and MSA-63 is an internal hydrophilic region that constitutes 50% of the protein and contains 3 major repeat motifs consisting of 5 residues: (V/S, G, E, Q, P/S), (P/L/S, G, E, H, A/L), and (S, E, H, G/A, S). In baboons, macaques, and humans, most of these repeats can be grouped into 3 larger adjacent repeat motifs each consisting of 25 residues. The repeat motifs contain several probable endoproteolytic cleavage sites, as will be shown below, thought to be responsible for the characteristic polymorphic pattern of primate SP-10 peptides when SP-10 is extracted in its native form from the acrosome (5, 13). These cleavages result in SP-10 peptides that contain the repeat region as their aminoterminal portion.

The terminal 78 amino acids of the primate SP-10 proteins demonstrated the highest interspecies conservation. The high degree of conservation in this region implies a similar function in all species. Additionally, contained within these 78 carboxyl amino acids are 10 cysteine residues that are completely conserved in all 4 species.

In summary, characterization of primate SP-10 cDNAs was undertaken in the anticipation of fertility trials using human SP-10 as a contraceptive vaccinogen in female baboons. We have demonstrated that human and baboon SP-10 are sufficiently homologous such that antibodies raised against human SP-10 in baboons may be predicted to recognize SP-10 on baboon or macaque sperm. The carboxyterminus of SP-10 demonstrated the highest interspecies conservation, and we suggest any vaccine that includes SP-10 as an immunogen should incorporate this region, as it likely contains functionally essential epitopes.

Isolation and Microsequencing of Native SP-10

Human sperm extracts, analyzed on 1- and 2-D Western blots with a monoclonal antibody to SP-10 (2), have shown heterogeneity of SP-10 peptides ranging from 17.5 to 34 kd with pIs from 4.9 to 5.3 (Fig. 22.5).

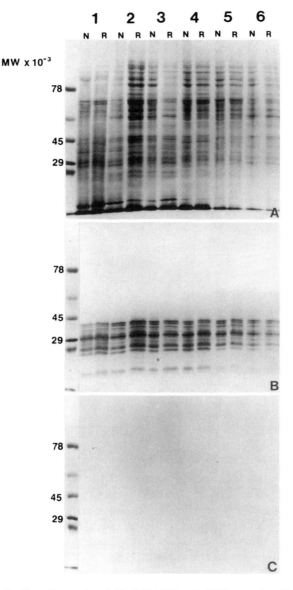

FIGURE 22.5. One-dimensional SDS-PAGE gel (10% acrylamide) nitrocellulose electroblot stained with amido black (*A*) and identical nitrocellulose sheet reacted with the MHS-10 mAb (*B*) or control IgG1 (*C*). Sperm extracts from 6 donors (1–6) contained B-mercaptoethanol (lanes marked R = reduced) or lacked this agent (N = nonreduced). Protein in the amount of 25 μg was run per lane. The pattern of SP-10 immunoreactive peptides is identical both between persons and in reduced and nonreduced extracts. Reprinted with permission from Herr, Klotz, Shannon, Wright, and Flickinger (13).

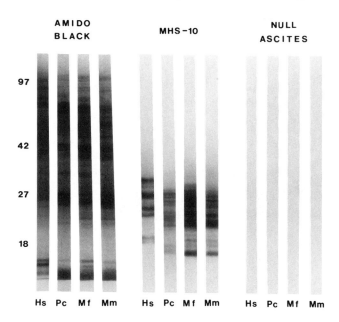

FIGURE 22.6. Immunoblot of human (Hs), *Papio cynocephalus* (Pc), *Macaca fascicularis* (Mf), and *Macaca mulatta* (Mm) sperm extracted with 1% SDS. Each lane was loaded with 10 µg of protein that was separated by SDS-PAGE and transferred to nitrocellulose. Lanes were stained with amido black, a 1:2000 dilution of MHS-10 ascites, or a 1:2000 dilution of null ascites as indicated. Lanes incubated with ascites were subsequently incubated with HRP-labeled goat antimouse IgG secondary antibody followed by 0.05% DAB and hydrogen peroxide. Reprinted with permission from Herr, Klotz, Shannon, Wright, and Flickinger (13).

Similar heterogeneity in immunoreactive SP-10 peptides has been observed (5) in sperm extracts from *Papio cyanocephalus*, *Macaca mulatta*, and *Macaca fascicularis* (Fig. 22.6). Although the entire SP-10 amino acid sequence of 265 amino acids (28.3 kd) has been deduced from sequencing SP-10 cDNAs (4), the nature of these multiple immunoreactive SP-10 peptide bands has been incompletely understood. A 3-step purification method for SP-10 peptides has been developed using monoclonal antibody affinity chromatography, reverse-phase HPLC, and preparative gel electrophoresis (13).

Silver Stain and Immunoblot of Purified Antigen

Human sperm preparations at various steps in the 3-step purification process are presented in Figure 22.7, where silver-stained protein bands are compared to an identical gel that was immunoblotted. A wide array

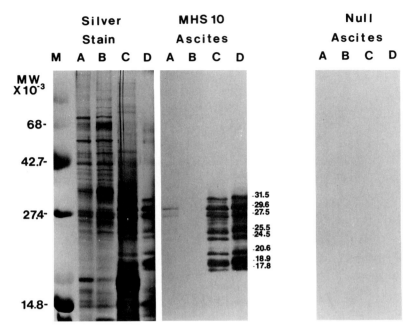

FIGURE 22.7. Demonstration of protein content and SP-10 immunoreactivity in fractions obtained at various steps in the purification of the SP-10 antigen. Shown are a silver-stained 10% gel and Western blots probed with the MHS-10 monoclonal antibody or control monoclonal antibody (*Null Ascites*). Each lane was loaded with 5-µg protein: (A) sperm extract (the starting material), (B) pH 8 wash of column containing proteins that did not bind to the MHS-10 mAb affinity column, (C) pH 2.4 bump of proteins bound to the MHS-10 mAb affinity column, and (D) purified SP-10 peak from the reverse-phase HPLC column. The enrichment of SP-10 peptides during purification is evident on the immunoblot by comparing starting and purified fractions. The immunoblot was performed at a concentration of primary antibody that permitted visualization of distinct, individual SP-10 bands in the purified fractions while avoiding overreacting these lanes so as to overstain the concentrated SP-10 peptides. Under these immunoblot conditions the SP-10 bands in the starting material were not all displayed, although their presence in the starting material was confirmed by immunoblotting at a higher concentration of primary antibody in other experiments and by the fact that purification enriched the full complement of immunoreactive SP-10 peptides. Molecular weight markers were bovine serum albumin (68 kd), ovalbumin (42.7 kd), carbonic anhydrase (27.4 kd), beta-lactoglobulin (18 kd), and lysozyme (14.8 kd). Reprinted with permission from Herr, Klotz, Shannon, Wright, and Flickinger (13).

of protein species was present in the starting sperm extract (lane A, silver stain). The immunoblot of the starting material lane loaded with 5 µg showed faint but detectable SP-10 immunoreactive bands (lane A, stained with MHS-10 ascites). As expected, most proteins present in the starting sperm extract passed through the MHS-10 affinity column (lane B, silver stain), and no immunoreactivity for SP-10 could be detected in this fraction (lane B, stained with MHS-10 ascites). Proteins that bound and were eluted from the MHS-10 affinity column are shown in lane C where a number of protein bands are observed on silver staining with dense bands prominent in the 15- to >30-kd range. The immunoblot of the affinity-purified antigen showed intensely staining immunoreactive peptides from 31 to 20 kd (lane C, Western blot). At least 12 distinct SP-10 protein bands were detected on the immunoblot of the affinity-purified proteins on this gel. Comparison of the immunoblot of the starting material (lane A) with lane C shows the significant enrichment of SP-10 in the affinity-purified preparation.

HPLC Purification of SP-10

In the next step the SP-10 preparation from the affinity purification was run over a reverse-phase HPLC column. Further purification of SP-10 peptides was evident (Fig. 22.7, lanes marked D and Fig. 22.8). (Figure 22.7, lane D, contains fractions *a* and *b* from Figure 22.8.) Figure 22.8 presents the protein elution profile (panel A) from the HPLC column and shows fractions from the elution profile that were silver stained (panel B), as well as the immunoblot of the corresponding fractions (panel C). Protein peaks containing SP-10 that were immunoreactive with the MHS-10 monoclonal antibody (Fig. 22.8C) were found to elute in the first major double peaks, marked *a* and *b*, at 19.8 and 20.1 min (Fig. 22.8A). Peak *a-b* was clearly separated from the majority of other proteins that eluted from 22 to 35 min. Peak *a* contained prominent SP-10 peptides at 29, 27, and 17–19 and a faint 31.5-kd band (which was present to a greater extent in peak *b*). The smaller peptides in the 17- to 19-kd range in peak *a* were not prominent in peak *b*. Peak *b* contained the bulk of the purified peptides, with the exception of those at 17–19 kd. The elution profile thus indicated that the lower-mass peptides in peak *a* have slightly different hydrophilic properties than the other SP-10 forms in peak *b*.

A silver stain of the eluted fractions from the HPLC column (Fig. 22.8B) that were not immunoreactive (Fig. 22.8C). These nonimmunoreactive proteins were present in the preparation eluted from the monoclonal antibody affinity column and are clearly separated from immunoreactive SP-10 by reverse-phase HPLC. While some of these proteins may bind nonspecifically to the Sepharose beads, it is possible that one or several of these nonimmunoreactive species may be proteins

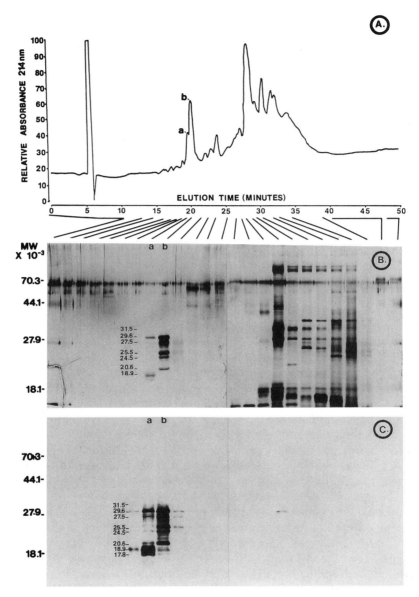

FIGURE 22.8. Elution profile (*A*), SDS-PAGE silver stain (*B*), and Western blot (*C*) of protein fractions from reverse-phase HPLC column purification of SP-10. *A*: In the elution profile of total protein (read at 214 nm) as a function of time (min), SP-10 appears highly hydrophilic since it elutes as the first major constituent, double peak a-b, separated from the majority of other proteins that eluted from 22 to 38 min. *B*: Fractions corresponding to indicated time points were electrophoresed and silver stained. Since the fractions were collected in

that complex with SP-10. In any case, these observations highlight the importance of the second HPLC step in refining SP-10's purification to eliminate contaminants that copurify in the eluant of the monoclonal antibody affinity column. A control blot was also run for the data presented in Figure 22.9 in which each lane was loaded with equal volumes of 20 µL (or 10%) of the fraction, and the blot was incubated with secondary antibody alone (data not shown). The absence of any reaction product confirmed that the staining that was evident with MHS-10 was not the result of a nonspecific or secondary antibody immunoreaction alone.

The purified SP-10 fraction (Figs. 22.8A and 22.8B) from the reverse-phase HPLC column is also shown in Figure 22.7, where a comparison may be made between the affinity column fraction (lane C) and the HPLC fraction (lane D). Enrichment of SP-10 peptides in the HPLC preparation compared to affinity column fraction is evident in the silver stained gel (Fig. 22.7, lane D). The immunoblot of the HPLC preparation shows recovery after HPLC of at least 12 SP-10 peptide bands, all of which were present after elution from the monoclonal antibody affinity column. Controls for these studies included the use of another monoclonal antibody, IgG1 (lane labeled *Null Ascites*), that showed no immunoreaction, verifying the specificity of the binding of the MHS-10 monoclonal antibody in the Western blots.

Preparative SDS-PAGE, Transfer to PVDF

As a third step in preparing individual immunoreactive SP-10 peptides for microsequencing, the fractions from the HPLC column were electrophoresed by preparative SDS-PAGE, the peptides were electrotransferred to PVDF membrane, and the membrane was cut into strips. Figure 22.9 shows a coomassie blue stain of the PVDF membrane containing those purified SP-10 peptides that were subsequently employed in peptide microsequencing, with the approximate molecular weight of the sequenced peptides noted on the figure's right margin.

varying volumes, each fraction was dried on a SpeedVac and redissolved in 200-µL ddH$_2$O, and 20 µL of each fraction was loaded per lane. The high degree of purity of SP-10 is evident in lanes marked a and b that corresponded to peaks a and b. C: Western blot of fractions collected from reverse-phase HPLC probed with MHS-10 monoclonal antibody ascites reveals that the double peak of SP-10 is the result of different forms of SP-10 having slightly different hydrophilic properties eluting at 19.86 and 20.09 min. Each lane was loaded with 5 µL of each fraction. Molecular weight markers were bovine serum albumin (68 kd), ovalbumin (42.7 kd), carbonic anhydrase (27.4 kd), beta-lactoglobulin (18 kd), and lysozyme (14.8 kd). Reprinted with permission from Herr, Klotz, Shannon, Wright, and Flickinger (13).

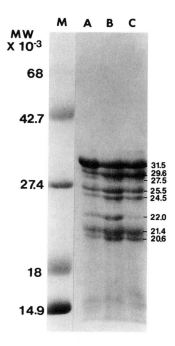

FIGURE 22.9. Coomassie blue-stained blot on PVDF membrane of purified SP-10 employed for microsequencing. Each lane, A, B, and C, represents approximately 40 μg of SP-10 prepared from a pool of 10–12 ejaculates. The approximate molecular weights of the SP-10 peptides sequenced are indicated on the right of the figure. Molecular weight markers were bovine serum albumin (68 kd), ovalbumin (42.7 kd), carbonic anhydrase (27.4 kd), beta-lactoglobulin (18 kd), and lysozyme (14.9 kd). Reprinted with permission from Herr, Klotz, Shannon, Wright, and Flickinger (13).

Microsequencing

Satisfactory N-terminal amino acid sequences were obtained for 8 SP-10 polypeptides. Table 22.1 compares the amino acid sequences obtained by microsequencing to the SP-10 amino acid sequence deduced from cDNAs. The number of N-terminal amino acid residues determined to correspond to the predicted sequence was 5 for the 21.4- and 27.5-kd polypeptides and 10 or more residues for the other 7 peptides (Table 22.1). SP-10 peptides with progressively lower apparent mass aligned farther toward the carboxyterminus. The SP-10 peptide of approximately 31.5 kd aligned at amino acid 78, the 29.5-kd peptide at amino acid 86, the 27.5-kd peptide at amino acid 96, the 25.5-kd peptide at amino acid 106, the 24.5-kd peptide at amino acid 112, the 22-kd peptide at amino acid 127, the 21.4-kd peptide at amino acid 137, and the 20.6-kd peptide at amino acid 140 (Table 22.2).

TABLE 22.1. Comparison of N-terminal amino acid sequences from isolated SP-10 peptides to the amino acid sequence deduced from cDNA cloning and sequencing.

M	N	R	F	L	L	L	M	S	L	Y	L	L	G	S	A	R	G	T	S	20
S	Q	P	N	E	L	S	G	S	I	D	H	Q	T	S	V	Q	Q	L	P	40
G	E	F	F	S	L	E	N	P	S	D	A	E	A	L	Y	E	T	S	S	60
G	L	N	T	L	S	E	H	G	S	S	E	H	G	S	S	K	H	T	V	80
															(31.5 kd)-	H	T	V-		
A	E	H	T	S	G	E	H	A	E	S	E	H	A	S	G	E	P	A	A	100
A	E	H	T	S	G	E	-(31.5 kd)													
		(29.6 kd)-	G	E	H	A	E	E	E	H	A	-(29.6 kd)								
								(27.5 kd)-	G	E	P	A	A							
T	E	H	A	E	G	E	H	T	V	G	E	Q	P	S	G	E	Q	P	S	120
		(25.5 kd)-	G	E	H	T	V	G	Q	Q	P	S	-(25.5 kd)							
				(24.5 kd)-	E	Q	P	S	G	E	Q	P	S-							
G	E	H	L	S	G	E	Q	P	L	S	E	L	E	S	G	E	Q	P	S	140
G	E	-(24.5 kd)																		
		(22.0 kd)-	E	Q	P	L	S	E	L	E	S	G	E	Q	P	S-				
					(21.4 kd)-	E	Q	P	S											
							(20.6 kd)-	S												
D	E	Q	P	S	G	E	H	G	S	G	E	Q	P	S	G	E	Q	A	S	160
D	-(22.0 kd)																			
D	-(21.4 kd)																			
D	E	Q	P	S	G	E	H	G	S	G	-(20.6 kd)									
G	E	Q	P	S	G	E	H	A	S	G	E	Q	A	S	G	A	P	I	S	180
S	T	S	T	G	T	I	L	N	C	Y	T	C	A	Y	M	N	D	Q	G	200
K	C	L	R	G	E	G	T	C	I	T	Q	N	S	Q	Q	C	M	L	K	220
K	I	F	E	G	G	K	L	Q	F	M	V	Q	G	C	E	N	M	C	P	240
S	M	N	L	F	S	H	G	T	R	M	Q	I	I	C	C	R	N	Q	S	260
F	C	N	K	I	TER															265

Amino acids are presented according to the single-letter code from data published in reference 6.
Source: Reprinted with permission from Herr, Klotz, Shannon, Wright, and Flickinger (13).

378   J.C. Herr et al.

TABLE 22.2. Summary of the apparent molecular weight (from SDS-PAGE) of the isolated SP-10 peptides, the N-termini at which they align on the deduced amino acid sequence, and the predicted mass of the peptide, provided no additional endoproteolytic cleavages or posttranslational modifications occur.

N-terminal amino acid	Apparent M_r by SDS-PAGE $\times 10^{-3}$ kd	Calculated $M_r \times 10^{-3}$ kd	Approx. mass (difference kd)
78	31.5	19.9	11.6
86	29.6	19.0	10.6
96	27.5	18.0	9.5
106	25.5	17.0	8.5
112	24.5	16.4	8.1
127	22.0	14.9	7.1
137	21.4	13.8	7.6
140	20.6	13.5	7.1

Source: Reprinted with permission from Herr, Klotz, Shannon, Wright, and Flickinger (13).

Assuming that the carboxyterminus of each of the SP-10 peptides was identical with that deduced from cDNA cloning (no further endoproteolytic or posttranslational modifications), the mass of the SP-10 peptides was calculated from the cDNA sequence. Comparison of these deduced masses to the apparent mass obtained by estimation from SDS-PAGE standards (Table 22.2) indicated that each peptide appeared with higher apparent mass on SDS-PAGE than predicted. These differences range from approximately 11 kd for the 31.5-kd peptide to 7 kd for the 20.6-kd peptide.

Two of the 77 amino acids identified by aminoterminal sequencing of purified SP-10 peptides did not match the amino acid sequences deduced from cDNAs—a frequency of 2.5% incorrect identification. These mismatches were a glutamic acid in the 29.6-kd peptide rather than a serine at amino acid 91 and a glutamine in the 25.5-kd peptide rather than a glutamic acid at amino acid 102. It is unclear if these differences represent actual misreadings at the time of amino acid sequencing or are due to substitutions representing genetic drift between the individual from whom the cDNA sequence was obtained and the pool of donors from which the SP-10 peptides were isolated and sequenced.

Predicted Peptide Linkages Cleaved

Examination of the amino acids proximal to the N-termini of each of the 8 sequenced SP-10 peptides revealed some interesting similarities (Table 22.3). An arginine lies adjacent to the N-terminal histidine of the 31.5-kd peptide; serines are adjacent to the 29.6- and 27.5-kd peptides' N-terminal glycine residues; glutamic acid lies adjacent to the 25.5-kd N-terminal glycine; glycine residues are adjacent to each of the N-terminal

TABLE 22.3. Predicted endoprotease cleavage sites in the SP-10 molecule determined by amino acid sequencing of 8 naturally occurring SP-10 peptides.

SP-10 peptide	Endoproteolytic site	Motif
31.5	GSSK-HTVA	
29.6	EHTS-GEHA	
27.5	EHAS-GEPA	EHXX-GEXX
25.5	EHAE-GEHT	
24.5	HTVG-EQPS	
22.0	HLSG-EQPL	XXXG-EQPX
21.4	LESG-EQPS	
19.7	GEQP-SDEQ	

Source: Reprinted with permission from Herr, Klotz, Shannon, Wright, and Flickinger (13).

glutamic acid residues of the 24.5-, 22-, and 21.4-kd peptides; and proline lies adjacent to the 19.7-kd peptide's N-terminal serine.

Significance of Results of Microsequencing SP-10

We began the purification and microsequencing of SP-10 peptides with alternative hypotheses that the MHS-10 monoclonal antibody was recognizing a common epitope on peptides that were (i) otherwise unrelated or (ii) derived from a single protein. The fact that the N-terminal amino acid sequences of 8 purified SP-10 peptides aligned on the SP-10 sequence derived from cDNA cloning indicated that the heterogeneity in immunoreactive SP-10 peptides observed on Western blots is due to the presence of peptides derived from a common precursor SP-10 protein.

We previously published a complete open reading frame for SP-10 (4) and concluded that we had sequenced authentic SP-10 because (i) a rabbit polyclonal antibody produced to a fusion protein consisting of 212 amino acids of SP-10 reacted with the identical series of multiple SP-10 peptides on Western blots, as did monoclonal antibody MHS-10; and (ii) the polyclonal antiserum precisely stained the acrosomal cap by immunofluorescence. The purification and microsequencing of SP-10 peptides that aligned on the SP-10 sequence deduced from cDNA cloning in the present study provide further proof that the correct cDNAs for SP-10 were previously cloned and sequenced.

The SP-10 purification by affinity chromatography on bound monoclonal antibody followed by HPLC gave an average (N = 3) recovery of 0.8% of the total starting sperm protein in the fraction designated SP-10. Recently, Dr. Shen in our group has calculated the amount of SP-10 in sperm extracts—utilizing capture ELISA and recombinant SP-10 as a standard—and determined that SP-10 represents approximately 1% of total sperm protein. The 0.8% recovery compares favorably with this determination and is of similar magnitude to recoveries obtained for the single-step monoclonal antibody affinity purifications of the guinea pig sperm antigens PH20 and PH30, where 0.24% and 0.1% of starting protein was obtained, respectively (14, 15).

Our data predict that at least in part the SP-10 polypeptide forms of variable mass observed on Western blots of SDS-PAGE occur as a result of hydrolysis by endoproteases. The multiple forms of SP-10 from 18 to 34 kd are observed in both reduced and nonreduced sperm extracts from all individuals we have tested, as well as in extracts collected into a cocktail of protease inhibitors (2). These multiple forms of SP-10 from 18 to 34 kd are also present in SDS extracts of testes and of sperm collected by ejaculation directly into SDS, a procedure that should minimize secondary proteolysis by enzymes in seminal plasma from accessory glands. After incubation at room temperature, the pattern of digestion changes little over 24 h, also suggesting that the digestion of SP-10 is not due to nonspecific proteolysis. Because of these observations we suggest that many of the polymorphic forms of SP-10 are generated intra-acrosomally and are present in the acrosome of normal sperm.

We hypothesize that the endoprotease cleaving between arginine 77 and histidine 78 is acrosin, the well-characterized trypsin-like serine protease with specificity for arginine and lysine bonds (16–18). Analysis of the amino acids immediately proximal to the microsequenced NH_2 termini of the SP-10 peptides leads us to propose further that the human acrosome contains at least 2 and possibly 4 additional endoproteases. To our knowledge, these endoproteases have not been described within the acrosome of any species. The first of these putative hydrolases is a endopeptidase possessing hydrolytic activity at serine-glycine residues. This endopeptidase would be responsible for creation of SP-10 polypeptides by hydrolysis at serines 85 and 95. The second predicted intra-acrosomal protease is an endopeptidase responsible for hydrolysis of SP-10 at glycines 111, 126, and 136. A third predicted endoprotease would cleave SP-10 at glutamic acid 105, while a prolyl endoprotease would act at proline 139.

While a maximum of 4 endoproteases other than acrosin is predicted from our data, 2 or 3 enzymes might suffice. The cleavage after proline 139 may be performed by prolyl endopeptidase, an enzyme found in several tissues (19, 27). Cleavage after serines 85 and 95 and after glycines 111, 126, and 136 may be due to proteinases specific for cleavage after those residues or by a proteinase that recognizes the motifs shown in Table 22.3, which presents the 4 amino acids flanking either side of the apparent cleavage sites of the SP-10 peptides. Three of the peptides (29.6, 27.5, and 25.5 kd) might be generated by an endoprotease acting within one motif of amino acid sequence (EHXS-GEXX), suggesting that the same endoprotease is responsible for the serine-glycine and the glutamic and glycine cleavages. Three other peptides (24.5, 22, and 21.4 kd) could be generated by another endoprotese cleaving at the glycine-glutamic acid bond within the common motif (XXXG-EQPX). Based upon a Chou-Fasman plot of the SP-10 sequence, it appears that the sites of digestion are in regions where the predicted secondary structure has either turns, coils, or some irregularity of helix; thus, a relatively nonspecific proteinase might also cleave at those sites. However, if the presence of the described motifs and nonhelical structure were sufficient for digestion, additional sites on the SP-10 sequence should have been cleaved and resolved on the Western blot. Thus, the importance of secondary structure at cleavage sites is unclear.

Whereas the confirmation of these newly predicted intra-acrosomal proteases awaits further isolation and characterization, their biological functions may include (i) a cascade-like activation of intra-acrosomal hydrolases, (ii) mediation of key steps in the events of capacitation, (iii) generation of active ligands between the inner acrosomal membrane and the egg investments (mediators of what has been called secondary binding [20]), and/or (iv) hydrolysis of constituents of the zona pellucida or corona radiata.

The multiple peptide forms of the intra-acrosomal protein SP-10 are reminiscent of the heterogeneity of other intra-acrosomal proteins that has been previously reported. Human acrosin exists as a zymogen of 52–55 kd on SDS-PAGE and is autoactivated to a 49-kd form followed by further conversion to several lower molecular mass forms ranging from 34 to 38 kd (21–23). Human sperminogen, another acrosomal trypsin-like hydrolase that acts on arginine peptide linkages, presents as a series of enzymatically active peptides from 32 to 36 kd on gelatin SDS-PAGE zymography (24, 25). Acrogranin, which appears as a single protein band with M_r 67,000 in guinea pig testis, consists of 4 bands in epididymal sperm extracts with M_r of 62,000, 51,000, 39,000, and 22,000 (26). It has been hypothesized that these acrogranin variants result from proteolytic processes that develop during epididymal maturation (25).

Acknowledgments. Supported by NIH Grants HD-23789, HD-29099, and GM-08401, and the R.W. Johnson Pharmaceutical Research Institute.

References

1. Kurth BE, Klotz K, Flickinger CJ, Herr JC. Localization of sperm antigen SP-10 during the six stages of the cycle of the seminiferous epithelium in man. Biol Reprod 1991;44:814–21.

2. Herr JC, Flickinger CJ, Homyk M, Klotz K, John E. Biochemical and morphological characterization of the intra-acrosomal antigen SP-10 from human sperm. Biol Reprod 1990;42:181–93.
3. Anderson DJ, Johnson PM, Jones WR, Griffen PD. Monoclonal antibodies to human trophoblast and sperm antigens: report of two WHO-sponsored workshops, June 30, 1986, Toronto, Canada. J Reprod Immunol 1987; 10:231–57.
4. Wright RM, John E, Klotz K, Flickinger CJ, Herr JC. Cloning and sequencing of cDNAs coding for the human intra-acrosomal antigen SP-10. Biol Reprod 1990;42:693–701.
5. Herr JC, Wright RM, John E, Foster J, Kays T, Flickinger CJ. Identification of human acrosomal antigen SP-10 in primates and pigs. Biol Reprod 1990;42:377–82.
6. Dubova-Mihailova M, Mollova M, Ivanova M, Kehayov I, Kyurkchiev S. Identification and characterization of human acrosomal antigen defined by a monoclonal antibody with blocking effect on in vitro fertilization. J Reprod Immunol 1991;19:251–68.
7. Kozac M. Compilation and analysis of sequences upstream from the translational start site in eukaryotic mRNAs. Nucleic Acids Res 1984;12:857–73.
8. Pless DD, Lennarz WJ. Enzymatic conversion of proteins to glycoproteins. Proc Natl Acad Sci USA 1977;74:134–8.
9. Hart GW, Brew K, Grant GA, Bradshaw RA, Lennarz WJ. Primary structural requirements for the enzymatic formation of the N-glycosidic bond in glycoprotein studies with natural and synthetic peptides. J Biol Chem 1979;254:9747–53.
10. von Heijne G. A new method for predicting signal sequence cleavage sites. Nucleic Acids Res 1986;14:4683–90.
11. Liu MS, Ruedi A, Lee CYG. Molecular and developmental studies of a sperm antigen recognized by HS-63 monoclonal antibody. Biol Reprod 1992;46:937–48.
12. Stevens VC, Powell JE, Lee AC, Griffin D. Antifertility effects of immunization of female baboons with C-terminal peptides of the β-subunit of human chorionic gonadotropin. Fertil Steril 1981;36:98–105.
13. Herr JC, Klotz K, Shannon J, Wright RM, Flickinger CJ. Purification and microsequencing of the intra-acrosomal protein SP-10: evidence that SP-10 heterogeneity results from endoproteolytic processes. Biol Reprod 1992;46:981–90.
14. Primakoff P, Hyatt H, Tredick-Klein J. Identification and purification of a sperm surface protein with a potential role in sperm-egg membrane fusion. J Cell Biol 1987;104:141–9.
15. Primakoff P, Cowan A, Hyatt H, Tredick-Klein J, Myles D. Purification of the guinea pig sperm PH20 antigen and detection of a site-specific endoproteolytic activity in sperm preparations that cleaves PH20 into two disulfide-linked fragments. Biol Reprod 1988;38:921–34.
16. Baba T, Kashiwabara S, Watanabe K, et al. Activation and maturation mechanisms of boar acrosin zymogen based on the deduced primary structure. J Biol Chem 1989;264:11920–7.

17. Polakoski KL, Parrish RF. Boar proacrosin. Purification and preliminary activation studies of proacrosin isolated from ejaculated boar sperm. J Biol Chem 1977;252:1888–94.
18. Parrish RF, Polakoski KL. Boar Mα-acrosine. Purification and characterization of the initial active enzyme resulting from the conversion of boar proacrosin to acrosin. J Biol Chem 1978;253:8428–32.
19. Orlowski M, Wilk E, Pearce S, Wilk S. Purification and properties of a prolyl endopeptidase from rabbit brain. J Neurochem 1979;33:461–9.
20. Saling PM. Mammalian sperm interaction with extracellular matrices of the egg. Oxford Rev Reprod Biol 1989;11:339–88.
21. Siegel MS, Polakoski KL. Evaluation of the human sperm proacrosin-acrosin system using gelatin-sodium dodecylsulfate-polyacrylamide gel electrophoresis. Biol Reprod 1985;32:713–20.
22. Siegel MS, Bechtold DS, Kopta CI, Polakoski KL. The rapid purification and partial characterization of human sperm proacrosin using an automated fast protein liquid chromatography (FPLC) system. Biochim Biophys Acta 1986;883:567–73.
23. Elce JS, McIntyre EJ. Acrosin: immunochemical demonstration of multiple forms generated from bovine and human proacrosin. Can J Biochem Cell Biol 1982;61:989–95.
24. Siegel MS, Polakoski KL. Human sperm proteinases: purification and characterization of proacrosin and a unique zymogen referred to as sperminogen. Biol Reprod;30(suppl 1):177.
25. Siegel MS, Bechtold DS, Willand JL, Polakoski KL. Partial purification and characterization of human sperminogen. Biol Reprod 1987;36:1063–8.
26. Anakwe OO, Gerton G. Acrosome biogenesis begins during meiosis: evidence from the synthesis and distribution of an acrosomal glycoprotein, acrogranin, during guinea pig spermatogenesis. Biol Reprod 1990;42:317–28.
27. Rosen J, Tomkinson B, Pettersson G, Zetterqvist O. A human serine endopeptidase, purified with respect to activity against a peptide with phosphoserine in the P1 position, is apparently identical with prolyl endopeptidase. J Biol Chem 1991;266:3827–34.
28. Freemerman AJ, Wright R, Flickinger CJ, Herr JC. Cloning and sequencing of baboon and cynomolgus monkey intra-acrosomal protein SP-10; homology with human SP-10 and a mouse sperm antigen (MSA-63). Mol Reprod Dev 1992.

Author Index

A
Adler, A., 246
Alikani, M., 246
Anderson, C., 246
Andrews, G.K., 195

B
Bass, K.E., 182
Bavister, B.D., 30, 110, 331
Berger, N.G., 331
Boatman, D.E., 30, 110, 331

C
Carson, D.D., 290
Chandrasekher, Y.A., 124
Cohen, J., 246
Cranfield, M.R., 331

D
Dahl, J.F., 46
Damsky, C.H., 182
Das, S.K., 195
Dey, S.K., 195
Donnelly, K.M., 169
Dukelow, W.R., 73

E
Enders, A.C., 145

F
Fazleabas, A.T., 169
Ferrara, T.A., 246

First, N.L., 317
Fisher, S.J., 182
Flanders, K.C., 195
Flickinger, C.J., 360
Foster, J., 360
Freemerman, A., 360

G
Gould, K.G., 46

H
Handyside, A.H., 349
Hearn, J.P., 158
Herr, J.C., 360
Hild-Petito, S., 169

I
Ialeggio, D.M., 331

J
Jacobs, A.L., 290
Jany, L., 279
Julian, J., 290

K
Kempske, S.E., 331
Khatchadourian, C., 279
Kissin, E., 246
Klotz, K., 360

Author Index

L

Leibfried-Rutledge, M.L., 317

M

Mavrogianis, P., 169
Menezo, Y., 279

O

Overstreet, J.W., 103

P

Paria, B.C., 195

R

Rall, W.F., 223
Rawlins, R.G., 308
Reing, A., 246
Rohde, L.H., 290
Roth, I., 182

S

Schaffer, N., 331
Schramm, R.D., 110
Seshagiri, P.B., 158
Shannon, J., 360
Smart, J., 331
Stouffer, R.L., 85, 124

T

Trounson, A., 3

V

VandeVoort, C.A., 103
Verhage, H.G., 169

W

Webley, G.E., 158
Wolf, D.P., 85, 124
Wright, R.M., 360

Z

Zelinski-Wooten, M.B., 124

Subject Index

Ablation, chromosome, 246
Abortion
  in gorilla, 59
  rate of, 353
  in squirrel monkey, 75
Acid rain, 312
Acrosin, 380
Acrosome, human, 380
Acrosome reaction, 90, 259, 360
  after ovulation, 37
  and calcium, 104, 105–106
  in macaques, 103
  in primates, 35
  rhesus monkey, 32
  and SZI, 250
  in vivo, 106
Activation, competence for, 321
Adhesion, 145–146
  cell, 148–156, 293
  pelvic, 50
Adhesion molecules, 185–187
Adrenoleukodystrophy, 354
Age
  appropriateness for AR, 62
  barrier reduction, 15
  and coculture, 283
  human maternal, 12, 269
Amenorrhea, lactational, 57
Amino acid
  comparisons of, 367
  metabolism, 11
  and SP-10, 364
Amniocentesis, 351
Amniotic fluid, human, 11
Amphibians, nuclear transfer, 319
Amplification, failure of, 355
Analysis, PCR, of SP-10, 364

Aneuploidy, 349
Angiogenesis, 175, 213
Animals
  domestic and cloning, 323
  transgenic, 325
Anomalies, of oocyte, 93
Antibodies, gonadotropin, 136. See also Refractoriness
Antigenicity, 73, 132
Antrum formation, 115
Apes, 159
Apposition, in primates, 145–146
AR. See Assisted reproduction
Arrest, third meiotic, 320
Arteries, spiral, 171
Arteriosclerosis, in chimpanzee, 56
Artificial insemination, 75, 223
  after cryopreservation, 232, 239
  and lion-tailed macaque, 333, 337
  success rate of, 60
Aspiration technique, 9
Assay
  hemizona, 36
  improvements in, 48
  multiple-sperm penetration, 35
Assisted breeding. See Assisted reproduction
Assisted fertilization, 64. See also Assisted reproduction, Artificial insemination
  drawbacks to, 258–261
  in humans, 255–258
Assisted hatching. See Hatching, assisted
Assisted reproduction, 313
  in female, 61
  future trends, 23–24
  in great apes, 46–65

Subject Index

and market for, 309
and nonhuman primates, 308
  protocols for, 62–64
  purpose of, 46
Asthenozoospermia, 252, 255, 257
Atenascin, 300
Attachment
  embryo, periimplantation, preimplantation, 161. *See also* Implantation
  human, 165
Autocrine effect, 94, 188
  and coculture, 280
  mediation of, 196
  regulation of, 175
Availability, of Old World monkeys, 86

Baboon, 154, 169–178
  anchoring stage, 155
  and assisted reproduction, 308
  chorionic gonadotropin, 368
  and cryopreservation, 237, 240
  epithelial invasion in, 149
  first IVF, 73, 75
  implantation, 145, 158
  proteins of, 169
  and SP-10, 360, 369
  trophoblast, 150, 152
  uterus, 171
Baby rate, take-home, 261
Baltimore Zoo, 332, 337
Banking, of gametes, 49, 223, 233, 235
Basal lamina, 294, 297
BCS. *See* Bovine calf serum
Biggers Whitten and Whitingham (BWW) culture, 104, 106
Binding site, 296–299
  and TGFβ, 210
Bioassay, of attachment hormone, 163
Biochemical analysis, 78–79
Biologist
  behavioral, 308
  reproductive, and species extinction, 313
Biomedicine, and live capture, 311
Biopsy
  cleavage-stage, 351
  polar body, 21
Biosynthesis, mucin, 298

Birth control, need for, 313
Blastocoel, formation of, 211
Blastocyst, 80, 113, 118, 164, 169, 198
  and assistive hatching, 274
  and coculture, 285
  and cryopreservation, 237
  culture of, 279
  formation of, 321
  human and HSPG, 294
  implantation, 145
  marmoset, 163
  and periimplantation, 195
  regulation of, 280
  rhesus monkey, 33, 94
  and superstimulation, 113
  and TGFβ, 201, 202, 211
Blastomere, 80, 264
  biopsy of, 246
  2-cell, 322
  as recipient, 320
  separation of, 274
Blood pressure, maternal, 186
Bonobo
  and artificial insemination, 61
  and productivity, 48
Boost, human, 5
Bovine
  and coculture, 280
  and IVF, 321
  oviduct and cumulus cells, 94
Bovine calf serum, 115
Bovine serum albumin (BSA), 104, 201
Brown, Louise, 309
Buffalo rat, liver cells, 94
Buserelin, 5

Caffeine, 87, 104, 107, 337, 344
Caffeine with cAMP, as acrosomal capacitor, 32
Calcium
  and acrosome reaction, 104
  and zona pellucida, in macaque, 105
Callithrix jacchus. *See* Marmoset
CAMA. *See* Computer-assisted motion analysis
Candidate selection, 246, 250
Capacitation, 381
  initiation, 103
  sperm, 85, 259

Capacitation (*cont.*)
  species specific, 103–104
Captive rearing, 332
Carbon dioxide, and embryo development, 118
Cattle, 49, 320, 321, 322, 323
  and coculture, 279
  and cryopreservation, 232, 237
  whole-cell fusion of, 318
cDNA, 298, 364, 369, 378
  SP-10, 362–363
Ceboidea, 159
Cell
  adhesion activity, 148–156
  during cryopreservation, 224–225
  cycle stage of, 324
  dehydration of, 227–228, 230
  differentiation of, 317
  gene-modified, 325
  marrow-derived, 184
  osmotic volume of, 226
  peri-inner region, 147
  stroma, 177, 195
  and TGFβ production, 209
  tumor invasion, 183
Cell fusion, trophoblast-epithelial, 149
Cercopithecoidea, 159
Cervical mucus, 103
CG. *See* Chorionic gonadotropin
Chilling, 240
Chilling injury, 237
Chimera, 275
Chimpanzee, 63
  and cryopreservation, 234
  implantation, 145
  maternal function, 48
  menstrual cycle in, 56
  similarity to humans, 46
  success of artificial insemination, 60
Chlamydia spp., 56, 60
Chondroitin sulfate proteoglycan (CSPG), 291, 299, 302
Chorionic gonadotropin (CG), 158, 159, 160, 162, 368
Chorion villus sampling (CVS), 351
Chromatin, oocyte, 322
  removal of, 319
Chromosomal painting, 353
Chromosome
  ablation, 246
  abnormalities, with microinjection, 18
  abnormalities of, 231, 325, 349
  analysis of, 78–79
  human, in superovulation, 12
  metaphase, 352
  in oocyte, and cryopreservation, 235
  PB2, 350
CITES. *See* Convention on International Trade in Endangered Species
Class I molecules, and cytotrophoblast invasion, 187–188
Cleavage
  after fertilization, 195
  third, and TGFβ, 203
Cleavage stage embryo. *See* Embryo
Clomid. *See* Clomiphene citrate
Clomiphene citrate, 7, 11, 56, 125
  and superovulation, 5
Clone, 170
Cloning, 310, 319, 320, 322
  and cryopreservation, 235
  of domestic animals, 323
  of SP-10, 362
  of TGFβ, 214
Cloprostenol, and marmoset, 77
CMRL medium, 115
Coculture
  autocrine effect, 280
  embryo and helper cells, 247
  human embryo, 282
  mechanism of, 284
  medium, 281–282
  oviduct and uterine cell, 280–281
Cold shock, 225, 240
  and chilling injury, 237
Collection
  for cryopreservation, 232
  oocyte and semen, 223–232
Compaction, 211
Compatibility, social, 62
Competitive selection. *See* Selection, competitive
Computer-assisted motion analysis (CAMA), 51, 63
Computer-assisted video image analysis system (CASMA), and rhesus monkey sperm motility, 87

Conservation
  and cryopreservation of endangered species, 223. *See also* Endangerment
  wildlife, 331
Contraception, in lion-tailed macaque, 344
Contraceptives, oral, for superovulation in human, 7
Convention on International Trade in Endangered Species (CITES), 47, 331
Cooling
  rapid, cryopreservation, 226. *See also* Freezing
  rates of, 90
  sensitivity to, 237
Corpus luteum, 158
Cortical granules, 239
Cryobiology, 223. *See also* Cryopreservation
Cryoinjury, 226, 234, 240
Cryopreservation, 64, 93, 280, 310, 325
  and blastocyst, 237
  of embryos, 18–21
  of human oocytes, 9, 233–234
  and live birth, 239
  methods of, 225–232
  of monkey sperm, 40
  of NHP, 234–235
  and population size, 49
  and rapid cooling, 226
  sperm, 90–91
  of squirrel monkey oocytes, 80
  survival rate, 240
  three phases of, 224–225
  of in vitro fertilization embryos, 62
Cryoprotection, 226
Cryoshock, 259
Crystallization, 229
CSPG. *See* Chondroitin sulfate proteoglycan
Culture
  conditions of, 40
  embryo, in rhesus monkey IVF, 32
  for IVM, 8
  long-term, 96
  media, effects of, 114–115
  medium, 164
  new media, 10
  variables of, 117–118
CVS. *See* Chorion villus sampling
Cycle, menstrual, 159, 333
Cyclic adenosine monophosphate (cAMP), and meiotic progression, 115
Cyclicity
  evidences of, 62
  normal menstrual, 136
  reproductive, and cryopreservation, 223
Cynomolgus macaque. *See* Cynomolgus monkey
Cynomolgus monkey, 85, 104, 125, 333
  and cryopreservation, 234, 237, 240
  frozen-thawed semen, 90
Cystic fibrosis, 355–356
  oocyte identification of, 21
Cytokines, 183, 187, 188–191
Cytology, vaginal, in squirrel monkey, 75
Cytolytic processes, in culture medium, 11
Cytoplasm
  during cyropreservation, 228
  and enucleation, 319
  in nuclear transfer, 325
  recipient, age and stage of, 321–322
Cytoplasmic injection, 18
Cytoplasmic maturation, 41
Cytoskeleton, 239
Cytotoxic processes, in culture medium, 11
Cytotrophoblast invasion, human, 182–193

dbcAMP. *See* Dibutyryl cyclic adenosine monophosphate
Decapacitation factor (DF), 87
Decidual cell, 170, 291
Decidual cell response, 299
Decidualization, 211
Defect
  genetic, detection of, 352–355
  single-gene, 355
Deforestation, 311, 312
Dehydration
  of cell, 227
  osmotic, 236

Deoxyadenosine, 259
Dermatan sulfate, 291
Densensitization, from hCG, 132
Developmental defect, and IVM oocytes, 110
Diabetes, mellitus type II, 136
Diagnosis
   of nonproductivity, 64
   preimplantation, of cystic fibrosis, 356
   prenatal, of X-linked disease, 354
Diakinesis/MI, MII, 112
Dibutyryl cyclic adenosine monophosphate (dbcAMP), 87, 104, 107, 337, 344
Differentiation, cell, 317
Dimethyl sulfoxide (DMSO), 18, 224, 231, 237–238
Disease
   genetic, 246
   and preimplantation diagnosis, 351–357
   research on, 310
Dizygotic twin births, human, 12
DMSO. See Dimethyl sulfoxide
DNA, 274, 350, 352
   analysis of, 354
   and defect diagnosis, 355
   probe of, 353
   recombinant technology, 137
   synthesis of, 324
DNA-construct, 246
Donor cell, 323–324
Dopamine receptor agonist (DRAg), 59
Down-regulation, of glycoproteins, 302
DRAg. See Dopamine receptor agonist
Duchenne's muscular dystrophy (DMD), 353
Dysmenorrhoea, membranous, 59

Edema, maternal, 186
EGF. See Exogenous growth factor
Egg
   follicular vs oviductal, 37
   hamster, 38
   mouse, 39
   quality and yield of, 111
   recovery of, 92
Electroejaculation, 77. See also Rectal probe electroejaculation, 86, 104, 336
Electroporation, 259, 319, 323
Embryo
   aneuploid, 349
   attachment competence, 301
   biopsy of, 351
   blastocyst, in rabbit, 113
   bovine, 322
   1–2 cell, 196
   2-cell, 114, 118, 201, 211, 317
   4-cell, 79, 201, 266, 275, 317
   6-cell, 79, 114
   8-cell, 114, 202, 211, 264, 317, 350, 355
   chromosomal abnormality, 350
   chromosomal study of, 78
   cleavage stage, 18, 110, 356
      early, 10, 93, 135
      nutritional requirements, 23
   coculture of, 279
   cryobiology of, 223–241
   and cryopreservation, 13, 18–21
   development, 78, 89
      after nuclear transfer, 319
      in culture, 10–12, 94
   early research, 309
   early transfer of, 279
   and endometrium, 169
   enucleated, 264
   genetic abnormalities of, 349
   implantation of, 158
   IVF, inadequacy of culture media, 30
   legal status of, 274
   loss of, 161
   in micromanipulation, 249
   mammalian, cryopreservation of 236–237
   metabolism of, 40
   morula, 94, 158
      in rabbit, 113
   mouse, 195–215, 196
      8-cell, 274
      and cryopreservation, 224
   mouse and assisted hatching, 266
   nonhuman primate, crypreservation of, 236–239
   of older women, 353
   osmotic volume, 231
   oviductal-stage transfer, 95

permeability of, 225, 236
preimplantation, 159, 211, 355
pronucleate, 262
recovery of, 75
residual, 31
rhesus monkey, transfer of, 33
and sex identification, 354
splitting of, 74
squirrel monkey, 74
with thin zona, 269
transfer of, 62, 70, 223. *See also* Implantation
  first successful, 74
  in lion-tailed macaque, 342
  and macaque, 333
viability of, 11, 265, 271
and vitrification, 229, 230
and zona hardening, 271
Embryogenesis, 85, 124, 211, 309, 313
postimplantation study of, 196
Embryonic, stem cell, 323, 325
Embryonic cell, 285
Embryotrophic effect, 280, 282
Endangered species. *See* Species, endangered
Endangerment, rate of change, 310
Endocrine development, abnormal, 50
Endocrine profile, 53
  lack of knowledge of, 310
Endometrial casting. *See* Dysmenorrhoea, membranous
Endometriosis, 59
Endometrium, 182
  maternal, 169
  uterine
    and embryo attachment, 290
    and embryo interaction, 169-178
Endoplasmic reticulum, 369
Endoprotease, 380
Energy, cell, 224
Enucleation
  ethics of, 274
  of oocyte, 319-320, 323
  and survival rate, 263
Environment
  concern for, 309. *See also* Habitat
  factors in infertility, 64
Epithelia, lumenal and tenascin production, 300
Epithelial cell, 205, 280, 281, 282, 294

maternal, 149
mouse uterine, 299
uterine surface, 296-298
Epithelium
  conversion of, 301
  endometrial, invasion of, 147-149
  lumenal, 290
  uterine, 147
Equilibruim freezing. *See* Freezing, equilibrium
Estradiol, 92, 113
  embryo uptake of, 79
  levels after superstimulation, 127
Estradiol-17β, 196
Estrogen, 170, 171, 195, 340-341
  human, 4, 6, 12
Estrual cycle, in squirrel monkey, 75
Ethanol, 320
Ethics
  and gene correction, 274
  and micromanipulation, 246, 273
Evaluation, 56-60
Exogenous growth factor (EGF), 164
Extinction. *See also* Species, endangered
  ultimate cause, 312
Extracellular fragments. *See* Fragments, extracellular
Eye, as immunologically privileged site, 187

F2α, 77
Facilitated reproduction. *See* Assisted reproduction
Fallopian tube, human, transfer site, 10
FCS. *See* Fetal calf serum
Feeder cell layers, human, 11
Female, evaluation of, 56
Fertility
  and cryopreservation, 233
  evaluation of, 49
  human, and assisted conception, 5
  and hyperactivity, 90
  NHP, after cryopreservation, 234-235
  suppression of, in marmoset, 76
Fertilization, 85, 124, 261, 313
  abnormal, 349
  and coculture, 283

Fertilization (*cont.*)
  dispermic, 350
  and microinjection technique, 17
  oviductal influences, 37
  polyspermic, 17
  process of, 39
  study of, 36
  xenogenous, 80
    in squirrel monkey, 74
  and zona characteristics, 271
Fetal calf serum, 94
Fetus, immunorejection, 187. *See also*
    Immunoreactivity
  and preeclampsia, 186
  triploid, 350
Fibrinoid, during anchoring, 155
Fibroblast growth factor, 212
Fibronectin, 176, 293, 296
First polar body (PB1), 110, 340
FISH. *See* Fluorescent method to detect
    in situ hybridization
Flare, human, 5
Fluid
  oviductal, 162
  uterine and oviductal, 169
Fluorescent method to detect in situ hybridization (FISH), 350, 353
Follicle
  development of, 61, 76
  diameter of, 115–117
  dominant, 124
  growth of, 5–6
    in squirrel monkey, 73
  multiple development of, 62
  preantral, 111
Follicle stimulating hormone, 128
Follicle stimulating hormone (FSH), 4, 6, 7, 73, 113, 124, 137, 266–267, 269, 333, 334, 338
  and preovulatory follicle, 125
  recombinant, 5
Follicle stimulation protocol, 91–92
Follicular growth, in squirrel monkey, 74
Follicular phase, human, 3
Follicular stimulation
  advantages of, 124. *See also* Ovulation
  NHP, 36
  protocols for, 125

research limitations, 135
Folliculogenesis, 354
Fragments, extracellular, 272
Freezing
  controlled slow, 226–228
  equilibrium, of embryos, 19
  intracellular, 229
  of mouse embryos, 341–342
  rapid, 225, 230–233. *See also* Cryopreservation
  slow, 225, 237
Frog, and cell totipotency, 317
FSH. *See* Folicle stimulating hormone
Fusion. *See also* Cell fusion
  myoblast, 213
  rabbit blastocyst, 322
  sperm-egg (human), 252
  whole-cell, 318

Gamete, cryobiology of, 223–241
Gamete fallopian tube transfer (GIFT), 10, 12, 15, 80
  human, 4
  for male factor infertility, 19
Gametogenesis, 124, 349
Gender, and micromanipulation, 246
Gene
  abnormality of, and micromanipulation, 246
  nucleotides of, 363
Gene correction, 274
Gene therapy, and micromanipulation, 273
Genetic analysis, preconceptive, 21
Genetic diversity, loss of, 311. *See also*
    Nuclear transfer
Genetic gain, 319
Genome, embryonic, activation of, 160
Genomic activation, 284
Germinal vesicle (GV), 110
  breakdown of, 112
Germ plasm, cryopreservation of, 239
Gibbon, menstrual cycle in, 56
GIFT. *See* Gamete fallopian tube
    transfer
Glands, 171
Global warming, 312
Glucose, 285
Glutamic acid, 378

Subject Index

Glutathione, 284, 285
Glycerol, 230
  for freezing, 223
Glycine, 378
Glycoconjugates, on blastocyst, 290
Glycogen, 285
Glycoprotein, 104, 302
Glycosaminoglycan, 291
  chains of, 292
GnRH, 5, 6, 7, 23, 62, 63, 132
Golgi apparatus, 369
Gonadotropin, 6, 23, 124, 125
  antibodies against, 136
  chorionic, 162–163
  and egg quality, 111
  exogenous, 4
  and oocyte maturation, 92
  pregnant mare serum, 32
Gorilla, 49
  abortion in, 59
  and cryopreservation, 234
  infertility in, 56
  menstrual cycle in, 56
  semen analysis in, 63
  success of artificial insemination, 60
Government, United States and population control, 313
Granulosa cell, 112
  and hCG, 132
Greenhouse effect, 312
Growth factor, 183, 196, 285
  trophoblast, 188
Growth hormone, 23
  human, 6–7
GV. *See* Germinal vesicle
Gylcerol, 90

Habitat
  destruction of, 311, 312
  preservation of, 308
  rainforest, loss of, 331
Ham's F10 medium, 282
Hamster, 118
  chromosomal embryo study, 78
  egg, as surrogates, 90
  and infertility study, 349
Hatching, assisted, 266–267
Hatching in humans, assisted, selective, 268–269

hCG. *See* Human chorionic gonadotropin
Heifer, and early transfer, 280
Hemorrhage, of implantation site, 155
Heparan sulfate-binding growth factor, 301–302
Heparan sulfate proteoglycan (HSPG), 291, 294, 296, 300, 302
Hereditary sensory motor neuron disease type II, 354
Herpes hominis type II, 56
Herpes virus, 319
Heteroduplex formation, 356
Heterokaryons, formation of, 149
Hexosamine, 291
Histidine, 78, 380
History, of assisted reproduction in great apes, 47
HIV, and cryopreservation, 233
HLA-G, 183, 187
hMG. *See* Human menopausal gonadotropin
Hominoidea, 159
Homo sapiens, 164–165. *See also* Human
Hormonal regulation, 298
Hormonal replacement, human, 13
Hormone, ovarian, 290
Hormone steroid, in embryo, 79. *See also* Steroid
HPLC, and SP-10, 373
HS. *See* Heparan sulfate proteoglycan
HSPG. *See* Heparan sulfate proteoglycan
HTF medium, 10
Human
  and chorionic gonadotropin, 164–165
  implantation in, 145, 147
  implantation site, 154
  preimplantation embryo, 159
  SP-10, 369
  and sperm cryopreservation, 233
  trophoblast differentiation, 152
Human chorionic gonadotropin (hCG), 4, 5, 6, 7, 32, 73, 79, 111, 132, 173, 188, 354
  and adhesion, 152
  in squirrel monkey, 76
  ovulatory initiator, 130
Human cord serum, 164

Human in vitro fertilization, 3-29
Human luteinizing hormone (LH), 132
Human menopausal gonadotropin (hMG), 5, 6, 7, 12, 125
  and cystic results, 76
Hyaluronate, 291
Hyaluronic acid, 296
Hybrid
  birth of, 343
  macaque and cryopreservation, 237
Hybridization, 353, 355
Hydrolases, intra-acrosomal, 381
Hyperactivation, 60, 337. See also Motility
  in rhesus monkey, 87
Hyperprolactinaemia, human, 4
Hypotaurine, and hamster eggs, 38
Hypothalamic-pituitary-ovarian axis, 124
Hypothalamo-pituitary, 53
Hypoxanthine, 115, 285
  as meiotic progression inhibitor, 115
Hypoxanthine phosphoribosyl transferase gene, 357
Hysterosalpingography, 64

Ice
  and cryopreservation, 226
  formation of, in cryopreservation, 19-20
  lack of, 228
ICM. See Inner cell mass
IGFBP-1. See Insulin-like growth factor binding protein
Immobilization, improved techniques of, 48
Immune fuction, maternal, 187
Immune system, maternal, 190
Immunologic sequelae, 86
Immunoreactivity, for SP-10, 369
Immunorejection, 213
Implantation, 124, 161, 165, 211, 313
  blastocyst, 145
  and coculture, 283
  delayed, 212
  human, 145, 147, 165
  and micromanipulation, 246
  multiple, 343
  physiology of, 158-167
  and protein synthesis, 175-177
  rate, 10, 260, 265
  and selective assisted hatching, 268
  site clean-up, 154
  timing of, 160-162
India, and rhesus monkey, 311
Indonesia, and live capture, 311
Infection, and infertility, 56
Infertility
  of apes, 56
  causes of, 103
  and cryopreservation, 223
  early IVF, 309
  human, treatment options, 3
  human male factor, 15
  idiopathic and GIFT, human, 4
  and micromanipulation, 255
  social effect on, 56
  and SZI, 250
Inflammatory cell, 213
Inheritance, 319
Inhibitor, protease, 380
Inner cell mass (ICM), 163, 195, 317
Insemination
  human, from IVM oocytes, 9
  microinjection, 16
  xenogenous, in rabbit, 236
Insulin, 164
Insulin-like growth factor binding protein (IGFBP), 7, 170-173, 175, 177
Integrins, 176
Interaction, 294
  embryo-uterine, and proteoglycans, 291-303
Interbirth interval, in great apes, 48
Interblastomere, 267
Interface, embryo-maternal, 159
In utero, human, transfer site, 10
Invasion
  cytotrophoblast, 182-193
  epithelial, 148
In vitro culture, rhesus monkey embryo, 93
In vitro fertilization (IVF), 15
  biotechnologies, 73, 80
  chronology, 41
  early success of, 74, 86
  human, 3, 158

and macaque, 333
in Old World Monkeys, 93
protocol
in marmoset, 77
in monkeys, 32–33
in squirrel monkey, 77
as quantitative measure, 41
Ionic component, of culture medium, 11
Ionophore, 320
Irradiation, and cryopreservation, 224

Karyoplast, 264, 319
Karyotyping, 353
Keratan sulfate, 291
Ketamine, HCl, 48
Ketaset, 48

β-Lactoglobulin, 174
Lacunae, in trophoblast, 152
Laminin, 293, 296
Laparoscopy, 64
Laparotomy, 339
Laser energy, and micromanipulation, 273
Lesch-Nyhan syndrome, 354, 357
Leuprolide, 5, 6
LH. *See* Luteinizing hormone, Human luteinizing hormone
LHRH agonist, 354
Ligands, 210
Linkages, peptide, 378
Lion-tailed macaque. *See* Macaque
Lipid peroxidation, in culture medium, 11
Lipoproteins, human CG, 162. *See also* Chorionic gonadotropin (CG)
Live capture, 311
Livestock, domestic, and cryopreservation, 223
Lumen, during implantation, 147
Lupron, 132
Luteal phase
adequacy of, in chimpanzee, 57
after stimulation, 134
inadequacy of, 62
in squirrel monkey, 76
Luteal stage, and IGBFP-1, 171

Luteinization, of granulosa cells, 132
Luteinizing hormone, surge, 4, 32, 112, 130, 137, 333
detrimental effects of, 125
human, 4, 5
length of, 135
various species, 132

Macaca mulatta, 163–164
Macaca silenus. See Macaque
Macaque, 154, 309
after LH treatment, 135
anchoring stage, 155
and assisted reproduction, 308
chorionic gonadotropin, 368
and cryopreservation, 90
epithelial invasion in, 149
follicle stimulation protocols, 125–130, 135–137
and gonadotropin dosage, 125
implantation, 145
and Indian government, 311
lion-tailed, 331, 343
contraception in, 346
propagation management, 331–346
normal values, 333–335
ovulatory events, 130–135
pig-tailed, 333
and SP-10, 360, 369
sperm capacitation in, 103–104
trophoblast differentiation, 150–152
usefulness of, 108
Male evaluation, 59
Mammal
and cryopreservation, 236
and nuclear transfer, 325
Marker, cellular, 112
Marmoset, 73, 75, 161, 163
adhesion study of, 147
chorionic gonadotropin, 368
and chorionic gonadotropin, 162
and cryopreservation, 237, 240
cycle in, 76
and embryo transfer, 79
epithelial invasion in, 149
implantation, 145, 158
meiosis in, 117
preimplantation embryo, 159

Marmoset (*cont.*)
  trophoblast invasion of, 150
Maternal vessels, invasion of, 149–152
Maturation, of oocytes, 92–93
MDBK cell, 282
Medium
  for coculture, 282
  for culture, oocyte incubation, 77
Meiosis
  after LH surge, 130
  marmoset, 117
  progression of, 112
  resumption, 111
Meiotic/mitotic transition, 320
Meiotic spindle, 239
Menopause, premature, 13
Menotropins, 49
Menstrual cycle
  artificial, in women, 13
  of chimpanzee, 47
  human, 4
  normal, in apes, 56
Menstruation, lack of, 75
Mesenchyme, and villi growth, 155
Metabolism, embryo, 285
Metalloproteinases, 184
  matrix-degrading, 183
Metaphase spindles, 110
Methionine, 284
Metrodin, 5, 62, 126
Mevalonate, 285
MHS-10, antibody to SP-10, 360
Mice
  4-cell nuclear transfer, 323
  and cryopreservation, 237
  G1 stage, 324
  and nuclear transfer, 317
  and pronuclear ova, 321
Microdrop insemination, 16
Microinjected fallopian transfer (MIFT), 17, 23
Microinjection, 16, 64, 80, 319
  cytoplasmic, 18
  single sperm, 247
Micromanipulation, 15–18, 64, 93, 310
  and assisted fertilization, 247
  candidates for, 246
  future applications of, 273–275
Microsequencing, 376–378
  and SP-10, 369, 379

Midluteal phase, human, 4
MIFT. *See* Microinjected fallopian transfer
Miscarriage, idiopathic, and placenta donation, 275
MIST. *See* Subzonal insertion (SZI)
Mitochondria, in nuclear transfer, 325
Mitogen, 176
Mitotic cell, 322
Mometrium, 182
Monkey. *See also* Marmoset, Rhesus monkey, Squirrel monkey, Tamarin, Macaque
  Old World, IVF-ET in, 85
  rhesus, 163–164
Morphology, of implantation, 145–156
Morula, 94, 113, 158, 212
  rhesus monkey, 163–164
  and TGFβ binding, 211
Morula cell, 323
Morulae, 280
  and cooling, 237
  and TGFβ, 202
Mosaicism, 350
Motherhood, definition of, 274
Motility
  and cryopreservation, 232, 234
  of sperm, 103. *See also* Hyperactivation
  sperm, and micromanipulation, 257
  of sperm, 344
Motion, rating scale for, 50
Mouse, 118
  and binding studies, 294
  cryopreservation of oocyte in, 235
  embryo, 196, 280, 294
  and enucleated study, 318
  and growth factor, 285
  intra-acrosomal antigen, 364
  lack of polyspermy, 262
  oocyte study of, 110
  and zona drilling, 248
  zona pellucida in, 104
mRNA, 190, 205, 363, 364
MSA-63, 364, 367
Multiple births. *See* Twins, Triplets
Mumps, in gorilla, 56
Muscle, cardiac, 213
Mycoplasma genitalium, 56, 59

Myogenesis, 213
Myometrium, uterine, 212–213

Natural cycle IVF, human candidates for, 4
Neonatal death loss, in squirrel monkey, 75
Neonate, and time with mother, 48
Nested primers, 356
New World primates, 73
  characteristics, 74
  implantation, 145
NHP. *See* Nonhuman primates
Nipple stimulation behavior (NSB), treatment and monitoring, 57–60
Nonhuman primates (NHP), 85
  and cryopreservation, 234
  and human reproduction research, 78
  IVF, value, 30–31
  oocyte biology in, 110
  threats to, 310
Nonproductivity, diagnosis of, 64
Nuclear changes, timing of, 112
Nuclear maturation, 41
Nuclear transfer, 310
  in mammals, 317–326
Nucleation, 226
Nuclei, tetraploid, 355
Necleoli, 264
Nucleotides, 363

Occlusion, oviductal, 50
Occult ovarian failure, human, 6
Offspring, of IVF rhesus monkeys, 33, 40
Old World primates, 158
  implantation, 145
Oligoasthenospermia, 16
Oligozoospermia, 252, 256, 257
Olive baboon, and superstimulation, 125
Oocyte
  activation of, 320
  aging, 322
  anomalies of, 93
  chromatin in, 322
  chromosome in, 235
  collection of, 91, 338
  Old World monkeys, 91–92
  rhesus monkey, 32
  competence of, 114
  cryopreservation of, 235–236
  and live birth, 239
  and culture media, 117
  donation of, human, 7
  early metaphase, 324
  enucleation of, 319, 323
  and environmental influences, 110
  freezing of, 74
  GV stage, 111
  human donation, 12–15
  hyperhaploid, 349
  management with lion-tailed macaque, 338
  maturation of, 2, 8, 34, 77, 92–93, 110
    inhibitor, 115
    in primates, 41
  penetration initiatiors, 103
  quality of, 112–113, 114
    and assisted fertilization, 259
    human, 15
  recovery of, 64
  recovery of, human, 5
  and superstimulation, 113–114
  in vitro fertilization, in squirrel monkey, 73
  in virto maturation of, 76
Oogenesis, 85
Oolemma, and sperm fusion, 250
Orangutan, 62
  and artificial insemination, 61
  menstrual cycle in, 56
Oregon Regional Primate Research Center, 85, 136
Organogenesis, 175
Osmosis, during cell freezing, 226
OT. *See* Oxytocin
Ovarian dysgenesis, human, 13
Ovarian failure, human, 4
Ovarian stimulation, 36
Ovarian suppression, 6
Overpopulation, and primates, 312
Ovids, 162
Oviduct, 211
  and paracrine effect, 280
Oviductal cell, 279, 281
Oviductal environment, 41
  influence of, 37–38

Oviduct cell, human, 11
Ovine, and coculture, 280
Ovulation
   and cryopreservation, 234
   early knowledge of, 73
   induction, 64. See also Follicular stimulation
   in squirrel monkey, 75
   and swelling phases, 57
Oxygen, consumption, in embryo, 79
Oxygen gas, and embryo development, 118
Oxytocin, 59
Ozone, depletion of, 312

Pan paniscus, 57
Pan troglodytes. See Chimpanzee
Papio anubis. See Baboon
Paracrine, 94, 171, 175, 188
   mediation of, 196
Paracrine effect, 279, 280, 285
Partial zona dissection, 249, 250, 251
   and human pregnancy rate, 267
Partial zonal drilling, 255–258
Partial zona opening (PZD), 16–17
Patient, selection of, 250, 252–255
Pattern, hormonal, 132
PB1. See First polar body
PDGF. See Platelet-derived growth factor
Pentapeptides, 364
Pentoxyfylline, 259
Peptide
   immunoreactivity of, 379
   linkages, 378
Pergolide, 57
Periimplantation, 298
   and mouse embryo, 195–215
Perineal swelling. See Swelling, perineal
Perivitelline area, 252, 262
Perivitelline space
   microinjection, 17
   and SZI, 247
Perlecan, expression of, 294
Permeability, and cryopreservation, 240
Phillipines, and live capture, 311
Phosphodiesterase inhibitors, 259
Phospholipase, membrane, 104

Phosphorylation, of sperm membranes, 103
Pig, 118
   and activation, 321
   and nuclear transfer, 320
Pig-tailed macaque. See Macaque, pig-tailed
Pituitary suppression, human, 6
Placenta
   chorioallantoic, 145
   development of 175, 176
   donation of, 275
   formation of, 155, 291
   hemochorial, formation of, 182
   human, 184
Placentation, 211
Plasmalemma, 262
Plasma membrane interaction, 23
Plasmin, 176
Plasminogen activators, 184
Platelet activating factor (PAF), 164
Platelet-derived growth factor, 164
Ploidy, normal, 324
PMSG. See Pregnant mare serum gonadotropin
Polar body, 1, 2, 350
Pollution, industrial, 312
Polycystic ovarian disease, 9
Polycystic ovarian syndrome, 4, 7
Polymerase chain reaction (PCR), 352
Polyploidy, 319, 350
Polyspermic fertilization, 17
Polyspermy, 252
   block to, 39
   correction of, 262–265
   and genetic abnormality, 350
   as oocyte defect, 110
   and PZD, 249
   and zona hardening, 271
Population, human, 312
Porcine, embryo of, 280
Postimplantation, and TGFβ, 205
Preeclampsia, 185
Pregnancy
   after micromanipulation, 261
   ectopic, 286
   from embryo transfer, 343
   extrauterine, 59
   human, and embryo quality, 11
   rate, 15–16

termination of 351
twin, 354
Pregnanediol (PdG), 57
Pregnant mare serum gonadotropin (PMS), 113, 125, 280, 338
Pregnant mare serum (PMS) and squirrel monkey, 76
Preimplantation, 78, 81, 349
  and genetic abnormalities, 349-357
Preincubation, of semen, 63
Preservation
  of genetic stock, 310
  of habitat, 308
Previllous stage, 152
Primagravidas, 185
Primary decidual zone (PDZ), 205
Primates
  New World, assisted reproduction in, 73
  transgenic, 80
Primatologist, concerns of, 308
Primer extension, mRNAs, 363
Profasi, 126
Progesterone, 57, 61, 113, 170, 171, 195, 340-341
  embryo uptake of, 79
  human, 5
  human use of, 12
  plasma, human, 3-4
  and squirrel monkey, 76
Prolactin, 4, 57
  and cells in culture, 113
Prolonged postpartum amenorrhea (pPA), 57
Pronuclear development, of human oocytes, 9
Pronuclear extraction, 246
Pronuclei, formation of, 252
Propagation management, of lion-tailed macaque, 331-346
Propanediol, 93
1,2-Propanediol, as cryoprotectant, 18
Protein, hCG, 162. See also Chorionic gonadotropin
  binding, 292
  Group I, 170
  Group II, 170
  hormonally regulated, 169
  placental, 14, 173
  removal, 103

synthesis of, in embryo, 79
utilization by embryo, 11
Proteinases
  cytotrophoblast, 185
  role of, 183-185
Proteinuria, maternal, 186
Proteoglycans, 214, 291
  as modulators, 291-303
Protocol
  monkey gametes, 31
  step-down/up, 125
Pseudopregnancy, 209
PTM. See Macaque, pig-tailed
Purification, of SP-10, 379-380
PVDF membrane, and SDS-PAGE, 375
PZD. See Partial zona dissection

Rabbit
  and activation, 321
  4-cell nuclear transfer, 323
  as embryo recipient, 80
  and fusion, 322
  G1 stage, 324
  and nuclear transfer, 320
  whole-cell fusion of, 318
Radiation, and cryopreservation, 224
Radicals, superoxide, in culture medium, 11
Radioimmunoassay, 162
Rain forest
  destruction of, 311
  loss of, 331
Rapid cooling. See Cooling, rapid
Rapid freezing. See Freezing, rapid and slow
Rate
  fertilization, 17, 236, 337, 338, 340
  pregnancy, 351
    in cattle, 325
    and coculture, 283
  success, 80, 93, 94-95, 233, 235, 236, 240, 265, 267, 269-270, 284, 309, 354
    in apes, 64
    cryopreservation, 230, 237
    in marmoset, 77
    and micromanipulation, 255, 261
    in PZD, 249
    in squirrel monkey, 73

RBP. *See* Retinol binding protein
Rearing, captive, 332
Receptors, 171
   adhesion, 185. *See also* Adhesion protein, 213
Recessive disease, 358
Recipient, lion-tailed macaque, 340
Recipient cell, 324
Rectal probe electroejaculation (RPE), 63. *See also* Electroejaculation, 334
Refractoriness, 136
   and follicular induction, 73
   lack of, in squirrel monkey, 76
Renal, dysfunction of, 186
Renou device, 32
Reproduction, in captivity, 332
Reproductive biologist. *See* Biologist
Reproductive biotechnology. *See* Assisted reproduction
Research, quality of, 31–32
Research needs, for nonhuman primate hormones, 135
Retardation, mental, 354
Retinitis pigmentosa, 354
Retinol binding protein (RBP), 170, 173–174, 175
   and baboon implantation, 177
Retroejaculation, electroejaculation, 335. *See also* Rectal probe ejaculation
Rhesus macaque. *See* Rhesus monkey
Rhesus monkey, 75, 85, 104, 125, 333
   blastocyst, 33, 94
   and chorionic gonadotropin, 162–164
   and coculture, 281
   and cryopreservation, 234, 237, 240
   and gonadotropins, 113
   implantation, 158
   and Indian government, 311
   IVF protocol, 32–33
   and LH surge, 135
   meiotic progression in, 116
   oocyte maturation, 114–115
   preimplantation embryo, 159
   response pattern, 127–130
   semen processing, 86
   in vitro culture, 246
RNA, 110, 189

synthesis of, 79
Rodents, 162
RPE. *See* Rectal probe electroejaculation

Saimiri sciureus. *See* Squirrel monkey
SDS-PAGE and PVDF membrane, 375
Seasonality
   and cryopreservation, 232
   in squirrel monkey, 75
Second polar body (PB2), 263, 350
Selection
   competitive, 49
   phenotypic, 325
Semen
   collection and storage of, 60
   collection of, 333
      in rhesus monkey, 32
   profile analysis (SPA), 63
   profile of, 18
   quality factors, 86
   rhesus monkey, processing, 86
Sequence analysis, of SP-10, 362
Sequence comparison
   cDNAs, 363
   significance of, 368
Serum
   human cord, 164
   for oocyte incubation, 77
   as supplement, 93
Sex identity, embryo, 21
Sheep
   and coculture, 279, 281–282
   whole-cell fusion of, 318
Shigella flexner, 59
Shrink-swell change, 228, 229
Site, expansion of, 152–155
Somatic cell coculture, 95, 96
SP-10, 360–383
   cloning, 368
   isolation of, 369
   microsequencing of, 379
Species
   endangered, 35, 41, 47
      and IVF, 30
   endangerment, 308
   and extinction, and assisted reproduction, 313
   yearly loss of, 312

Subject Index

Species Survival Plan/Program (SSP), 47, 331
Sperm
  abnormalities of. See Asthenozoospermia, Oligozoospermia, Teratozoospermia
  abnormality, and microinjection, 64
  capacitation of, 34-35, 74, 86-90
    in macaques, 103
  changes in, 103
  collection of, 75, 77, 343
  cryopreservation, 40
  cryopreservation of, 90-91, 225, 236
  epididymal, 37, 77
  frozen, and coculture, 283
  human
    cryopreservation of, 233-234
    fertilizing ability of 258-261
    subzonal insertion of, 250
    and zona pellucida, 107
  mammalian, and cryopreservation, 232
  management of, in lion-tailed macaque, 335
  motility of, 51, 60
    and fertilizing ability, 34-35
  NHP, cryopreservation of, 234-235
  nuclei, 110
  penetration assay, 52
  as pronuclei, 321
  recovery of 337
    in macaque, 106
  treatment of, 32
  washing procedures, 87
  zona pellucida interaction, 104
Sperm penetration assay, 90
Splicing, alternative, 364, 368
Splitting, embryo, 310
Squirrel monkey, 75, 309
  and cryopreservation, 236
  fertilization rate, 80
  and in vitro fertilization protocol, 77
SSP. See Species Survival Plan/Program
Stage matching, for nuclear transfer, 324-325
Steriod, 298
  and in vivo development, 114
Stimulation, follicular. See Follicular stimulation
Storage, of cryopreserved cell, 224
Stroma, 213, 299
Stromal extracellular matrix, and HSPG, 299
Stromal invasion, 149-152
β-Subunit of human chorionic gonadotropin (β-hCG), 368
Subzonal insertion (SZI), 246, 251
  advantages of, 256
  and polyspermy, 262
Suckling, as contraceptive effect, 57
Sucrose, 230
Suidae, 162
Superovulation, 3-7, 310, 354
  early efforts, 85-86
  human, 9
Superstimulation
  and egg yield, 111
  and oocytes, 113-114
Surrogacy, 64
SUZI. See Subzonal insertion (SZI)
Swelling, 237
  ano-genital, 57, 61
  and cyclicity, 62
  osmotic, 225
  patterns of, 62
  perineal, 47, 61, 63
Swell-shrink change, 230
Swimming speed, of sperm, hyperactivation, 60. See also Motility
Swine
  4-cell nuclear transfer, 323. See also Pig
  whole-cell fusion of, 318
Synchronization
  human donor and recipient, 13
  of menstrual cycle, 63
Syncytiotrophoblast, 171
Syncytium, human, 151
Syndecan, expression of, 294
SZI. See Subzonal insertion

TALP, 94, 115, 118
TALP-HEPES, 86
Tamarin, cotton-top, 159, 161
Taurine, 284
Telazol, 48
Teratozoospermia, 250, 252, 256, 257
TEST. See Tubal embryo stage transfer
Testosterone, 334

TES-Tris, 90
Thawing
  after cryopreservation, 228
  survival rate, 240
Tiletamine zolezapam, 48
Tissue
  stromal, 291
  of trophoblast, 152
Totipotency, 160, 317
  in amphibians, 325
  of donor nucleus, 323
Toxic compounds, and coculture, 284
Toxicity, to cryoprotectant, 225
Transcriptase-polymerase chain reaction method (RT-PCR), 189
Transfection, 323
Transfer
  at blastocyst stage, 286
  embryo, 94–95
    in New World species, 79
  intrafallopian, 342
  nuclear, 318
Transferrin, 164
Transforming growth factor β, 196, 205
Transgenic animal, 111, 118
Transgenic primate, 80
Transplantation, nuclear, 273
Trapping, 312
Trimester
  and cell study of, 184
  and endometrial changes, 176
Triplets. See also Twins
  cattle, 9
  in marmosets, 161
Triploidy, 79
Trisomies, 353
Troops, and breeding, 345
Trophectoderm, 275, 294
Trophoblast, 195, 285, 294, 317
  adhesion of, 147
  invasion of, 186
Trophoblastic cell, 279
Trophoblast, syncytial, 94
Trophectoderm cell, 351
Tubal embryo stage transfer (TEST), 10, 15
Tubal occlusion, 4, 16
Twinning
  artifical, and micromanipulation, 273
  in marmoset, 74, 80
Twin ovulations, human, 12
Twins. See also Triplets
  in marmoset, 161
Tyrode's solution, 266, 269, 271
  and zona drilling, 248

Ultrasonography, and follicular response, 91
Ultrasound, 64
  in follicular recovery, 76
United Nations, 313
Ureaplasmas, 56, 59
Uridine incorporation, 79
Urinary hormone assay, 64
Uronic acid, 291
Uterine receptivity, 11–15
Uterus
  and embryo, 290
  nonpregnant and protein synthesis, 170
  and TGFβ, 203

Vaccine
  and SP-10, 360
  testing of, 311
Vascular contact. See Implantation
Vascular smooth muscle cell, 213
Vasculature, maternal, 177
Vero cell, 94, 282
  human, 11
Veterinarian, interests of, 309
Viability
  during cryopreservation, 224
  of embryo, 265
  loss of, 279
  sperm and motion characteristics, 50
Villi, primary, formation of, 155
Vitamin E, 285
Vitrification, 225, 228–230, 231. See also Cryopreservation
Vitronectin, 296

Warming, after cryopreservation, 224, 228
Waymouth's and Ham's F12, 118

Whitten's medium, 201
Wisconsin Regional Primate Research Center, 340
World Health Organization, 313, 360

X-linked disease, 350, 353

Yerkes colony, 57, 59

Zgyotes, as recipient, 320
ZIFT. *See* Tubal embryo stage transfer
Zona
   ablation, 274
   hardening, 93, 271
   human, drilling and dissection, 248–250
   premature hardening, in human oocytes, 9
   thinning of, 269
Zona drilling, 246, 262, 266, 268
   limitations, 17. *See also* Subzonal insertion
Zona pellucida, 34, 36, 94, 107, 263, 265, 301
   glycoprotein receptor on, 104
   homologous, in macaque sperm interaction, 104
   human, 16
   and macaque sperm, 104–105
   manipulation of, 247, 249
   receptor sites, 103
   salt-stored, 35
   thickness and selective hatching, 268
   and Tyrode's solution, 271
Zoo, and live capture, 311
ZP1-3. *See* Glycoprotein
Zygote, 110, 351–352
   and cryopreservation, 237
   dispermic, 262
   human, 263
   and pronuclear injection, 274, 275
   and superstimulation, 113
Zygote intrafallopian transfer (ZIFT), 15